Diatoms Biotechnology

This book covers biotechnology and applications of diatoms including various applications such as the use of diatom biogenic silica in drug delivery and their cultivation for wastewater remediation. It contains case studies in diatom biorefinery and provides information about the bioactive compounds, antioxidants, and lipids from diatoms. It also discusses the latest innovations in the genetic engineering of diatoms.

Key Features:

- Discusses the applications of diatoms in drug delivery
- Covers the application and cultivation of diatoms for wastewater bioremediation
- Describes the innovations and gaps in genetic engineering of diatoms
- Throws light on the lipid content, antioxidant content, and various bioactive compounds present in diatoms
- Includes extraction technologies in diatoms

The book is meant for academicians and microbiologists, biotechnologists, and marine biologists.

Diatoms Biotechnology

Edited by
Archana Tiwari
Diatom Research Laboratory, Amity Institute of Biotechnology, Amity University, Noida, India

Ashok Pandey
Centre for Innovation and Translational Research, CSIR-Indian Institute of Toxicology Research, Lucknow, India, and Sustainability Cluster, School of Engineering, University of Petroleum and Energy Studies, Dehradun, India

Pau Loke Show
Department of Chemical and Environmental Engineering, Faculty of Science and Engineering, University of Nottingham Malaysia, Malaysia

Binod Parameswaran
Microbial Processes and Technology Division, CSIR-National Institute for Interdisciplinary Science and Technology, Trivandrum, India

CRC Press
Taylor & Francis Group
Boca Raton London New York

CRC Press is an imprint of the
Taylor & Francis Group, an **informa** business

Designed cover image: © Shutterstock ID 2232399515, Elif Bayraktar

First edition published 2024
by CRC Press
6000 Broken Sound Parkway NW, Suite 300, Boca Raton, FL 33487-2742

and by CRC Press
4 Park Square, Milton Park, Abingdon, Oxon, OX14 4RN

CRC Press is an imprint of Taylor & Francis Group, LLC

ISBN: 9781032562179 (hbk)
ISBN: 9781032566450 (pbk)
ISBN: 9781003436553 (ebk)

DOI: 10.1201/ 9781003436553

Typeset in Times
by Newgen Publishing UK

Contents

Preface

This book is a compilation of the biotechnological applications of diatoms. The biochemical composition of the diatom biomass includes a myriad of high-value compounds such as omega 3 fatty acids, DHA, EPA, polysaccharides, and phenols, to name but a few. Each of these metabolites has tremendous nutraceutical value and they find applications as sustainable aquafeeds. The antimicrobial efficacy of diatoms against bacteria and pathogenic fungi opens broad horizons in the pharmaceutical industry. The challenges in the path of sustainable diatom biorefinery are also elaborated in this book. The recent advancements in diatom research are also discussed, along with the pertinent research gaps.

The role of diatom biogenic silica for drug-delivery applications has been highlighted in Chapter 1. The frustules of diatoms are rich in silica, and they hold immense potential as drug-delivery systems. Chapter 2 elaborates the diverse bioactive compounds from diatoms including lipids, pigments, proteins, carbohydrates, and many more. The genetic engineering of diatoms, including innovations and future perspectives, is discussed in Chapter 3. The lipids from diatoms are highly valuable as biofuels and therapeutic molecules and Chapter 4 focuses on these lipids. The photonic materials from diatoms are described in some details in Chapter 5. Case studies in diatom biorefinery are explained in Chapter 6. The extraction technologies from diatoms have been elaborated in Chapter 7, while antioxidative compounds from diatoms are described in Chapter 8. The wastewater remediation potential of diatoms is highlighted in Chapter 9, wherein the bioremediation efficacy of diverse diatoms in different types of wastewaters is explained at length. Chapter 10 deals with the cultivation of diatoms for wastewater remediation along with the opportunities and challenges they present.

We acknowledge the authors for their scientific contributions and the reviewers for their valuable feedback in improving the chapters of this book. We are thankful to the publishers and Dr. Madhurima Kahali for constant support.

Editors
Archana Tiwari
Ashok Pandey
Pau Loke Show
Binod Parameswaran

Editor Biographies

Archana Tiwari is a professor at Amity Institute of Biotechnology, Amity University, Noida, India. She obtained a Ph.D. degree from Allahabad, and is a Gold medalist and distinction holder in Botany. Her key research interests include phycoprospecting diatoms for wastewater remediation and high-value products. She has been working on algal biotechnology for two decades and her research work has been published in 85 international journals with high impact factors. She has also authored 10 books and several chapters in publications by Springer, Wiley Blackwell, and Elsevier on algae and their applications.

Ashok Pandey is a distinguished scientist at CSIR-Indian Institute for Toxicology Research, Lucknow, India, and Honorary Executive Director at the Centre for Energy and Environmental Sustainability—India. His major research and technological development interests are in industrial and environmental biotechnology, which span from biomass to fuels and chemicals, waste to wealth/energy, industrial enzymes, solid-state fermentation, etc. His major research and technological development interests are industrial & environmental biotechnology and energy biosciences, focusing on biomass to biofuels & chemicals, waste to wealth & energy, etc. He has 16 patents, 120 books, >1000 papers and book chapters, etc. with *h* index of 131 and >74,000 citations.

Pau Loke Show is a professor in Biochemical Engineering from the University of Nottingham, Malaysia. His research interests cover upstream to downstream bioprocessing, microalgae technology, and the circular economy. Professor Show has published over 550 research papers and more than 200 contributions to international congresses, three books, and a number of patents that are currently applied in the industry. Prof. Show has received numerous prestigious academic awards from national and international organizations. He is now serving as an editor-in-chief at *Current Nutrition & Food Science*, Editor at *Scientific Report, Biocatalysis and Agricultural Biotechnology*, associate editor at *Bioengineered; Current Biochemical Engineering*, and editorial board member at several other journals.

Binod Parameswaran is a principal scientist in the Microbial Processes and Technology Division of CSIR-National Institute for Interdisciplinary Science and Technology, Trivandrum, India. He obtained his Ph.D. in Biotechnology from the University of Kerala, Thiruvananthapuram, India. He worked as a post-doctoral fellow at the Korea Institute of Energy Research, Daejeon, South Korea. He joined the CSIR-National Institute for Interdisciplinary Science and Technology, Thiruvananthapuram, India as a scientist. He has more than 140 publications to his credit. His name is listed in the world's top two percent of scientists as per a study by Stanford University and Elsevier carried out in the year 2020. He is a recipient of several awards and fellowships including the Young Scientist Award from International Forum on Industrial Bioprocesses (IFIBiop), France, Kerala State Young Scientist Award from Kerala State Council for Science, Technology, and Environment, Prof. S. B. Chincholkar Memorial Award of the Biotech Research Society, India, Elsevier Impactful Research Award, Elsevier Renewable Energy Best Paper Award, Visiting Fellowship, EPFL, Switzerland, Marie Curie Fellow, and others. He is fellow of the International Society for Energy, Environment and Sustainability (ISEES). He is a National Honorary Advisory Board Member of the Centre for Energy and Environmental Sustainability (CEES), India, and Central Office Executive of the Biotech Research Society, India.

Contributors

Kul Chandra Adhikari
School of Molecular and Life Sciences
Curtin University
Bentley, Western Australia, Australia

Rewati Raman Bhattarai
School of Molecular and Life Sciences
Curtin University
Bentley, Western Australia, Australia

Kit Wayne Chew
School of Chemistry, Chemical Engineering and Biotechnology.
Nanyang Technological University
Singapore

Wen Yi Chia
Department of Chemical and Environmental Engineering, Faculty of Science and Engineering
University of Nottingham Malaysia
Selangor, Malaysia

Tjandra Chrismadha
Research Center for Limnology and Water Resources
National Research and Innovation Agency
Bogor, Indonesia

Tasneem Fatma
Department of Biosciences
Jamia Millia Islamia
New Delhi, India

G. Flora
Department of Botany
St. Mary's College (Autonomous)
Thoothukudi, Tamil Nadu, India

Hao Yi Gay
School of Energy and Chemical Engineering
Xiamen University Malaysia
Selangor, Malaysia

Ee Von Goh
Future Food Malaysia & School of Biosciences
University of Nottingham Malaysia
Selangor, Malaysia

Nur Akmalia Hidayati
Research Center for Applied Microbiology
The National Research and Innovation Agency (BRIN)
Jakarta, Indonesia

Pannaga Pavan Jutur
Omics of Algae Group and DBT-ICGEB Centre for Advanced Bioenergy Research, Industrial Biotechnology
International Centre for Genetic Engineering and Biotechnology
New Delhi, India

Nida Jamil Khan
Department of Biosciences
Jamia Millia Islamia
New Delhi, India

Fong Fong Liew
Department of Oral Biology & Biomedical Sciences, Faculty of Dentistry
MAHSA University, Saujana Putra Campus
Selangor, Malaysia

Heli Siti Halimatul Munawaroh
Universitas Pendidikan
Bandung, Indonesia

Asha Arumugam Nesamma
Omics of Algae Group and DBT-ICGEB Centre for Advanced Bioenergy Research, Industrial Biotechnology
International Centre for Genetic Engineering and Biotechnology
New Delhi, India

Wee Jun Ong
School of Energy and Chemical Engineering
Xiamen University Malaysia
Selangor, Malaysia

Der Jiun Ooi
Department of Oral Biology & Biomedical Sciences, Faculty of Dentistry
MAHSA University, Saujana Putra Campus
Selangor, Malaysia

Win Nee Phong
School of Molecular and Life Sciences
Curtin University
Bentley, Western Australia, Australia

Riska Nur Pratiwi
Universitas Pendidikan
Bandung, Indonesia

T.V. Ramachandra
Energy & Wetlands Research Group, Centre for Ecological Sciences
and
Centre for Sustainable Technologies
and
Centre for Infrastructure, Sustainable Transportation and Urban Planning
Indian Institute of Science
Bangalore, Karnataka, India

Roshan Regmi
CSIRO Agriculture and Food, Waite Campus
Urrbrae, South Australia

Mohammed Rehmanji
Omics of Algae Group and DBT-ICGEB Centre for Advanced Bioenergy Research, Industrial Biotechnology
International Centre for Genetic Engineering and Biotechnology
New Delhi, India
and
Department of Biosciences
Jamia Millia Islamia
New Delhi, India
and
Camborne School of Mines and Environment and Sustainability Institute
University of Exeter
Penryn, Cornwall, UK

G. Saranya
Energy & Wetlands Research Group
Centre for Ecological Sciences
and
Centre for Sustainable Technologies
Indian Institute of Science
Bangalore, Karnataka, India

Awalina Satya
Research Center for Limnology and Water Resources
National Research and Innovation Agency
Bogor, Indonesia

Azalea Dyah Maysarah Satya
Department of Chemical and Environmental Engineering, Faculty of Science and Engineering
University of Nottingham Malaysia
Selangor, Malaysia

Ika Atman Satya
Research Center for Computation
National Research and Innovation Agency, (BRIN)
Komplek Cisitu, Bandung, Indonesia

Abhishek Saxena
Diatom Research Laboratory, Amity Institute of Biotechnology
Amity University
Noida, Uttar Pradesh, India

Pau Loke Show
Department of Chemical and Environmental Engineering, Faculty of Science and Engineering
University of Nottingham Malaysia
Selangor, Malaysia

G. Mary Stephy
Department of Botany
St. Mary's College (Autonomous)
Tamil Nadu, India

Doris Ying Ying Tang
Department of Chemical and Environmental Engineering, Faculty of Science and Engineering
University of Nottingham Malaysia
Selangor, Malaysia

Archana Tiwari
Diatom Research Laboratory, Amity Institute of Biotechnology
Amity University
Noida, Uttar Pradesh, India

Ashokkumar Veeramuthu
Center of Excellence in Catalysis for Bioenergy and Renewable
Chemicals (CBRC), Faculty of Science
Chulalongkorn University
Bangkok, Thailand
and
Center for Transdisciplinary Research, Department of Pharmacology, Saveetha

Dental College, Saveetha Institute of Medical and Technical Sciences
Saveetha University
Chennai, India

Gou Rean Wong
Department of Oral Biology & Biomedical Sciences, Faculty of Dentistry
MAHSA University, Saujana Putra Campus
Selangor, Malaysia

Sara Kazemi Yazdi
Department of Chemical and Environmental Engineering, Faculty of Science and Engineering
University of Nottingham Malaysia
Selangor, Malaysia

1 Diatom Biogenic Silica for Drug-Delivery Applications

Abhishek Saxena and Archana Tiwari
Diatom Research Laboratory, Amity Institute of Biotechnology,
Amity University, Noida, Uttar Pradesh, India

1.1 INTRODUCTION

A pharmaceutical formulation known as a "carrier" serves as a drug-delivery system when a drug is successfully delivered to the target site, providing the body of the patient with therapeutic effects (Phogat et al., 2021). Drug-delivery methods made possible by nanotechnology have increased the effectiveness and performance of therapeutic drugs to improve patients' survival and quality of life. Because of their decreased toxicity, improved therapeutic efficacy, drug solubility, and controlled and sustained drug release to the target site, nano-form drug-delivery systems are preferable to classical/conventional drug-delivery methods (Uthappa et al., 2018). Due to their enormous surface area to volume ratio and permeable absorbent network, porous structures have recently attracted a lot of attention. As a result, porous materials that are organized in a hierarchy and systematically are very common in research fields (Yang et al., 2016). In the Earth's crust, silicon (27.2%) is the most common element after oxygen (45.5%) (www.doubtnut.com/question-answer-chemistry/silicon-is-the-second most-abundant-element-272-in-the-earths-after-oxygen-455-carbon-silicon german33102763).

Numerous industries use silicon and its oxide forms because of their unique physical and chemical characteristics. Applications related to biomedicine are proliferating and expanding quickly in many fields (Karaman and Kettiger, 2018). Silica-based nanoparticles are used as excipients in the pharmaceutical industry. Mesoporous silica nanoparticles have recently been developed as a drug-delivery system due to their porous character, which enables them to have a sizable drug-loading capacity that is insoluble in water (Karaman and Kettiger, 2018). Because of its outstanding rigid nature, high biocompatibility, broad surface area, and drug loading capacity, it is generally recognized in the treatment of cancer, dermal formulations, and cartilage regeneration, among other applications (Phogat et al., 2021). Researchers have primarily used silica-based drug-delivery systems that already exist, such as the Santa Barbara Amorphous-15 (SBA-15) and Mobile Composition of Matter No. 41 (MCM-41) with pore sizes of 2–50 nm. However, they have some drawbacks, including the use of synthetic pathways, high cost, the release of harmful compounds into the environment, and the need for highly skilled labor (Katiyar et al., 2006; Pagar et al., 2018; Phogat et al., 2021). In this context, the development of natural-based silica materials that can successfully replace synthetic silica is critical. Diatom biosilica, also known as frustules, is a 3-D nanoarchitecture material that varies from synthetic silica in that it is created by a biological self-assembling process that is ecologically stable, capable of self-healing, and displays characteristics that adapt to environmental changes. It is being studied for various applications in nanofabrication, life sciences, separation routes, catalysis, and so on (Saxena et al., 2022). The advantages, difficulties, and potential applications of diatom biosilica as an effective drug-delivery system are briefly discussed in this chapter.

DOI: 10.1201/9781003436553-1

1.2 TYPES OF DRUG-DELIVERY SYSTEMS

A drug-delivery system is a process of administering an active pharmaceutical ingredient (API) to the target site to have an effective and efficacious therapeutic effect on the body (Kumar et al., 2022). To improve product effectiveness, patient convenience, and compliance, it aims to modify the drug release profile and pharmacokinetic characteristics. It also aims to deliver the medication at a rate determined by the body's needs during the treatment period and direct the active ingredient to the site of action. A medicine's effectiveness can be significantly impacted by how it is delivered. By developing a variety of drug-delivery methods, it is now possible to more effectively control the pharmacokinetics, pharmacodynamics, toxicity, immunogenicity, and effectiveness of drug-delivery systems (DDSs) (Kumar et al., 2022). Inactivation by the gastric juice, metabolism before reaching the target cell, and first-pass metabolism in the lung, liver, and small intestine are only a few of the problems of the conventional drug-delivery system. There are far too many negative responses. It is a costly, prolonged, multi-dose therapy. This can be overcome by increasing the rate of drug administration, reducing biodegradation, using time-release drugs, and using site-specific targeting. Only injectable drugs should be administered orally; they should be inexpensive, strong, time-released, or self-triggering formulations. By figuring out the optimum way to administer a specific medicinal molecule, one can "optimize" how well medicine works in the body (Bhagwat and Vaidhya, 2013).

Drugs can enter the body through a variety of channels (Figure 1.1) which are categorized according to their "starting point," or the place where the medication is administered. Each route has a unique set of benefits and drawbacks.

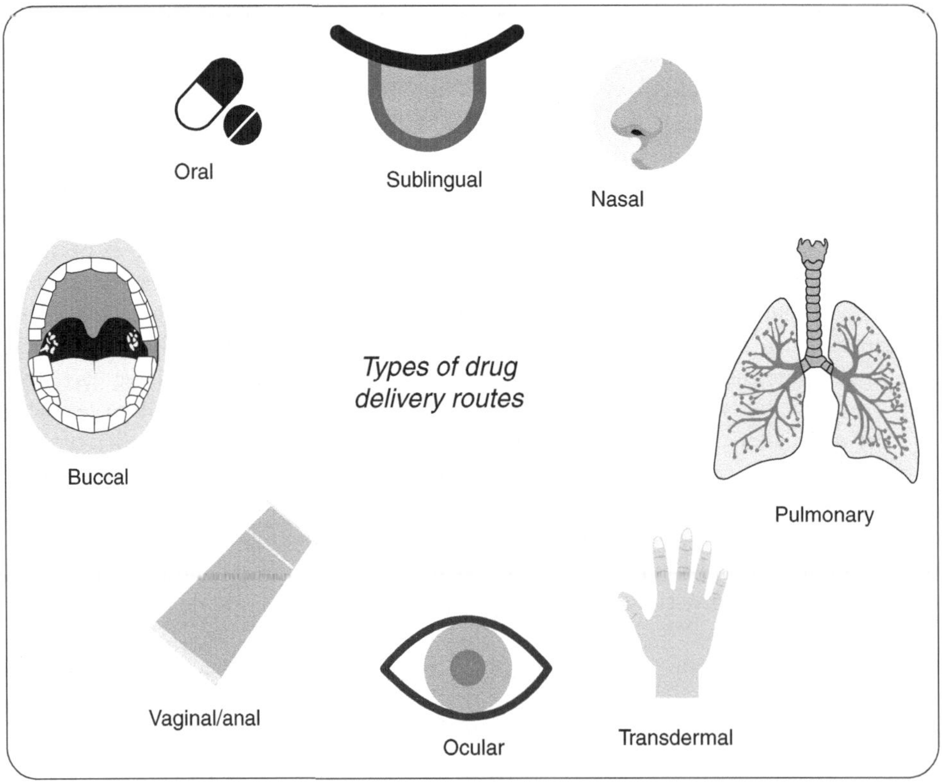

FIGURE 1.1 An overview of various drug-delivery routes.

1.3 BUCCAL DRUG DELIVERY

The distribution of drugs through the buccal mucosa (cheek lining) is referred to as "buccal drug delivery" (Hao and Heng, 2003). While this mode of delivery eliminates first-pass effects, including quick drug uptake and liver conversion into inert molecules, it does present a "difficult" barrier to drug absorption, particularly for larger biopharmaceuticals. Delivery is currently restricted due to the ease with which small molecules with lipophilic properties can penetrate the membrane (Smart, 2005). For extended-release drug administration (in which the drug is delivered in a controlled manner over a long period), the buccal route is frequently used; formulations that may adhere to the mucosa are typically favored. Several formulations have been developed for buccal delivery, including pills, gels, lozenges, and patches (Smart, 2005).

1.4 NASAL DRUG DELIVERY

Nasal drug delivery refers to the process of administering a drug through the nasal cavity. Local illnesses of the upper respiratory tract, such as allergic rhinitis and nasal congestion, are frequently treated using nasal spray medications (Illum, 2012). However, this administration strategy was used for systemic distribution of small-molecule drugs, such as the migraine therapy olmatriptan (Zomig®), for instance, when a quick onset is required (Zomig, 2019). Due to the nasal mucosa's high level of vascularization, first-pass metabolism, as with buccal administration, can be avoided and rapid delivery to the systemic blood circulation is possible. Drugs can be administered nasally using liquid or (less frequently) powder forms (Kublik and Vidgren, 1998).

1.5 OCULAR DRUG DELIVERY

Ocular drug administration has emerged as a challenging topic for drug-delivery experts due to the eye's distinctive structure and physiology, where both static, dynamic, and metabolic ocular barriers hinder medicine absorption through the eye. Drug delivery to particular eye regions can be accomplished using several different methods. By identifying specific efflux and inflow transporters in the eye and modifying pharmaceuticals to target these transporters, researchers have been able to overcome challenges associated with administering drugs to ocular tissues (Gaudana et al., 2010).

1.6 ORAL DRUG DELIVERY

Oral drug delivery is by far the most popular and frequently chosen route of drug administration due to its non-invasiveness, ease of use, cost-effectiveness, and the gastrointestinal tract's (GIT's) high absorption capacity (Inamuddin, 2018). The water solubility of the therapeutic molecule in the GI system should be assessed to see if modifications are required to boost bioavailability to ensure the effectiveness of oral drug administration. Oral drug delivery has drawbacks when it comes to its suitability for specific patient populations, such as children, the elderly, and individuals with cognitive impairment (Inamuddin, 2018).

1.7 PULMONARY DRUG DELIVERY

The administration of a drug through the mouth and into the airways is referred to as pulmonary drug delivery. Local lung conditions can be effectively treated with inhaled medications. Due to the large absorptive surface area and highly permeable membrane of the alveolar region, pulmonary medication delivery has also been investigated as a potential route of administration for systemic disorders in more recent years. Additionally, dietary challenges and interpatient metabolic diversity had little impact on pulmonary delivery (Labiris and Dolovich, 2003).

1.8 SUBLINGUAL DRUG DELIVERY

When a drug is administered under the tongue, it is absorbed into the circulation via the ventral surface of the tongue and the floor of the mouth. This process is known as sublingual drug delivery. Because of the rapid rate of sublingual absorption, the impact starts right away. Additionally, hepatic first-pass metabolism is avoided with this delivery strategy. It is not desirable for it to interfere with conversing, eating, or drinking. Additionally, due to vasoconstriction of the arteries, smoking decreases absorption and, as a result, the effectiveness of the medication; therefore, usage in smokers is not advised (Narang and Sharma, 2011).

1.9 TRANSDERMAL DRUG DELIVERY

Transdermal drug delivery involves applying a formulation to healthy skin to distribute a drug systemically (Alkilani et al., 2015). The stratum corneum receives the drug first, followed by the deeper epidermis and dermis, and then dermal microcirculation, where it is finally absorbed systemically (Donnelly, 2012). Two important benefits are its non-invasiveness and adaptability for patients who are unconscious or throwing up.

1.10 VAGINAL/ANAL DRUG DELIVERY

Vaginal/anal pharmaceutical administration routes provide a speedier onset of effect and improved absorption when compared to the oral route. Laxatives and sedatives are examples of rectal drugs that have local or systemic effects (e.g., analgesics when other routes are contraindicated). Drugs are administered via the vaginal route, which avoids first-pass metabolism and is unaffected by stomach side effects (Alexander et al., 2004). For the delivery of hormones and the treatment of women's health issues, the vaginal route is frequently investigated. Alternatives to vaginal formulations include hydrogels, tablets, pessaries, and suppositories (Sahoo et al., 2013).

1.10.1 Targeted Drug Delivery

The ability to focus the administration of a medical substance to its intended target, increasing effectiveness and reducing side effects, makes targeted drug-delivery systems particularly interesting. Targeted drug delivery is a technique for administering a drug so that it is concentrated much more in some areas of the body than in others. The two general categories of a drug's targeting mechanism are (a) passive targeting and (b) active targeting.

1.11 DRUG-DELIVERY VEHICLES

Once inside the body, drug molecules with low bioavailability require defense against enzymatic and acid-catalyzed degradation. Due to insufficient bioavailability, the pharmaceutical industry rejects about 40% of novel active pharmaceutical ingredients (APIs) (Svenson, 2004). To address this, carrier systems that improve bioavailability and protect drug molecules from degradation have been created. In this section, we will look at three different methods of carrier-based distribution.

1.12 NANOPARTICLES

Sizes of nanoparticles range from 100 to 500 nm (Rizvi and Saleh, 2018). By altering their characteristics, it is possible to boost bioavailability, decrease clearance, and increase stability, making them ideal "carriers" for delivering a particular treatment to its particular target area. Nanoparticles are more soluble and, hence, have enhanced bioavailability due to their small size and larger surface area. Their ability to cross the blood–brain barrier (BBB), enter the pulmonary system, and pass through endothelial cell tight connections increases their appeal as drug transporters (Rizvi and Saleh, 2018). Three common ways for nanoparticles to enter the body are through injection, inhalation, or ingestion. It is essential to first determine how the body will react to the exterior particulate matter to develop the "ideal" nanoparticle-based delivery method. By altering the surface properties of the particle, this problem can be solved. If the body perceives a nanoparticle as "foreign," the body's normal immune response will cause it to be expelled from the body. Plasma protein binding, which typically boosts clearance, is blocked by adding polymer complexes to the surface. For several diseases, including cancer, neurological disorders, and acquired immune deficiency syndrome (AIDS), nanoparticles have been studied as drug-delivery systems (Rizvi and Saleh, 2018).

1.13 LIPOSOMES

Sphere-shaped vesicles known as liposomes are constructed of one or more phospholipid bilayers (Akbarzadeh, 2013). These vesicles provide a desirable delivery strategy due to their adaptable biochemical and physiochemical properties that make it simple to modify them (Sercombe et al., 2015). Liposomes are good drug carriers because of their rare capacity to contain both lipophilic and hydrophilic molecules. They can self-assemble, carry substantial pharmacological loads, and are biocompatible, among other advantages. They are "pharmacologically inactive," which means they have a low level of toxicity because they are made of natural phospholipids. There are four different categories of liposomes: conventional, theranostic, PEGylated, and ligand targeted. It is important to note that whereas liposomes can be easily modified, this also enhances the likelihood of detection and clearance (Sercombe et al., 2015).

1.14 MICROSPHERES

Microspheres are tiny (1–1000 μm) spherical particles made of linear polymers (Varde and Pack, 2004) that are defined as "free-flowing powdered drug-delivery devices" (Alli et al., 2013). They offer various benefits when employed as a drug-delivery system, including being biodegradable, biocompatible, and easily administered. They can be used to deliver drugs both locally and specifically. The polymer matrix of the microspheres is permeable to medicine. Microspheres can be created using both synthetic and natural polymers (Bhagwat and Vaidhya, 2013). Based on whether they are biodegradable, synthetic polymers are further divided into several groups. The longevity of non-biodegradable microspheres in the body increases the possibility of harm over time. Because they don't pose the same risk, biodegradable polymers are better suited for parenteral applications (Bhagwat and Vaidhya, 2013). Resealed erythrocytes, niosomes, and monoclonal antibodies are examples of carrier-based approaches.

1.15 THE PLACE OF DIATOMS IN THE PHARMACEUTICAL INDUSTRY

The global economy is changing toward green and sustainable development in order to tackle health concerns and climate change issues. Science has recently been startled by the commercial application of diatoms since they provide enormous promise for various industrial applications. They can survive in adverse environments because of their diverse pore architectures and distinct cell walls, making them an attractive alternative for industrial products. Diatoms have all of the characteristics required to become an alternate raw material for pharmaceuticals. However, despite the fact that diatoms grow at a rate of almost 132 MT dry diatoms ha^{-1} per year, little attention has been paid to them (Wang and Seibert, 2017). According to the Tufts Centre for the Study of Medicine Development, the cost of bringing a new medicine to market is around 2.6 billion dollars (DiMasi et al., 2016). Furthermore, existing drug-delivery technologies have low solubility, poor biodistribution, low selectivity, early degradation, and undesirable pharmacokinetics (Aw et al.,

2011a,b). As a consequence of these constraints, pharmaceutical companies and scientists have been developing novel drug-delivery systems that improve bioavailability while reducing unwanted side effects associated with the performance of existing treatments. Without a question, current biomedical technology has improved patient health, but human society is seeking ways to increase quality of life by employing ecologically friendly tactics. As a result, boosting diatom research in the pharmaceutical sector by extracting the most from diatom frustules is the hope and other option for prolonged and regulated drug release to the target organs. The pore size of diatom frustules' 3D nanoarchitecture spans from nanometers to micrometers. The complex frustule properties, such as high specific surface area, strong thermal stability, high biocompatibility, and changeable surface morphology, have caught the attention of researchers for their potential use in gene and drug delivery even though liposomes, nanogels, and carbon nanotubes are well-known drug-delivery tools (Chandrasekaran et al., 2014; Cicco et al., 2015; Ragni et al., 2017). These qualities are sufficient to research different silica-based materials with lower costs for the biomedical industry (Mishra et al., 2017; Terracciano et al., 2018). Depending on the environment and silicon absorption efficiency, the homogeneity, space, and complexity of diatom frustules can vary (Knight et al., 2016). The role of silicic acid transporters, silica transport, polymerization of silica monomers, and other aspects of the biosilicification process have all been thoroughly documented, which can be utilized by the healthcare and nanotechnology industries (Martin-Jézéquel et al., 2000; Knight et al., 2016). Several diatom species, including *Thalassiosira weissflogii* (Aw et al., 2011a) and *Coscinodiscus concinnus* (Gnanamoorthy et al., 2014), are potential drug carriers because of their amorphous nature and morphology. Additionally, many studies have shown that diatom microcapsules are effective delivery systems for water-soluble and poorly soluble medications that can be employed for implantable as well as oral applications (Aw et al., 2011a; Ragni et al., 2017). Due to their distinctive qualities, such as uniform pore structure, chemical inertness and biocompatibility, non-toxicity, ease of transport, filtering efficiency, and selective drug delivery, diatoms are a suitable model for drug-delivery applications in the pharmaceutical industry, as shown in Figure 1.2. Consequently, previous research has paved the way for innovative drug-delivery systems based on renewable materials.

1.16 THE DIATOM FRUSTULES

Diatoms are unicellular (single-celled) photosynthetic organisms that exhibit a normal eukaryote structure within their silica walls. They have complicated skeletons that can be elongated, with bilateral plane symmetry, or spherical, with a radial symmetry. Many diatoms are asymmetrical, but they always fit into one of these two groups. A diatom's skeleton, or frustule, is formed of exceedingly pure silica covered with an organic coating. This skeleton is separated into two sections, the epitheca covers the hypotheca like a box lid or Petri dish lid. Both sections of a frustule can be severely perforated. Pennate diatoms have a large slit, known as the raphe, that runs down the long axis. The live diatom secretes mucilage through the raphe, which it can use to cling to a substrate or move by gliding over it (Kooistra and Göran Pohl, 2005). Figure 1.3 shows a schematic representation of the valve of both centric and pinnate diatoms.

The valves are often the components with the most elaborate biomimetic designs. The valve face often has a thicker midrib or center section from which ribs extend out to the outside rim. Most species have a single layer for their valves, but some, especially larger species, have chambered valves comprised of an outer and an inner layer linked, for instance, by a honeycomb matrix. A small aperture or a tiny mesh of pores is frequently used to connect the inner layer to the interior. Large species usually use thicker ribs, buttresses, flying buttresses, cross beams, and other architectural features to strengthen their internal valves. Frustules must be resilient enough to withstand a herbivore's onslaught (Hamm et al., 2003). However, this is not its only use; it also acts as a means of moving objects. Between each rib, there are rows of pores that permit communication with the environment. Near the valve apices, many species have fields of closely spaced pores. These pore

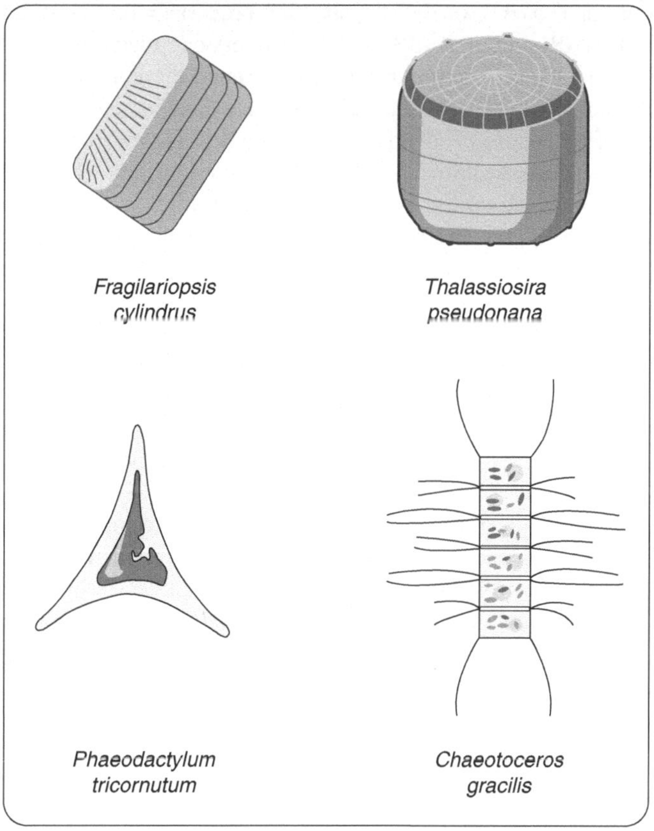

FIGURE 1.2 Types of diatoms utilized in pharmaceutical applications.

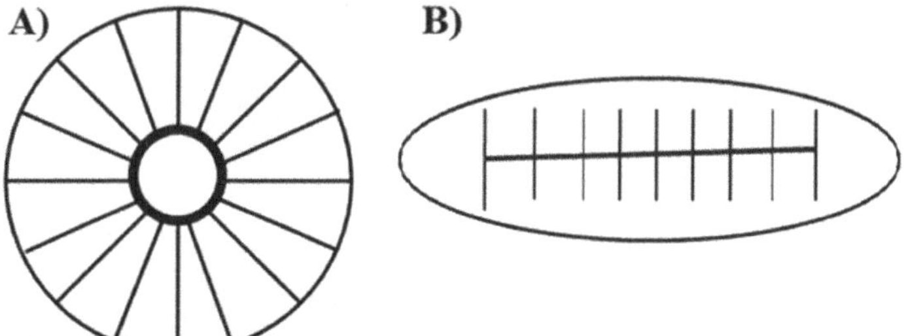

FIGURE 1.3 Schematic organization of: (A) a valve of a centric diatom in valve view; (B) a valve of a pinnate diatom.

fields generate gooey polysaccharides, which, when combined, form mucilage attachment pads. These fields are usually very dense and enclosed by support structures since they may endanger structural integrity.

Most diatoms have tubular processes that run through the valve face. Labiate processes (also known as rimoportulae), which most likely excrete and/or absorb organic molecules, are the most prevalent. A labiate is a term for the tube's side-folded internal end, which resembles a lip. Another type of tub found only in *Thalassiosirales*, the strutting process (= fultoportula), is involved in the excretion of chitin fibers (Round et al., 1990). These fibers can reach into the surrounding space or connect sister cells in a chain. The minuscule struts around these tubes anchor the tube into the surrounding cell wall and serve as structural reinforcements against the tensile stress of the chitin fibers. The third sort of procedure is called a raphe, which consists of two slits that enable active cell movement over the substrate utilizing a low-cost traction system (Edgar and Pickett-Heaps, 1984). The slits often border the midrib and are ">" shaped in sections through the cell wall. The valve's structural fault is the raphe. Several species have a collection of bridge-like structures called fibulae that span the raphe on the inside, supporting the valve, and preventing it from breaking at this weakness. Additionally, the ends of the slots are frequently bent sideways or formed like drops to reduce breaking.

Frustule components have been preserved in remarkable detail over millions of years (e.g., Gersonde and Harwood, 1990; Harwood et al., 2004; Suto, 2006; Ishii et al., 2011). Multipolar centric first appears in the Jurassic, then pennate in the Late Cretaceous, and finally raphid pennate in the Paleogene. Radial centric first appear in the Jurassic (Rabosky and Sorhannus, 2009). This sequence supports conclusions drawn from an evolutionary analysis of the diversity already present. Since several cell walls, designs, decorations, and linking techniques to form chains do not exist in the existing diversity, the fossil record is crucial for understanding their architecture. For instance, the historic diatom structures presented by Round et al. (1990) and Gersonde and Harwood (1990). The biological niches of these ancient diatoms, including the methods of chain formation, can be statistically recreated and inferred based on what is known about analogous but unrelated modern diatoms. The girdle bands and valves are frequently preserved in such great ultrastructural detail.

1.17 PURIFICATION AND FUNCTIONALIZATION OF FRUSTULES

The microparticles of naturally occurring diatoms contain contaminants such as organic material, clay, terrigenous particles, and volcanic gas as a result of aging conditions and the environment. All pollutants must be eliminated to obtain diatoms in their purest state, also known as "isolated silica micro-shell" (Yang et al., 2016). A study found that using a crusher to crush the diatoms and piranha solution to remove contaminants would yield pure diatoms (Rosi et al., 2004). This is the method that is most frequently used among the several techniques for harvesting and purifying diatoms. By removing organic layers and enhancing the surface's hydroxyl groups, the piranha solution enhances surface modification (Karakoy et al., 2014). This technique has several advantages, including simple filtration, eco-friendliness, and a quick procedure (Rosi et al., 2004). This approach may also be used to purify water by extracting and eliminating harmful metal ions. Figure 1.4 shows the purification, functionalization, and drug loading in a diatom cell.

Gordon and Drum (1994) were the first to investigate the usage of diatom frustules for a variety of applications in nanotechnology. Diatoms are the most abundant category of algae species, accounting for around one-fifth of the global total (Phogat et al., 2021). They are an ideal option for desired applications due to their self-assembled complex 3D porous structure (Rea and Stefano, 2019). Diatom frustules are highly valued as an alternative to manufactured silica because they are composed of amorphous silica that has a significant number of hydroxyl groups (–OH) bonded to their surface due to the fact that diatoms form the frustules in an aquatic environment and they constantly remain there (Phogat et al., 2021). These microshells can be chemically altered, for example, by modifying a glass slide used in immunoassays. Purification of frustules before functionalization is strongly advised for biomedical applications to ensure that the surface is biocompatible and safe. Many organic and inorganic substances (Al_2O_3, FeO_3, $CaCO_3$, CaO, K_2O, Na_2O, MgO) from the environment hide frustules in nature. To make the surface pure and usable, these chemicals must be eliminated before any surface alteration (Tramontano et al.,

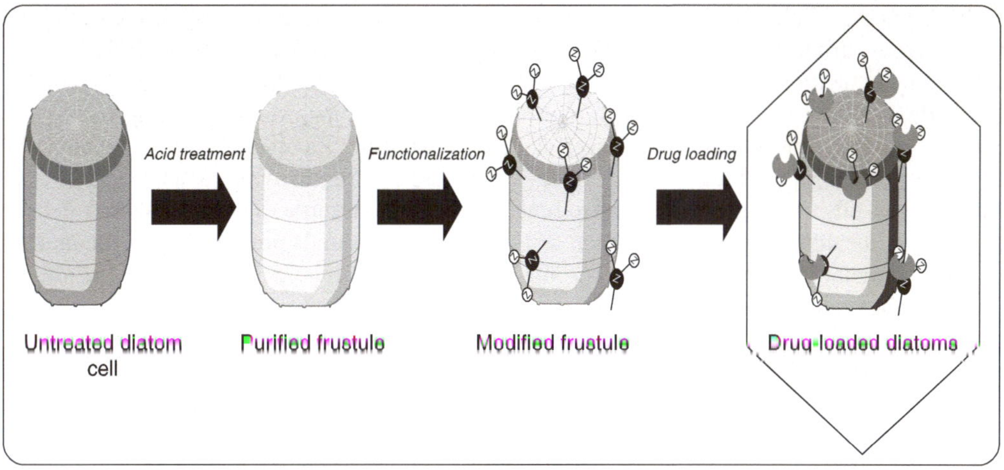

FIGURE 1.4 Schematic overview of purification, functionalization, and drug loading in a diatom cell.

2020). Diatomite, a fossilized diatom skeleton, and living diatom frustules were treated with acid, i.e. piranha solution (2 M H_2SO_4, 10% H_2O_2), and then rinsed many times with distilled water to eliminate any residues of sulfuric acid before use. The composition of frustules changes during purification, exposing the highly reactive silanol groups (Si-OH), which aid in the binding of biomolecules to biosilica walls. Hence, purification is highly desirable before functionalization and is widely utilized in drug-delivery applications (Utthappa et al., 2018; Phogat et al., 2021). It is possible to further engineer the immobilization strategy of diatom frustules, resulting in modified orientation and yield.

1.18 DIATOMS AS EFFICIENT DRUG-DELIVERY CARRIERS

Due to their abundance and amorphous nature, specific surface area, modifiable surface, low density, mechanical resistance, hollow and large internal space with nano-scale porosity, high chemical inertness, excellent optical and photonic properties, thermal stability, high permeability, and non-toxicity, diatoms are effective drug-delivery systems (Khan, 2017; Maher et al., 2018). Because of their nanostructured layout, they may be employed in many applications, including liquid filtration, DNA purification, immunoprecipitation, photonics, and drug administration (Terracciano et al., 2018). Because of their specific traits and outstanding structural morphology, species such as *Aulacoseira* sp., *Coscinodiscus concinnus*, *Thalassiosira weissflogii*, *Thalassiosira pseudonana*, and *Nitzschia* sp. have been identified as possible drug nanocarriers (Karakoy et al., 2014; Maher et al., 2018). Yang and colleagues used *Aulacoseira* sp. diatoms in drug-delivery applications (Uthappa et al., 2018). Iron oxide nanoparticles that self-assemble and accumulate cationic material onto anionic material when electrostatically driven were used to modify the surface of a diatom (Uthappa et al., 2018). The indometacin drug-loading capacity was 0.28 g/g, with a burst release (6 h) of around 60% and the remaining drug released over two weeks. The sluggish drug release was caused by the drug being held in the unoccupied interior structures of the diatom micropores, as shown by zero-order kinetics (Karakoy et al., 2014). Due to the weak connections between the drug molecules and the silica surface, indometacin showed a 65–70% burst release within 2 hours without surface change (Uthappa et al., 2018). Numerous in-depth studies on surface modification utilizing the drugs indometacin and gentamicin, which are hydrophobic and hydrophilic, respectively, have been conducted. It was discovered that hydrophilic surfaces, rather than hydrophobic surfaces, are more prone to drug-loading augmentation and regulated release of pharmaceuticals (Mishra et al., 2018). Hydrophilic surfaces like epoxy-rich glycidoxypropyltrimethoxysilane (GPTMS), 3-trimethoxysilane, and amino-rich APTES increase the loading of hydrophobic drugs. Polyhydroxyalkanoates (16-PHA), OTS, and polyethylene glycol (PEG)-silane, three diatoms with hydrophobic surfaces, had a reduced drug-loading capacity and quicker drug release (Mishra et al., 2018). Multiple functional groups led to a shift in release profiles of 6–15 days and a 15–24% increase in drug-loading capacity (Yang et al., 2016). Various other strategies were used to functionalize diatom surfaces for desired drug-delivery applications, including organosilanes such as N-(3-(trimethoxysilyl) propyl) ethylenediamine, 3-aminopropyltriethoxysilane (APTES), 16-phosphonic-hexadecanoic acid phosphonic acids, and 2-carboxyethoxyphosphonic acid (Aw et al., 2012). Organosilanes regulate the hydrophilic and hydrophobic properties of diatom surfaces. A significant regulated release of the drug was found with greater drug loading (Zaharudin et al., 2020). With indometacin, APTES-DE was investigated with a greater drug-loading capacity of 28.5 percent for GO-DE (diatomaceous earth) and a lower drug-loading capacity of roughly 15.8 percent for the control sample. The drug release profile with DE modified with APTES at pH 3.5 exhibited a 14-day sustained release profile, followed by a 37-day prolonged release profile (Kumeria et al., 2013; Bariana et al., 2013; Dhas et al., 2019). Persistent drug release was caused by hydroxyl group, hydrophobic interactions, hydrogen bonding, and linkages (Maher et al., 2019). For an hour (Khan, 2017), mesoporous materials, such as SBA-15, demonstrated poor control over the continuous release of indometacin to the target location. The APTES-DE complex has a decreased drug-loading capacity due to weak van der Waals forces of contact (Cha et al., 2014). This demonstrates that GO-modified diatoms provide continuous drug release, which is a critical element in drug-delivery applications. The cytotoxicity of pure frustules and APTES functionalized frustules against H1355, epidermoid malignant cells of the lung (Yang et al., 2016), revealed that functionalized frustules are particularly efficient against cancer cell lines. Using a fluorescent probe, a strong conjugation between the drug molecule and the frustules was generated, resulting in high and efficient cellular absorption (Uthappa et al., 2018). This has cleared the way for novel anticancer or antitumor compounds in the future. Magnetized diatoms, which are diatoms that have been functionalized with iron oxide, have also been shown to be effective against malignant cells in an animal model using a fluorescent probe (ZW800 Dye) (Khan, 2017; Aw et al., 2012). They also demonstrated that magnetized diatom frustules are biodegradable and have little toxicity. Diatomite with a wavelength of 450 nm was effectively used as a nanocarrier for the transport of small interfering RNA (siRNA) within H1355, human epidermoid malignant cells. Before this, it was shown that nanoparticles such as liposomes and gold are incapable of crossing the cell membrane (Rudzinski et al., 2010). Diatomite and siRNA were biologically conjugated to silence harmful genes. After 72 hours of exposure to 300 g/mL diatomite, minor cytotoxicity was reported, confirming the absorption of nanoparticles by malignant cells. This type of conjugation might potentially be used in gene silencing to reduce the expression of target proteins in malignant cells (Lang et al., 2013). Chitosan-polyguluronate (particle size 110–430 nm) conjugation with siRNA is used for gene silencing in HELA cells and HEK293FT with minimal toxicity (Rudzinski et al., 2010; Todd et al., 2014).

To create a solidified oral drug, solid self-emulsifying phospholipid suspension (SSEPS) in mesoporous diatoms was used for the first time (Mishra et al., 2017). Carbamazepine (CBZ), a water-insoluble medicine, was absorbed into diatoms using two methods: (i) the drug was gently combined into the drug, and (ii) ethanol evaporation—the superior approach. In this investigation, the medication was loaded under a liquid condition, which sped up the pace of dissolution because the sample had to enter the diatoms' pores through the partial adsorption procedure (Yang et al., 2016). It was shown that SSEPS accelerated the water-insoluble medication's disintegration. Streptomycin, a hydrophilic medication, was used by Gnanamoorthy et al. (2014) in oral drug-delivery applications, using the marine diatom species *Coscinodiscus concinnus* as a nano-drug carrier. The dispersion was biphasic, with a first burst release in 6 hours and second-phase drug releases from the inner diatom pores for around 7 days. The drug loading efficiency was 33% (Uthappa et al., 2018; Gnanamoorthy et al., 2014). Diatomite was functionalized with aluminum sulfate [$Al_2(SO_4)_3$] in an aqueous solution under stirring conditions, then progressively added to a base sodium hydroxide (NaOH) to get a homogeneous mixture, which was filtered and dried. Finally, with sodium diclofenac, a high drug absorption of around 250 mg/g was observed (Trofimov et al., 2019; Janićijevića et al., 2015). For the first 8 hours, just 18% of the medication was released, followed by 37 percent release after 8 hours. In comparison to the control, 45 percent of the medication was released within the first 8 hours and was sustained after 8 hours. The "non-fickian model" (Yang et al., 2016) was used to show drug diffusion. The loading of ciprofloxacin in *Thalassiosira weissflogii* with 2,2,6,6,-tetramethylpiperidine-1-oxyl radical (TEMPO), a potent reactive oxygen scavenger, was studied in bone cell development (Cicco et al., 2015). The first drug release was quick, and the late release demonstrated sustained release for 7 days in a row, which is a useful therapy in dental surgery or orthopedics, thus boosting bone cell adhesion, increasing cell proliferation, and assisting in the prevention of inflammation (Uthappa et al., 2018). *Thalassiosira pseudonana* was selected to destroy malignant cells by attaching to the streptococcal protein G. In this study, IgG was clamped on the diatom's surface to actively link the cell's targeting antibodies (Delalat et al., 2015). This specifically targeted and destroyed neuroblastoma and B-lymphoma tumor cell lines. Diatoms were functionalized with APTES in a phosphate saline buffer at 7.4 pH for the loading of Sorafenib, a low water-soluble medication, and the drug release profile was investigated. This improves toxicity as well as cellular absorption. Diatomaceae could be used to enhance chemotherapy. Jawalkote et al. (2015) used two techniques to treat the diatom *Nitzschia* with iron oxide microparticles for curcumin delivery, both curcumin on diatom microshells (CDM) in situ and curcumin-loaded magnetic diatom micro shells (CMDM)-ferrofluid. Since CMDM-ferrofluid produced better results than CDM in situ, it was proven to be a better strategy and may now be considered a cutting-edge choice for more practical applications in magnetic hypothermia. The numerous diatom species used in medication delivery applications are listed in Table 1.1.

1.19 BENEFITS OF DIATOM BIOSILICA OVER TRADITIONAL DRUG-DELIVERY SYSTEMS

To address common therapeutic problems, several mesoporous silica particles have been created and exploited in medicinal applications. The synthesis of porous nanomaterials requires both bottom-up strategies, such as self-assembling driven chemistry, electrochemical etching, and sol–gel synthesis, as well as top-down strategies, such as standard or electron beam lithography, nano printing, template molding, and atomic layer deposition (Rea et al., 2017). Economically speaking, all of these techniques need expensive tools and materials as well as highly skilled labor, making mass production unappealing. Because diatoms are easily available, affordable, and environmentally friendly materials, employing diatom biosilica is advantageous over using conventional drug carriers (Gordon et al., 2009). Invading every watery habitat, diatoms are single-celled eukaryotic photosynthetic algae that are crucial to the persistence and development of life on Earth. The cell walls of diatoms, also known as frustules, are the most spectacular example of 3-D nanoarchitecture seen in nature (Mann, 1999). There are about 200,000 estimated species and 200 live genera of diatoms, all of which have distinctive sizes and morphologies (ranging from 2 μm to 200 mm) (Saxena et al., 2021). Based on frustule symmetry, diatoms are divided into two groups: pennates, which are bipolar, symmetrical, and have a longitudinal axis parallel to the plane of symmetry, and centric, which are radially symmetrical on an axis that passes through the center of the cell (Guiry, 2012). Diatom frustules are typically bipartite structures with two overlapping valves termed thecae, regardless of shape. Sometimes compared to a Petri dish or pillbox, the epitheca (top section) and hypotheca (bottom half), are characterized by a series of connecting bands (girdle bands or cingula) (Hildebrand et al., 2008; Battarbee et al., 2001). Pore sizes and

TABLE 1.1
List of Various Diatom Species, Their Functionalization, and Drug Delivery Efficiency

S. no.	Diatom	Functionality	Drug	References
1.	*Aulacoseira* sp.	Dopamine-iron oxide; hydrophobic and hydrophilic silane (organosilanes and phosphonic acid); graphene oxide; SEEPS (solid self-emulsifying phospholipid suspension); oligo(ethylene glycol)methacrylate	Indometacin, gentamicin, mesalamine, pednisone, carbamazepine, levofloxacin	(Bariana et al., 2013; Zhang et al., 2013; Delalat et al., 2015; Losic et al., 2010; Aw et al., 2012; Kumeria et al., 2013; Milovic et al., 2014; Vasani et al., 2015)
2.	*Coscinodiscus* sp.	NA	Streptomycin	(Gnanamoorthy et al., 2014)
3.	Diatomite	Aluminum sulfate hexadecahydrate	Diclofenac sodium	(Janicijevic et al., 2014; Janicijevic et al., 2015)
4.	*Thalassiosira weissfloggi*	APTESTEMPO (2,6,6,tetramethylpiperidine Noxy)	Ciprofloxacin	(Cicco et al., 2015)
5.	Calcined Diatomite	APTES-PEGCPP; APTES Sulpho GMBS	Sorafenib; siRNA	(Terracciano et al., 2015; Rea et al., 2014)
6.	*Nitzschia* sp.	CMDM–F and CMDM-I	Curcumin	(Javalkote et al., 2015)
7.	*Thalassiosira weissfloggi*	NA	Curcumin	(Saxena et al., 2022)

patterns span from nanometers to micrometers and are arranged hierarchically in the cribellum, cribrum, and foramen, among other porous structures (Losic et al., 2007). In terms of the number of silicified structures they can produce, and the amount of biogenic silica produced globally, these unicellular algae are the most prevalent species involved in the biosilicification process (Poulsen et al., 2003). Silica deposition vesicles (SDVs) are particular intracellular compartments where biosilicification occurs. Certain silicic acid transporter proteins move soluble orthosilicic acid $Si(OH)_4$—a type of silicon—from the environment into the cytoplasm of the cell (SITs). Ten transmembrane domains make up these proteins, which may bind $Si(OH)_4$ directly through glutamine sections to allow it to enter SDVs. In the SDV reaction vessel, $Si(OH)_4$ is transformed into a silica network (SiO_2) by a combination of long-chain polyamines (LCPAs), silaffin proteins, and polyanionic silacidin peptides (Kröger et al., 2002). Diatoms are simple to produce in large quantities since their growth requires sunlight and inorganic salts as nutrition (Bowler et al., 2010). Diatomite, which was created by a million years of the fossilization process of dead algae and is being mined, is a less expensive source of diatom silica. The development of synthetic methods to produce new silica-based materials under mild reaction conditions has also been prompted by numerous investigations into the biomineralization process of diatom silica nanofabrication (Kröger et al., 1999). Diatom biosilica-related proteins and other chemicals have been carefully identified and characterized. In particular, silaniffs and silacidins have been extracted from the cell walls of diatoms and successfully used to create in vitro silica nanostructures such as spheres, rods, and platelets (Sumper et al., 2004).

1.20 CHALLENGES AND OPPORTUNITIES

The pharmaceutical industries have several obstacles in the creation and operation of medicine that works on a target organ. For example, it requires a high level of skill in its manufacture, administration, and storage, which can result in the sudden removal from the targeted system, at the target site accumulation of the drug can steer the production of toxins, it can spark an unwanted immune response that can be damaging to intravenous-directed carrier systems, it can cause redistribution and dispersal of the drugs, and it requires particularly advanced technology for its formulation (Jain, 2008). There is no defined pattern for the optimal release and variation in the sustained release can lead to confusion amongst physicians; the sustained drug-delivery system involves chemical alteration of drugs that can modify their properties such as solubility, dispensation, and pharmacokinetics (Muro, 2012). Diatom frustules are a natural biocompatible biomaterial with high promise in pharmaceutical sectors in the future. They are gaining popularity among nanotechnologists due to their exceptional 3D surfaces. Diatoms have been recommended as the next-generation biomaterial for smart drug-delivery vehicles because of their highly organized structure and form, vast surface area, great mechanical resilience, biodegradability, and biocompatibility (Delasoie et al., 2019). A suitable functionalization technique enables the transport of an insoluble medication to the target region without precipitation. Pharmaceutical companies must focus on increasing medicine bioavailability by researching the non-toxicity properties of biosilica for a variety of medicinal applications. Indeed, this will be accomplished by doing more research on animal models, which will undoubtedly pave the way for the prevention of serious diseases (Delasoie et al., 2019).

1.21 FUTURE DIRECTIONS

Diatoms are the primary producers of oxygen and biomass, thus they make a significant contribution to the environment (Li et al., 2017). They hold great therapeutic potential in fields such as cancer treatment, dermal formulation, cartilage regeneration, cell labeling, biosensors, bio-photonics, and drug delivery (Bharti et al., 2015). The evolution of form and the genome responsible for its development is a crucial biological question (Almqvist et al., 2001). This intracellular investigation will both show off diatoms' enormous potential and widen their range of uses as a very effective drug-carrier system for medical purposes (Brewer et al., 2007). Due to its excellent biocompatibility for gene transfer and simplicity of customization, diatom silica has been suggested as the best option for treatments. It can also provide a completely new program for magnetic imaging and non-viral gene delivery (Brewer et al., 2007).

Diatomite that has been fossilized and is available from the mining industry can be used to make nanoparticles including silica microcapsules (Diatoms of North America, 2020). It promises to be a strong and stable substitute for synthetically generated silica for drug administration due to its distinctive features, low cost, and biocompatibility (Venkatesan, 2015). They are the best delivery system for medications and drug formulations in the therapeutic industry, according to numerous studies. In contrast to conventional drug-delivery methods, these well-implemented systems only require on-site release and provide sustained drug release for a longer time (Lee et al., 2015). A more clinical study is required because it has recently been demonstrated that diatom frustules can be a potent choice for the steady release of medications. A good drug-delivery system should break down into non-toxic parts so that the body can swiftly eliminate it (Liu et al., 2016). Diatom provides new biomaterial routes with a competitive edge over produced silica material thanks to its naturally driven, self-assembling 3D silica cell wall (Uthappa et al., 2018). As a result, it could serve as a brand-new, inexpensive scaffold for a variety of biological applications. The process of turning diatom silica into biodegradable nano- or microcarriers raises the possibility that diatom biosilica will one day serve as a smart drug-delivery system (Uthappa et al., 2018). It is necessary to conduct more studies on the biological solution as a novel pharmaceutical drug carrier for clinical applications, including effectiveness and biocompatibility using in vivo conditions employing animal and pre-clinical models. The use of diatomaceous earth silica material in food and agriculture has received approval from the United States Department of Agriculture (USDA), the Environmental Protection Agency (EPA), and the Food and Drug Administration (FDA), but the pharmaceutical industry and medical practices still need formal regulatory approval. Diatom biosilica as a biomaterial has a possible future as a carrier for medicine delivery in therapeutic applications because of its biological nature (Maher et al., 2018).

1.22 CONCLUSIONS

This chapter focuses on the applications of natural diatoms, specifically how they may be utilized to enhance medicine delivery. They have repeatedly demonstrated that they are superior medication transporters. The pore size of the material may be reduced by face tuning, resulting in extended drug release. According to an in vivo toxicity study, neither the initial diatom nor the diatom changed inorganically caused any harmful responses or fatalities during the observation period. A powerful tool for the development and growth of therapeutic operating processes is biosilica, which is made from diatoms. The potential for significant developments in nanotechnology and biological research is increasing with each study on biosilica or diatomite. It is expected that shortly, new drug-carrying vehicles will be developed by concentrating on various surface functions that are compatible with one another. We have summarized the most recent and effective technologies of drug deliveries with glass-walled microbes. With the current and future innovations devoted to diatoms, it is expected that in the future it shall be possible to fully maximize their potential in the pharmaceutical industry.

ACKNOWLEDGMENTS

We sincerely thank the Department of Biotechnology (DBT), India (scientific research project BT/ PR 15650/ AAQ/3/815/2016) for providing financial support.

REFERENCES

Akbarzadeh, A., Rezaei-Sadabady, R., Davaran, S., Joo, S., Zarghami, N., Hanifehpour, Y. et al. (2013) Liposome: classification, preparation, and applications. *Nanoscale Research Letters* 8. doi: 10.1186/1556-276x-8-102

Alexander, N., Baker, E., Kaptein, M., Karck, U., Miller, L., Zampaglione, E. (2004) Why consider vaginal drug administration? *Fertility and Sterility* 82:1–12. doi: 10.1016/j.fertnstert.2004.01.025

Alkilani, A., McCrudden, M., Donnelly, R. (2015) Transdermal drug delivery: Innovative pharmaceutical developments based on disruption of the barrier properties of the stratum corneum. *Pharmaceutics* 7:438–470. doi: 10.3390/pharmaceutics7040438

Alli, S., Vemula, S., Veerareddy, P. (2013) Role of microspheres in drug delivery-an overview. *Drug Delivery Letters* 3:191–199. doi: 10.2174/22103031113039990003

Almqvist, N., Delamo, Y., Smith, B.L., Thomson, N.H., Bartholds, A., Lal, R., Brzezinski, M., Hansma, P.K. (2001) Micromechanical and structural properties of a pennate diatom investigated by atomic force microscopy. *Journal of Microscopy* 202: 518–532. https://doi.org/10.1046/j.1365-2818.2001.00887.x

Aw, M.S., Simovic, S., Addai-Mensah, J., Losic, D. (2011a). Polymeric micelles in porous and nanotubular implants as a new system for extended delivery of poorly soluble drugs. *Journal of Materials Chemistry* 21: 7082–7089. doi: 10.1039/c0jm04307a

Aw, M.S., Simovic, S., Addai-Mensah, J., Losic, D. (2011b) Silica microcapsules from diatoms as new carrier for delivery of therapeutics. *Nanomedicine (London)* 6: 1159–1173. doi: 10.2217/nnm.11.29

Aw, M.S., Bariana, M., Yu, Y., Addai-Mensah, J., Losic, D. (2013) Surface-functionalized diatom microcapsules for drug delivery of water-insoluble drugs. *Journal of Biomaterials Applications* 28: 163–174. https://doi.org/10.1177/0885328212441846

Aw, M.S., Simovic, S., Yu, Y., Mensah, J.A., Losic, D. (2012) Porous silica microshells from diatoms as biocarrier for drug delivery applications. *Powder Technology* 223: 52–58. https://doi.org/10.1016/j.powtec.2011.04.023

Bariana, M., Aw, M.S., Kurkuri, M., Losic, D. (2013) Tuning drug loading and release properties of diatom silica microparticles by surface modifications. *International Journal of Pharmaceutics* 443: 230–241. https://doi.org/10.1016/j.ijpharm.2012.12.012

Battarbee, R.W., Jones, V.J., Flower, R.J., Cameron, N.G. (2001) Diatoms. In: *Developments in Paleoenvironmental Research*. Springer: Heidelberg, Germany, 2001; Volume 3, p. 134.

Bhagwat, R.R., Vaidhya, I.S. (2013) Novel drug delivery systems: An overview. *International Journal of Pharmaceutical Sciences and Research* 4: 970–982. doi: 10.13040/ijpsr.0975-8232.4(3).970-82

Bharti, C., Nagaich, U., Pal, A.K., Gulati, N. (2015) Mesoporous silica nanoparticles in target Drug delivery system: a review. *International Journal of Pharmaceutical Investigation* 5: 124–133. https:// doi.org/ 10.4103/2230-973X.160844

Bowler, C., De Martino, A., Falciatore, A. (2010) Diatom cell division in an environmental context. *Current Opinion in Plant Biology* 13: 623–630.

Brewer, M., Zhang, T., Dong, W., Rutherford, M., Tian, Z.R. (2007) Future approaches of nanomedicine in clinical science. *Medical Clinics of North America* 91: 963–1016. https://doi. org/10.1016/ j.mcna.2007.05.006

Chandrasekaran, S., Sweetman, M.J., Kant, K., Skinner, W., Losic, D., Nann, T., et al. (2014). Silicon diatom frustules as nanostructured photoelectrodes. *Chemical Communications* 50: 10441–10444.

Chao, J.T., Biggs, M.J.P., Pandit, A.S. (2014) Diatoms: a biotemplating approach to fabricating drug delivery reservoirs. *Expert Opinion, Drug Delivery* 11: 1687–1695. https://doi.org/10.1517/17425247.2014.935336

Cicco, S.R., Vona, D., De Giglio, E., Cometa, S., Mattioli-Belmonte, M., Palumbo, F., et al. (2015). Chemically modified diatoms biosilica for bone cell growth with combined drug-delivery and antioxidant properties. *ChemPlusChem* 80: 1104–1112.

Delalat, B., Sheppard, V.C., Rasi Ghaemi, S., Rao, S., Prestidge, C.A., McPhee, G., Rogers, M.L., Donoghue, J.F., Pillay, V., Johns, T.G., Kroger, N., Voelcker, N.H. (2015) Targeted drug delivery using genetically engineered diatom biosilica. *Nature Communications* 6: 8791.

Delasoie, J., Zobi, F. (2019) Natural diatom biosilica as microshuttles in drug delivery systems. *Pharmaceutics* 11:537. https://doi.org/10.3390/ pharmaceutics11100537

Dhas, N., Parekha, K., Pandey, A., Kudarha, R., Mutalik, S., Mehta, T. (2019) Two dimensional carbon based nanocomposites as multimodal therapeutic and diagnostic platform: a biomedical and toxicological perspective. *Journal of Controlled Release* 308: 130–161. https://doi.org/10.1016/j.jconrel.2019.07.016

Diatoms of North America. *What are Diatoms?* https://diatoms.org/what-are–diatoms/, 2020 (accessed 4 April 2020).

DiMasi, J.A., Grabowski, H.G., Hansen, R.W. (2016). Innovation in the pharmaceutical industry: new estimates of R&D costs. *Journal of Health Economics* 47: 20–33.

Donnelly, R. (2012). *Microneedle-Mediated Transdermal and Intradermal Drug Delivery*. Chichester, UK: Wiley-Blackwell.

Edgar, L.A., Pickett-Heaps, J.D. (1984) Diatom locomotion. In: Round F.E., Chapman D.J. (eds.) *Progress in Psychological Research*, vol 3. Biopress, UK, pp. 47–88.

Gaudana, R., Ananthula, H., Parenky, A., Mitra, A. (2010) Ocular drug delivery. *The AAPS Journal* 12: 348–360. doi: 10.1208/s12248-010-9183-3

Gersonde, R., Harwood, DM. (1990) Lower Cretaceous diatoms from ODP Leg 113 site 693 (Weddell Sea) Part 1: Vegetative cells. *Proceedings of the ODP Science Results* 113:3 65–402.

Gnanamoorthy, P., Anandhan, S., Prabhu, V.A. (2014) Natural nanoporous silica frustules from marine diatom as a biocarrier for drug delivery. *Journal of Porous Materials* 21: 789–796. https://doi.org/10.1007/s10934-014-9827-2

Gordon, R., Drum, R.W. (1994) The chemical basis of diatom morphogenesis. *International Journal of Cytology* 150: 243–372.

Gordon, R., Losic, D., Tiffany, M.A., Nagy, S.S., Sterrenburg, F.A. (2009) The glass menagerie: Diatoms for novel applications in nanotechnology. *Trends in Biotechnology* 27: 116–127.

Guiry, MD. (2012) How many species of algae are there? *Journal of Phycology* 48: 1057–1063.

Hamm, C.E., Merkel, R., Springer, O. et al. (2003) Architecture and material properties of diatom shells provide effective mechanical protection. *Nature* 421: 841–843.

Hao, J., Heng, P. (2003) Buccal delivery systems. *Drug Development and Industrial Pharmacy* 29:821–832. doi: 10.1081/ddc-120024178

Harwood, D.M., Chang, K.H., Nikolaev, V.A. (2004) Late Jurassic to earliest Cretaceous diatoms from Jasong Synthem, Southern Korea: Evidence for a terrestrial origin. In: *Abstracts of the 18th International Diatom Symposium*. Międzyzdroje, 2–7 September 2004.

Hildebrand, M., Doktycz, M.J., Allison, D.P. (2008) Application of AFM in understanding biomineral formation in diatoms. *Pflügers Archiv European Journal of Physiology* 456: 127–137.

Illum, L. (2012) Nasal drug delivery—Recent developments and future prospects. *Journal of Controlled Release* 161: 254–263. doi: 10.1016/j.jconrel.2012.01.024

Inamuddin, Asiri, A., Mohammad, A. (2018) *Applications of Nanocomposite Materials in Drug Delivery* (pp. 509–573). Woodhead Publishing.

Ishii, K.-I., Iwataki, M., Matsuoka, K., Imai, I. (2011) Proposal of identification criteria for resting spores of Chaetoceros species (Bacillariophyceae) from a temperate coastal sea. *Phycologia* 50: 351–362.

Jain, K.K. (2008) Drug delivery systems–an overview. In: Jain K.K. (ed.), *Drug Delivery Systems. Methods in Molecular Biology*, Vol. 437. Human Press. https://doi.org/10.1007/978-1-59745-210-6_1.

Janicijevic, J., Krajisnik, D., Calija, B., Dobricic, V., Dakovic, A., Krstic, J., Markovic, M., Milic, J. (2014) Inorganically modified diatomite as a potential prolonged-release drug carrier. *Materials Science & Engineering, C* 42: 412–420.

Janićijevića, J., Krajišnika, D., Calijaa, B., Vasiljevićb, BN., Dobričićc, V., Dakovićd, A., Antonijeviće, M.D., Milić, J. (2015) Modified local diatomite as potential functional drug carrier-A model study for diclofenac sodium. *International Journal of Pharmaceutics* 496: 466–474. https://doi.org/10.1016/j.ijpharm.2015.10.047

Javalkote, V.S., Pandey, A.P., Puranik, P.R., Deshmukh, P.K. (2015) Magnetically responsive siliceous frustules for efficient chemotherapy. *Materials Science and Engineering C* 50: 107–116. https://doi.org/10.1016/j.msec.2015.01.079

Karakoy, M., Gultepe, E., Pandey, S., Khashab, M.A., Gracias, D.H. (2014) Silane surface modification for improved bioadhesion of esophageal stents. *Applied Surface Science* 311: 684–689. https://doi.org/10.1016/j.apsusc.2014.05.136

Karaman, D.S., Kettiger, H. (2018) Chapter 1 – Silica-based nanoparticles as drug delivery systems: Chances and challenges. In: Grumezescu, A.M. (ed.) *Inorganic Frameworks as Smart Nanomedicines*. William Andrew Publishing. Pages 1–40.

Katiyar, A., Yadav, S.K., Smirniotis, P.G., Pinto, N.G. (2006) Synthesis of ordered large pore SBA-15 spherical particles for adsorption of biomolecule. *Journal of Chromatography, A* 1122: 13–20. https://doi.org/10.1016/j.chroma.2006.04.055

Khan, M.J. (2017) The novel drug delivery system. *World Journal of Pharmacy and Pharmaceutical Sciences* 6: 477–487. https://doi.org/10.20959/wjpps20177-9607

Knight, M.J., Senior, L., Nancolas, B., Ratcliffe, S., Curnow, P. (2016) Direct evidence of the molecular basis for biological silicon transport. *Nature Communications* 7: 11926.

Kooistra, W.H.C.F., Pohl, G. (2015) Diatom Frustule Morphology and its Biomimetic Applications in Architecture and Industrial Design. *Evolution of Lightweight Structures* 75–102. https://doi.org/10.1007/978-94-017-9398-8_5

Kröger, N., Lorenz, S., Brunner, E., Sumper, M. (2002) Self-assembly of highly phosphorylated silaffins and their function in biosilica morphogenesis. *Science* 298: 584–586.

Kröger, N., Rainer, D., Manfred, S. (1999) Polycationic peptides from diatom biosilica that direct silica nanosphere formation. *Science* 286: 1129–1132.

Kublik, H., Vidgren, M. (1998) Nasal delivery systems and their effect on deposition and absorption. *Advanced Drug Delivery Reviews* 29: 157–177. doi: 10.1016/s0169-409x(97)00067-7

Kumar, V., Bansal, V., Madhavan, A., Kumar, M., Sindhu, R., Awasthi, M.K., Binod, P., Saran, S. (2022) Active pharmaceutical ingredient (API) chemicals: a critical review of current biotechnological approaches. *Bioengineered* 13(2): 4309–4327.

Kumeria, T., Bariana, M., Altalhi, T., Kurkuri, M., Gibson, C.T., Yang, W., Losic, D. (2013) Graphene oxide decorated diatom silica particles as new nano-hybrids: towards smart natural drug microcarriers. *Journal of Materials Chemistry B* 45: 6302–6311. https:// doi.org/10.1039/C3TB21051

Labiris, N, Dolovich, M. (2003) Pulmonary drug delivery. Part I: Physiological factors affecting therapeutic effectiveness of aerosolized medications. *British Journal of Clinical Pharmacology* 56: 588–599. doi: 10.1046/j.1365-2125.2003.01892.x

Lang, Y., Monte, F.D., Collins, L., Rodriguez, B.J., Thompson, K., Dockery, P., Finn, D.P., Pandit, A. (2013) Functionalization of the living diatom Thalassiosira weissflogii with thiol moieties. *Nature Communications* 4: 2683, https://doi.org/10.1038/ ncomms3683

Lee, B.K., Yun, Y.H., Park, K. (2015) Smart nanoparticles for drug delivery: boundaries and opportunities. *Chemical Engineering Science* 125: 158–164, https://doi.org/10.1016/j. ces.2014.06.042

Li, X.L., Tao, L., Tiwari, A., Marella, T.K., Li, R., Li, G. (2017) Optimization of growth conditions and fatty acid analysis for three freshwater diatom isolates. *Phycology Results* 65: 177–187. https://doi.org/ 10.1111/pre.12174

Liu, D., Yang, F., Xiong, F., Gu, N. (2016) The smart drug delivery system and its clinical potential. *Theranostics* 6: 1306–1323. https://doi.org/10.7150/ thno.14858

Losic, D., Pillar, R.J., Dilger, T., Mitchell, J.G., Voelcker, N.H. (2007) Atomic force microscopy (AFM) characterisation of the porous silica nanostructure of two centric diatoms. *Journal of Porous Materials* 14: 61–69.

Losic, D., Yu, Y., Aw, M.S., Simovic, S., Thierry, B., Addai-Mensah, J. (2010) Surface functionalisation of diatoms with dopamine modified iron-oxide nanoparticles: toward magnetically guided drug microcarriers with biologically derived morphologies. *Chemical Communications* 46: 6323–6325.

Maher, S., Aw, M.S., Losic, D. (2019) Diatom silica for biomedical applications. In: J. Seckbach, R. Gordon (Eds.), *Diatoms: Fundamentals and Applications*, pp. 511–536. https://doi.org/10.1002/9781119370 741.ch20

Maher, S., Kumeria, T., Aw, MS., Losic, D. (2018) Diatom silica for biomedical applications: recent progress and advances. *Advanced Healthcare Materials* 7: 1800552. https://doi. org/10.1002/adhm.201800552

Mann, D.G. (1999) The species concept in diatoms. *Phycologia* 38: 437–495.

Martin-Jézéquel, V., Hildebrand, M., Brzezinski, M.A. (2000) Silicon metabolism in diatoms: implications for growth. *Journal of Phycology* 36: 821–840.

Milovic, M., Simovic, S., Losic, D., Dashevskiy, A., Ibric, S. (2014) Solid self-emulsifying phospholipid suspension (SSEPS) with diatom as a drug carrier. *European Journal of Pharmaceutical Sciences* 63: 226–232.

Mishra, M., Arukha, A.P., Bashir, T., Yadav, D., Prasad, G.B.K.S. (2017) All new faces of diatoms: potential source of nanomaterials and beyond. *Frontiers in Microbiology* 8: 1239. https://doi.org/10.3389/ fmicb.2017.01239

Mishra, V., Bansal, K.K., Verma, A., Yadav, N., Sourav, T., Sudhakar, K., Rosenholm, J.M. (2018) Solid lipid nanoparticles: emerging colloidal nano drug delivery systems. *Pharmaceutics* 10: 191. https://doi.org/ 10.3390/ pharmaceutics10040191

Muro, S. (2012) Challenges in design and characterization of ligand-targeted drug delivery systems. *Journal of Controlled Release* 164: 125–137. https://doi.org/10.1016/j. jconrel.2012.05.052

Narang, N., Sharma, J. (2011) Sublingual mucosa as a route for systemic drug delivery. *International Journal of Pharmacy and Pharmaceutical Sciences* 3: 1822.

Pagar, O.B., Nagare, H.S., Autade, R.R., Chine, Y.M., Narode, P.R., Sanklecha, Y. (2018) Mesoporous silica: a review. *International Journal of Pharmaceutics and Drug Analysis* 6: 1–12.

Phogat, S., Saxena, A., Kapoor, N., Aggarwal, C., Tiwari, A. (2021) Diatom mediated smart drug delivery system. *Journal of Drug Delivery Science and Technology* 63: 102433.

Poulsen, N., Sumper, M., Kroger, N. (2003) Biosilica formation in diatoms: Characterization of native silaffin-2 and its role in silica morphogenesis. *Proceedings of the National Academy of Sciences USA* 100: 12075–12080.

Rabosky, D.L., Sorhannus, U. (2009) Diversity dynamics of marine planktonic diatoms across the Cenozoic. *Nature* 457: 183–186

Ragni, R., Cicco, S., Vona, D., Leone, G., Farinola, G.M. (2017). Biosilica from diatoms microalgae: smart materials from bio-medicine to photonics. *Journal of Materials Results* 32: 279–291.

Rea, I., De Stefano, L. (2019) Recent advances on diatom-based biosensors. *Sensors* 19: 5208.

Rea, I., Martucci, N.M., De Stefano, L., Ruggiero, I., Terracciano, M., Dardano, P., Migliaccio, N., Arcari, P., Tate, R., Rendina, I., Lamberti, A. (2014) Diatomite biosilica nanocarriers for siRNA transport inside cancer cells. *Biochimica et Biophysica Acta* 1840: 3393–3403.

Rea, I., Terracciano, M., De Stefano, L. (2017) Synthetic vs natural: Diatoms bioderived porous materials for the next generation of healthcare nanodevices. *Advances in Healthcare Materials* 6: 1601125.

Rizvi, S., Saleh, A. (2018) Applications of nanoparticle systems in drug delivery technology. *Saudi Pharmaceutical Journal* 26: 64–70. doi: 10.1016/j.jsps.2017.10.012

Rosi, N.L., Mirkin, C.A., Thaxton, C.S. (2004) Control of nanoparticle assembly by using DNA-modified diatom templates. *Angewandte Chemie International Edition* 43: 5500–5503.

Round, F.E., Crawford, R.M., Mann, D.G. (1990) *The Diatoms. Biology and Morphology of the Genera.* Cambridge University Press, Cambridge.

Rudzinski, W.E., Aminabhavi, T.M. (2010) Chitosan as a carrier for targeted delivery of small interfering RNA. *International Journal of Pharmaceutics* 399: 1–11. https://doi.org/10.1016/j. ijpharm.2010.08.022

Sahoo C.K., et al. (2013) Intra vaginal drug delivery system: An overview. *American Journal of Advances in Drug Delivery* 2013: 43–55.

Saxena, A., Dutta, A., Kapoor, N., Kumar, A., Tiwari A. (2022) Envisaging marine diatom *Thalassiosira weissflogii* as a smart drug delivery system for insoluble drugs. *Journal of Drug Delivery Science and Technology* 68: 102983.

Saxena, A., Tiwari A., Kaushik, R., Iqbal, H.M.N., Parra-Saldívar, R. (2021) Diatoms recovery from wastewater: Overview from an ecological and economic perspective. *Journal of Water Process Engineering* 39: 101705.

Sercombe, L., Veerati, T., Moheimani, F., Wu, S., Sood, A., Hua, S. (2015) Advances and challenges of liposome assisted drug delivery. *Frontiers in Pharmacology* 2015: 6. doi: 10.3389/fphar.2015.00286

Smart, J. (2005) Buccal drug delivery. *Expert Opinion, Drug Delivery* 2: 507–517. doi: 10.1517/17425247.2.3.507

Sumper, M., Kröger, N. (2004) Silica formation in diatoms: The function of long-chain polyamines and silaffins. *Journal of Materials Chemistry* 14: 2059–2065.

Suto, I. (2006) The explosive diversification of the diatom genus *Chaetoceros* across the Eocene/Oligocene and Oligocene/Miocene boundaries in the Norwegian Sea. *Marine Micropaleontology* 58: 259–269.

Svenson S. (2004). *Carrier-based Drug Delivery.* Washington, D.C.: American Chemical Society.

Terracciano, M., De Stefano, L., Rea, I. (2018) Diatoms green nanotechnology for biosilica-based drug delivery systems. *Pharmaceutics* 10: 242.

Terracciano, M., Shahbazi, M.A., Correia, A., Rea, I., Lamberti, A., Stefano, L.D., Santos, H.A. (2015) Surface bioengineering of diatomite based nanovectors for efficient intracellular uptake and drug delivery. *Nanoscale* 7: 20063–20074. https:// doi.org/10.1039/c5nr05173h

Terracciano, M., Stefano, L.D., Rea, I. (2018) Diatoms green nanotechnology for biosilica-based drug delivery systems. *Pharmaceutics* 10: 242. https://doi.org/ 10.3390/pharmaceutics10040242

Tiwari, A., Melchor-Martínez, EM., Saxena, A., Kapoor, N., Singh, K., Saldarriaga–Hern´andez, S.C., Parra-Saldívar, R., Iqbal, H.M.N. (2021) Therapeutic attributes and applied aspects of biological macromolecules (polypeptides, fucoxanthin, sterols, fatty acids, polysaccharides, and polyphenols) from diatoms–a review. *International Journal of Biological Macromolecules* 171: 398–413.

Todd, T., Zhen, Z., Tang, W., Chen, H., Wang, G., Chuang, Y.J., Deaton, K., Pan, Z., Xie, J. (2014) Iron oxide nanoparticle encapsulated diatoms for magnetic delivery of small molecules to tumors. *Nanoscale* 6: 2073–2076. https://doi.org/10.1039/ c3nr05623f

Tramontano, C., Chianese, G., Terracciano, M., de Stefano, L., Rea, I. (2020) Nanostructured biosilica of diatoms: From water world to biomedical applications. *Applied Sciences* 10: 6811. doi:10.3390/app10196811

Trofimov, A.A., Pawlicki, A.A., Borodinov, N., Mandal, S., Mathews, T.J., Hildebrand, M., Ziatdinov, M.A., Hausladen, K.A., Urbanowicz, P.K., Steed, C.A., Ievlev, A.V., Belianinov, A., Michener, J.K., Vasudevan, R., Ovchinnikova, O.S. (2019) Deep data analytics for genetic engineering of diatoms linking genotype to phenotype via machine learning. *npj Computational Materials* 5: 67. https://doi.org/10.1038/s41524-019-0202-3

Uthappa, U.T., Brahmkhatri, V., Sriram, G., Jung, H.Y., Yu, J., Kurkuri, N., Aminabhavi, T.M., Altalhi, T., Neelgund, G.M., Kurkuri, M.D. (2018) Nature engineered diatom biosilica as drug delivery systems. *Journal of Controlled Release* 281: 70–83. https://doi.org/10.1016/j.jconrel.2018.05.013

Varde, N., Pack, D. (2004) Microspheres for controlled release drug delivery. *Expert Opinion on Biological Therapy* 4: 35–51. doi: 10.1517/14712598.4.1.35

Vasani, R.B., Losic, D., Cavallaro, A., Voelcker, N.H. (2015) Fabrication of stimulus responsive diatom biosilica microcapsules for antibiotic drug delivery. *Journal of Materials Chemistry B* 3: 4325–4329.

Venkatesan, J., Lowe, B., Kim, SK. (2015) Application of diatom biosilica in drug delivery. In: S.K. Kim (Ed.), *Handbook of Marine Microalgae* (pp. 245–252). Academic Press. https://doi.org/10.1016/B978-0-12-800776-1.00015-7

Wang, J.-K., Seibert, M. (2017) Prospects for commercial production of diatoms. *Biotechnology for Biofuels* 10: 16.

Yang, X.Y., Chen, L.H., Li, L., Rooke, J.C., Sanchez, C., Su, B.L. (2016) Hierarchically porous materials: synthesis strategies and structure design. *Chemical Society Reviews* 46: 481–558.

Yang, X.Y., Chen, L.H., Li, Y., Rooke, J.C., Sanchez, C., Su, B.L. (2016) Hierarchically porous materials: synthesis strategies and structure design. *Chemical Society Reviews* 46: 481–558. https://doi.org/10.1039/C6CS00829A

Zaharudin, N.S., Isa, E.D.M., Ahmad, H., Rahman, M.B.A., Jumbri, K. (2020) Functionalized mesoporous silica nanoparticles templated by pyridinium ionic liquid for hydrophilic and hydrophobic drug release application. *Journal of the Saudi Chemical Society* 24: 289–302. https://doi.org/10.1016/j.jscs.2020.01.003

Zhang, H., Shahbazi, M.A., Makila, E.M., da Silva, T.H., Reis, R.L., Salonen, J.J., Hirvonen, J.T., Santos, H.A. (2013) Diatom silica microparticles for sustained release and permeation enhancement following oral delivery of prednisone and mesalamine. *Biomaterials* 34: 9210–9219.

Zomig 5mg Nasal Spray–Patient Information Leaflet (PIL)–(eMC). (2019). Accessed 19 July 2019: www.medicines.org.uk/emc/product/6712/pil

2 Bioactive Compounds from Diatoms

Win Nee Phong,[1] Roshan Regmi,[2] Kul Chandra Adhikari,[1] Rewati Raman Bhattarai,[1] and Pau Loke Show[3]
[1]School of Molecular and Life Sciences, Curtin University, Bentley, Western Australia, Australia
[2]CSIRO Agriculture and Food, Waite Campus, Urrbrae, South Australia
[3]Department of Chemical and Environmental Engineering, Faculty of Science and Engineering, University of Nottingham Malaysia, Selangor, Malaysia

2.1 INTRODUCTION

Microalgae are unicellular, eukaryotic microscopic organisms widely found in marine and freshwater (Spolaore, et al., 2006). Microalgae have a wide array of applications including carbon sequestration (Sethi, et al., 2020), wastewater treatment (Tiwari & Marella, 2019), animal and human nutrition (Rhodes et al., 2017), biofertilizers (Grzesik, et al., 2015), and bioenergy (Lakatos et al., 2019). Diatoms are an important industrial group of microalgae with the ability to produce different potential bioactive compounds. Bioactive compounds are nutritional compounds found in small amounts in plants, animal products, and specific lipid-rich foods. Because of their properties, various researches and trials have been conducted to estimate their benefit and identify and optimize their extraction from different sources (Tiwari et al., 2021). The diatom taxon is one of the most diverse groups of microalgae, with a contribution of almost 25 % of the Earth's primary productivity (Scala & Bowler, 2001). Because of their ability to survive in extreme conditions and the presence of different pore structures in the cell wall, diatoms are ideal for producing bioactive compounds. In addition, the highly ornamented siliceous cell, commonly called a frustule, is a unique characteristic of diatoms (Smetacek, 1985). It helps in sessile adhesion, preventing drying, and forming biofilms and colonies (Svetličić et al., 2013).

Diatoms have been reported to consist of novel metabolic pathways known as the origin of valuable bioactive compounds for a wide range of applications in food, cosmetics, medical, pharmaceutical, bioremediation, and nanotechnology. Given the ability of diatoms to produce different bioactive compounds like proteins, carbohydrates, fatty acids, vitamins, carbohydrates, and pigments, they are researched extensively worldwide. Fucoxanthin, the primary carotenoid of diatoms, is used mainly as an antioxidant, anti-inflammatory, anti-diabetes, and anti-cancer drug (Peng et al., 2011). Some polysaccharides such as laminarin and chryosolaminarin have been reported to be used in aquaculture (Fleurence, 2021; Lebeau & Robert, 2003; Lincoln et al., 1990). Some sulfated polysaccharides have also been reported to be used in the cosmetic industry as a protective agent against oxidative stress to ward off the effect of free radicals and reactive chemical species (Pouvreau et al., 2008; Yu et al., 2018). Some species of diatoms have been genetically engineered to produce essential proteins such as human IgG antibodies (Hempel & Maier, 2012). Moreover, diatoms derived from photosynthetic and photoprotective pigments are of great importance for food, cosmetic, and pharmaceutical products (Ambati et al., 2019).

 DOI: 10.1201/9781003436553-2

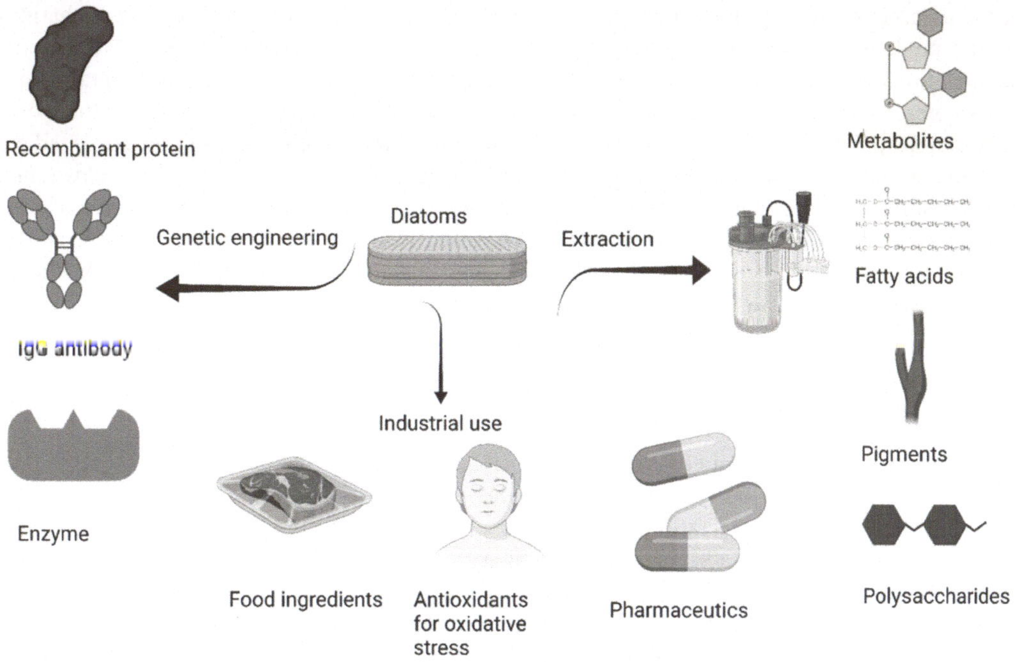

FIGURE 2.1 Use of diatoms for multiple purposes.

Accompanied by next-generation sequencing, the prospect of industrial application of diatoms has increased significantly with the availability of complete genomes for some species like *Thalassiosira pseudonana* (Mock et al., 2008), *T. oceanica* (Kilpert & Podsiadlowski, 2006), *Fragilariopsis cylindrus* (Mock et al., 2017), *Fistulifera solaris* (Tang & Bi, 2016), and *Synedra acus* (Brembu et al., 2011). Diatoms commonly comprise three different genomes: the nuclear genome, the plastid genome, and the mitochondrial genome. Complete genome and transcriptomic data enhance the discovery of new pathways for bioactive compounds. Diatoms are widely studied for their biotechnological applications and are currently used as bioreactors to produce important pharmaceutical compounds and secondary metabolites. It is worth mentioning that a closely associated microbiome plays a crucial role in shaping the growth, aggregation, and secretion of polymeric compounds. With the bacterial community, benthic diatoms are the dominant organisms in marine biofilm communities that colonize almost every biotic and abiotic surface in the marine environment. Therefore, a diatom/bacterial community model is a critical driver for the types of bioactive compounds secreted by diatoms (Amin et al., 2012). This chapter discusses different categories of bioactive compounds from diatoms, their cultivation techniques, potential benefits, challenges to their use and production, and future perspectives (Figure 2.1).

2.2 TYPES OF BIOACTIVE COMPOUNDS

2.2.1 Polysaccharides

Diatoms are among the primary producers in the marine ecosystem; a large portion of organic material produced in diatoms are polysaccharides, ranging between 10–80% (Haug & Myklestad, 1976; Hussein & Abdullah, 2020). In diatoms, cell wall silica is composed of organic matter with proteins, lipids, and polysaccharides. Some experimental studies suggest that polysaccharide content is more pronounced than protein and lipids (Reimann et al., 1966) in diatoms; however, X-ray photoelectron spectroscopy measurement in cell surfaces of *T. pseudonana* suggests polysaccharides are not predominant (Tesson et al., 2009). Therefore, quantifying polysaccharides from diatoms requires precise sequential extraction techniques (Gügi et al., 2015). Polysaccharide-related bioactive compounds from different species of diatoms have been reported to exhibit various functions. For example, polysaccharides from *P. tricornutum* have anti-inflammatory, antiadhesive, free radical scavenging activity (Chen et al., 2011; Guzmán-Murillo et al., 2007), *Cylinrotheca closterium* polysaccharides possess anti-inflammatory activity (Pletikapić et al., 2011), and polysaccharides from *Navicula salinarium* exert antiviral activity (Staats, et al., 1999). Similarly, polysaccharides of *Isochrysis* sp. possess nutraceutical, antiallergic, and anti-inflammatory activity (Xiao & Zheng, 2016). In this chapter, two major polysaccharides, chrysolaminarin and chitin, are explained in detail in their structure and applications.

2.2.2 Chrysolaminarin

Chrysolaminarin is a significant type of polysaccharide produced by diatoms (Chiovitti et al., 2004). Chrysolaminarin has previously been known as leucosin and is a linear polymer of β-(1–3)- and β-(1–6)-linked glucose units in a ratio of 11:1 (Beattie et al., 1961). The production of polysaccharides such as chrysolaminarin is attributed to the primary energy and carbon reserve during the light phase, which is used during the dark phase when cells prepare for the next growth cycle (Caballero et al., 2016; Kamalanathan et al., 2019). Although the production of polysaccharides from diatoms has been shown to be increased during favorable conditions, some active polysaccharides are produced more in unfavorable conditions (Kamalanathan et al., 2019). In addition to the carbon reserve, the polysaccharides secreted by the diatoms help shape the surrounding microbial community by providing them with valuable carbon sources required for their metabolism (Amin et al., 2012; Torres-Monroy & Ullrich, 2018). The role of polysaccharides in different diatoms has been studied previously under different conditions. The inhibition of chrysolaminarin in oil-sensitive diatoms from *T. pseudomona* negatively impacts its growth and intracellular monosaccharide accumulation. In addition, the inhibition of chrysolaminarin impacts the microbiome community in free-living fractions but not in the phycosphere (Kamalanathan et al., 2019). Bioactive activities of chrysolaminarin from different diatom species have been used for various purposes. For example, *Odontella aurita*-extracted chryolaminarin possesses strong hydroxyl radical scavenging activity, which can be used in human health and nutrition (Xia et al., 2014). *Syndera acus* chryolaminarin has been shown to repress the growth of colon tumor cells. In aquaculture, the antioxidant activity of chrysolaminarin from *P. malhamensis* promotes fin regeneration in zebrafish (Ma et al., 2021; Kusaikin et al., 2010).

Polysaccharides of diatoms generally comprise insoluble monosaccharides like fucose, galactose, mannose, glucose, glucuronic acid, and a minor amount of rhamnose and arabinose. Mannose is the most dominant fraction in most strains. However, fucose is dominant in *Nitzchia brevirostris*, and glucose is a significant monosaccharide in *Melosira granulata*, *Cyclotella stelligera*, *Nitzchia curvilineata*, and *Amphora salina* (Hecky et al., 1973; Tesson & Hildebrand, 2013). Details of diatom-specific oligosaccharides and polysaccharides have been reviewed (Gügi et al., 2015).

2.3 CHITIN

Chitin [β-(1-4)-poly-N-acetyl-D-glucosamine] is the second most available polysaccharide in nature after cellulose and has been well characterized in only a few eukaryotic lineages (Durkin et al., 2009). Although chitin does not produce bioactive compounds directly, its derivative component, chitosan, is widely used as a biomaterial in the health industry. Chitosan is produced with enzymatic and chemical modifications of chitin (deacetylation) (Elieh-Ali-Komi & Hamblin, 2016). Among hundreds of diatom genera, two have been reported to produce long fibers of chitin that protrude through their silica cell walls. Gene clusters for the putative synthesis of chitin synthases have been identified from *T. pseudomona* (Durkin et al., 2009). Chitin synthase is a key enzyme in the chitin synthesis pathway, which uses UDP-N-acetylglucosamine to produce chitin polymer. The unique characteristics of chitosan, such as biocompatibility, biodegradability, biofilm formation ability, and non-toxicity, make it a promising candidate for use in different industrial applications, mainly in biomedical applications such as tissue engineering (Jayakumar et al., 2010; Suh & Matthew, 2000; Venkatesan et al., 2014). The antibacterial activity of chitosan has also been reported (Kumar, 2000) and it has been used as scaffolds for wound-healing purposes (Menon et al., 2010). The modification of chitosan with other copolymers and nanoparticles has been used in antiaging cosmetics (Morganti et al., 2012). Antitumor activity of chitosan has been reported both *in vitro* and *in vivo* (Tokoro et al., 1988). This activity is attributed to enhanced lymphokine production, which triggers the proliferation of cytolytic T-lymphocytes (Cheung et al., 2015). Additionally, chitosan has shown promising application as a vaccine adjuvant in experimental studies in mice and guinea pigs (McNeela et al., 2000; Mills et al., 2003). Chitosan-coated diatom silica was used to control hemorrhage in military and civilian trauma cases (Feng et al., 2016).

2.4 PROTEIN AND BIOACTIVE PEPTIDES

Food-sourced proteins constitute a significant source of protein hydrolysates that contain bioactive peptides (Abeer et al., 2021). The terms "protein hydrolysates" and "bioactive peptides" are often used interchangeably (Chalamaiah et al., 2019; Chalamaiah, Yu, & Wu, 2018). Protein hydrolysates are a complex mixture of peptides of different chain lengths and free amino acids produced biologically or chemically via the breakdown of proteins from hydrolysis (Tiwari et al., 2021). Protein hydrolysates are substrates for the subsequent production of bioactive peptides (Abeer et al., 2021). Bioactive peptides generally include both oligopeptides and polypeptides. Oligopeptides are short protein fragments that are usually 2–20 amino acids in length. Polypeptides have a longer chain consisting of over 20 amino acids (Karami & Akbari-Adergani, 2019; Sathya et al., 2021). Depending on their amino acid composition, sequence, and structure, these peptides may exert different biological activities that modulate critical physiological functions in the human body (Chakrabarti et al., 2018; Chakrabarti et al., 2014; Sánchez & Vázquez, 2017). Due to their potent biological activities that positively impact human health, there has been a growing interest in applying food-derived bioactive peptides as functional foods or nutraceuticals in recent decades (Chakrabarti et al., 2018; Chakrabarti et al., 2014; Phong, et al., 2018).

Numerous food proteins from animal or plant origins, such as soy, oat, pulses, eggs, milk, meat, and fish, have been exploited as sources of bioactive peptide production (Chakrabarti et al., 2018). Nonetheless, exploiting these traditional food proteins for bioactive peptide production is challenging, especially in the current context of a rapidly growing population demanding proteins as food sources (Colgrave et al., 2021). For this reason, interest in other sustainable sources of food-derived bioactive peptides to meet future demand is continuing to grow and gain momentum.

The microalgae such as diatoms can be considered a commercial and sustainable source of protein for bioactive peptide production due to their rapid growth rate, higher productivity compared to conventional crops, and their culture conditions which can be optimized for maximal production of the biomolecules of interest, thus guaranteeing a steady and continuous supply of protein (Batista et al., 2013; Chakrabarti et al., 2018; Colgrave et al., 2021; Phong et al., 2017). Furthermore, the protein content of diatoms can compete quantitatively and qualitatively with traditional protein sources. In terms of quantity, several species of diatoms have been reported to possess very high protein concentrations. For example, *Amphora*, *Navicula*, and *Cymbella* cultured in Conway medium were reported to contain at least 400 g/kg of protein on a dry weight basis (Khatoon et al., 2009), which is comparable to soybeans (41%) (Yaklich, 2001). In terms of quality, the amino acid profile of diatoms is comparable to traditional dietary protein sources and contains essential amino acids that the human body cannot synthesize (Brown & Jeffrey, 1995; Eilertsen et al., 2022; Kolmakova & Kolmakov, 2019). For example, the percentage of arginine, histidine, isoleucine, leucine, and valine in *Cylindrotheca fusiformis* (Brown & Jeffrey, 1995) and Lincoln variety of soybean were similar (Van Etten et al., 1959). All these characteristics make diatoms more competitive compared to traditional protein sources. However, the exploitation of diatoms for protein production remains minimal (Barkia et al., 2019; Fu et al., 2021).

The challenge with protein production from diatoms is that diatoms carry unpleasant odors, and non-protein components such as chlorophylls may affect food products' taste and visual impact, hampering their application in the food industry (Saari, et al., 2019; Fu et al., 2021). Moreover, the overall process of protein production from diatoms is quite complicated. Therefore, the efficient and sustainable mass cultivation of biomass and protein extraction is required to ensure the successful application of diatoms in food manufacturing. However, according to Svenning et al. (2020), the efficient extraction from diatom biomass may be less challenging as the silica-based cell walls of the microalgae diatoms are generally more fragile compared to the rigid cell walls of some other green microalgal strains such as *Chlorella* sp., which can lower the extraction efficiency of intracellular proteins (Saari, et al., 2019; Phong et al., 2018).

Peptides from diatoms, also known as natural peptides, have demonstrated diverse biological activities with positive effects on human health, such as radical scavenging to combat oxidative stress, anti-hypertension, and anti-inflammatory (Alzahrani et al., 2018; Barkia et al., 2019). In addition, the bioactive peptides could reduce immune-mediated secondary side effects compared to synthetic supplements that may have side effects on the defense mechanisms and cause cell stress in the human body (Barkia et al., 2019). Nonetheless, the exploitation of diatoms as the source of bioactive peptides for functional ingredients manufacturing remains undervalued (Barkia et al., 2019). Among the diatom species, only a few species, such as *Phaeodactylum* sp., *Navicula* sp., and *Nitzschia* sp., have been extensively studied as a source of active polypeptides (Barkia et al., 2019). *Navicula incerta* was reported to contain around 46.5–51.7% of protein content, demonstrating a good source for peptides (Tiwari et al., 2021).

Hydrolysis of peptide bonds is vital in liberating the potent peptides, which has been shown to demonstrate better bioactivity of protein hydrolysates and peptides compared to their parent proteins (Udenigwe & Aluko, 2012). Previous research showed that the bioactive proteins and hydrolysate bioactive peptides derived from diatoms could serve as natural antioxidants. They have demonstrated good antioxidant properties *in vitro* and in cell cultures. For example, two purified bioactive peptides with amino acid sequences of Val-Glu-Val-Leu-Pro-Pro-Ala-Glu-Leu and Pro-Gly-Trp-Asn-Gln-Trp-Phe-Leu obtained from *Navicula incerta* by enzymatic hydrolysis demonstrated potent antioxidant properties in HepG2/CYP2E1 cells (Kang et al., 2012). Similarly, the protein hydrolysates isolated from *Nitzschia laevis* after enzymatic hydrolysis were reported to exert strong *in vitro* antioxidant activities (Alzahrani et al., 2018). This finding was in agreement with previous studies which evaluated the potential of bioactive peptides derived from indigenous marine diatoms such as *Bellerochea* sp. and *Nitzschia* sp. In this study, the protein extracts were hydrolyzed by proteases before being assessed for their angiotensin-converting enzyme (ACE)-inhibitory, antioxidant properties *in vitro*, and antihypertensive effects *in vivo*. It was found that the resultant protein hydrolysates showed potent antioxidant and ACE-inhibitory properties, showcasing the potential benefits of using diatom-derived bioactive peptides as nutraceuticals or functional foods (Barkia et al., 2019).

2.5 PIGMENTS

Most of the commercial pigments available currently are synthesized chemically, and only a small portion is of natural origin, such as plants or algae (Patel et al., 2022). However, consumer preferences for natural resources over synthetic ones have dramatically increased (Mezzomo & Ferreira, 2016).

One of the significant sources of natural pigments is diatoms. Diatom-derived pigments act as antioxidants, anti-cancer, anti-inflammatory, anti-angiogenic, anti-obesity, and neuroprotective agents, which can be used in different industries such as food, nutraceuticals, pharmaceuticals, aquaculture, and cosmetics (Kuczynska et al., 2015).

Diatoms typically contain two photosynthetic pigments: chlorophylls and carotenoids (Kuczynska et al., 2015). The pigment composition and content of diatoms vary from species to species and can be significantly influenced by environmental conditions. Diatom cells accumulate more pigments under various stress conditions such as temperature, low and high irradiance, salinity, and nutrient deficiency culture conditions (Ambati et al., 2019; McClure et al., 2018; Xia et al., 2013). The primary photosynthetic pigments of diatoms are chlorophylls and carotenoids, including β-carotene, xanthophylls, namely fucoxanthin, diatoxanthin, and diadinoxanthin, as well as blue pigments such as marennine and marennine-like pigments (Kuczynska et al., 2015; Mal et al., 2021; Nymark et al., 2009).

2.6 CHLOROPHYLLS

Chlorophylls are primary green pigments in diatoms that play an essential role in light-harvesting (Bayu et al., 2020; Kuczynska et al., 2015). Several kinds of chlorophylls are found in photosynthetic organisms; however, only two forms occur in diatoms: chlorophyll *a* and chlorophyll *c* (Kuczynska et al., 2015; Mal et al., 2021). In diatoms, chlorophyll *a* involves photochemical energy conversion, whereas chlorophyll *c* participates in photosynthesis as an accessory pigment (Kuczynska et al., 2015). Chlorophyll has been used as a colorant in the food and cosmetic industries. Chlorophyll production from natural resources is more expensive than artificial colorings. Diatoms proliferate, and their biomass productivity is enhanced, under controlled cultivation systems. Due to their sustainability of production and renewable nature compared to other natural resources, diatoms could be an economical source for chlorophyll production (Ambati et al., 2019; Dufossé et al., 2005; McClure et al., 2018; Xia et al., 2013).

Chlorophylls exhibit stability in their natural environment conditions. However, changes in temperature and pH during extraction and processing may cause structural modifications and undesirable changes in the chromatic properties of chlorophylls. Therefore, chlorophylls must undergo a chemical modification involving substituting the central hydrogen atoms with copper or zinc ions to improve their food colorant stability (Hosikian et al., 2010; Viera et al., 2019). Along with its use as a natural coloring agent, chlorophyll has drawn significant attention in the pharmaceutical industry as a functional ingredient due to its potential human health benefits. As a result, this bioactive compound has been consumed as a nutraceutical agent for health benefits, providing antioxidant, anti-inflammatory, antimutagenic, and antimicrobial properties (Wu et al., 2021).

2.7 CAROTENOIDS

Carotenoids are fat-soluble tetraterpene pigments that usually exhibit yellow, orange, or red color (Maoka, 2020). Based on their chemical structure, carotenoids can be further categorized into two classes: carotenes (hydrocarbon) and xanthophylls (their oxygenated derivatives) (Kuczynska et al., 2015). Fucoxanthin, diadinoxanthin, diatoxanthin, violaxanthin, antheraxanthin, and zeaxanthin are classified as the xanthophyll class of carotenoids as they contain a single oxygen atom in their structure. On the other hand, carotenoids which are purely hydrocarbons and have no oxygen, such as *β*-carotene, are classified as carotenes or carotenoids (Kuczynska et al., 2015; Maoka, 2020). Carotenoids are mainly involved in the photo-protection of cells from oxidative damage; however, fucoxanthin also plays a part in light-harvesting in diatoms (Bayu et al., 2020; Kuczynska et al., 2015; Maoka, 2020). Carotenoids have a wide range of beneficial functions in human health and have been used in dietary supplements, food colorings, animal feeds, and cosmetics (Bayu et al.,

2020; Wu et al., 2021). Human cells cannot synthesize carotenoids; therefore, they must rely on dietary intake sources for these essential nutrients (Wu et al., 2021). Due to the high productivity of carotenoids in diatoms compared to other resources including microalgae and macroalgae, diatoms can be a commercial source of carotenoids in the human diet (Bayu et al., 2020).

2.8 Β-CAROTENE

β-Carotene is an essential source of provitamin A carotenoid for humans as it can be converted to vitamin A (Gurmu et al., 2014). A diet rich in β-carotene promotes eye health and prevents night blindness. β-Carotene is crucial for the proper functioning of the retina (Bayu et al., 2020). It has been described to exert various critical biological activities such as anti-cardiovascular diseases, anti-carcinogenic, anti-angiogenic, and anti-macular degeneration (Guruvayoorappan & Kuttan, 2007; Saini et al., 2015; Sathasivam & Ki, 2018).

The anti-angiogenic activity of β-carotene has been investigated *in vivo* and *in vitro*. The anti-angiogenic activity of β-carotene is due to its ability to affect serum cytokine levels and prevent the activation and nuclear translocation of transcription factors (Guruvayoorappan & Kuttan, 2007). β-Carotene also has beneficial effects on bone health. For example, β-carotene intake was positively correlated with a higher lumbar spine bone mineral density while studying 189 postmenopausal Korean women (Kim et al., 2016). This corroborates with the study by Matsumoto, Tousen, and Ishimi (2018), which suggested a β-carotene-containing diet preserves bone health, preventing osteoporosis. In addition, β-carotene plays a vital role in the food industry as an attractive natural food colorant (Saini et al., 2015). Due to its provitamin and antioxidant properties, consuming β-carotene at appropriate levels provides several potential health benefits.

β-Carotene also plays a role in protecting the skin against photo-aging due to its antioxidant activity. Dietary β-carotene improved facial wrinkles and elasticity significantly and protected skin from ultraviolet-induced DNA damage (Cho et al., 2010). Therefore, 30 mg/day of β-carotene supplementation was recommended to prevent and repair photo-aging (Cho et al., 2010).

2.9 FUCOXANTHIN

Fucoxanthin ($C_{42}H_{58}O_6$) is one of the most abundant and attractive carotenoids in diatoms (Bayu et al., 2020). Due to the high quantity of fucoxanthin, diatoms typically exhibit a golden-brown color. This brown pigment molecule has a very high commercial value globally. Due to fucoxanthin's unique molecular structure, a wide variety of remarkable its biological properties are applied in human health, such as anti-obesity, antidiabetic, antioxidant, anti-inflammatory, anti-cancer, anti-angiogenic, antimicrobial, neuroprotective, and photoprotective activities as well as exhibiting its protective effects on the liver, blood vessels of the brain, eyes, bones, and skin (Pangestuti & Kim, 2011; Sachindra et al., 2007; Seth et al., 2021). Peng et al. (2011) suggested that the antioxidant activity of fucoxanthin can be compared to α-tocopherol (a type of vitamin E). Fucoxanthin can be found in marine brown seaweed (*Phaeophyceae*) and *Bacillariophyta* diatoms (Peng et al., 2011; Seth et al., 2021; Xia et al., 2013). Marine seaweeds such as *Eisenia bicyclis*, *Laminaria japonica*, and *Undaria pinnatifida* are popular sources of fucoxanthin (Kanazawa et al., 2008; Méresse et al., 2020) which has been exploited for use as a food, nutritional, and medicinal ingredient (Méresse et al., 2020). Nonetheless, the low content of fucoxanthin in marine seaweeds (ca. 0.2 mg per g of fresh weight), coupled with expensive and inefficient extraction methods, have impeded its large-scale production (Kanazawa et al., 2008; Méresse et al., 2020; Seth et al., 2021). Conversely, marine diatoms have a high content of fucoxanthin and multiply. Fucoxanthin production in diatoms can be enhanced under controlled cultivation systems without seasonal fluctuations, thereby exhibiting more significant potential for their use on a commercial scale compared to marine seaweeds (Seth et al., 2021; Steinrücken et al., 2018). Diatoms such as *Cylindrotheca closterium* (Nekvapil et al.,

2022), *Skeletonema costatum* (Nekvapil et al., 2022), *Phaeodactylum tricornutum* (Gómez-Loredo et al., 2016; Kim et al., 2012; McClure et al., 2018), *Chaetoceros* sp. (Bhattacharjya et al., 2020), *Isochyris galbana* (Bayu et al., 2020; Gómez-Loredo et al., 2016), and *Odontella aurita* (Xia et al., 2018) are well known for the production of fucoxanthin. Among the diatom species, only a few diatom species such as *Phaeodactylum tricornutum*, *Cyclotella cryptic*, and *Odontella aurita* have been explored for their commercial viability in fucoxanthin production (Seth et al., 2021; Xia et al., 2013). According to an experiment conducted by Kim et al. (2012), *P. tricornutum* demonstrated high economic potential as a promising source of fucoxanthin for various commercial applications because of the easy extraction and high content of fucoxanthin. It has been reported that fucoxanthin in diatoms was at least ten times more abundant than that in the present commercial fucoxanthin sources in any of the edible brown seaweeds of *Eisenia bicylis*, *Laminaria japonica*, *Undaria pinnatifida*, *Sargassum* sp., *Hijkia fusiformis*, and *Fucus* sp. (Bayu et al., 2020; Gómez-Loredo et al., 2016; Ishika et al., 2017; Poojary et al., 2016; Stonik & Stonik, 2015; Xia et al., 2013).

2.10 ANTIOXIDANT AND ANTIPROLIFERATIVE ACTIVITIES

Fucoxanthin is a powerful antioxidant reported to show substantial scavenging activities against different free radicals such as 2,2-diphenyl-1-picrylhydrazyl (DPPH), hydroxyl radical, superoxide anion, hydrogen peroxide, 2,2'-azino-bis(3-ethylbenzothiazoline-6-sulfonic acid (ABTS), and singlet oxygen, which has been demonstrated in several studies (Liu et al., 2011; Lourenço-Lopes et al., 2021; Neumann et al., 2019; Ristivojević et al., 2021). The molecular structure of fucoxanthin is different from that of other carotenoids such as β-carotene (Peng et al., 2011), which is characterized by an unstable allenic bond and oxygenic functional groups such as hydroxyl, epoxy, carboxyl, and carbonyl groups in the polyene hydrocarbon chain. These chemical properties make fucoxanthin an excellent candidate for its antioxidant properties (Guvatova et al., 2020; Miyashita et al., 2020; Peng et al., 2011; Xia et al., 2013). Neumann et al. (2019) investigated antioxidant and antiproliferative activities of fucoxanthin derived from *Phaeodactylum tricornutum in vitro*. Based on the ferric-reducing antioxidant power (FRAP) assay, there was no significant difference between the antioxidant effects of fucoxanthin derived from *P. tricornutum* and astaxanthin, a commercial carotenoid derived from the red algae *Haematococcus pluvialis* (Neumann et al., 2019; Lorenz & Cysewski, 2000; Vigani et al., 2015). Both fucoxanthin and astaxanthin demonstrated intense antioxidant activity in the FRAP assay compared to β-carotene (Neumann et al., 2019). It was found that fucoxanthin could inhibit the oxidative burst in human polymorphonuclear leukocytes (PMLs), scavenge radicals, and increase the glutathione (GSH) to glutathione disulfide (GSSG) ratio (a marker for oxidative stress) (Neumann et al., 2019). The antiproliferative effect of fucoxanthin has also been reported on several cancer cell lines, such as human leukemia cell lines, K562 and TK6 (Almeida et al., 2018), breast cancer cell line (Karkhane Yousefi et al., 2018), human glioblastoma cell line (Lopes et al., 2020), gastric adenocarcinoma SGC-7901 and BGC-823 cells (Yu et al., 2018). These findings suggest the potential benefits of using fucoxanthin in nutraceutical and pharmaceutical applications. Therefore, fucoxanthin could be considered a novel therapeutic agent to support traditional cancer treatment (Neumann et al., 2019) and enhances human health.

2.11 ANTI-HYPERLIPIDEMIC AND ANTI-OBESITY POTENTIALS

Fucoxanthin was found to possess anti-obesity and anti-hyperlipidemic potentials, as demonstrated in earlier *in vitro* and *in vivo* studies (Gille et al., 2019; Koo et al., 2019). The anti-obesity effect of fucoxanthin-rich *P. tricornutum* extract on lipid accumulation in 3T3-L1 adipocytes and high-fat diet-induced obese mice has been investigated by Koo et al. (2019). In addition, *Phaeodactylum* extract was found to employ anti-obesity effects by promoting lipolysis and inhibiting adipocytic lipogenesis (Koo et al., 2019). Likewise, fucoxanthin-rich *P. tricornutum* extract has also been found to ameliorate the effects of diet-induced obesity in C57BL/6J mice as has also been reported by Gille et al. (2019). The findings indicate that fucoxanthin derived from diatoms can be used as a promising candidate for developing anti-obesity foods and healthy functional foods.

2.12 ANTIDIABETIC ACTIVITY

The use of fucoxanthin in treating diabetes has been reported in previous studies (Kong et al., 2019; Oliyaei et al., 2021). For example, the fucoxanthin-rich extract from brown algae *Laminaria japonica* was found to decrease diabetic signs in streptozotocin-nicotinamide-induced diabetic rats (Kong et al., 2019). The *in vivo* antidiabetic effect of fucoxanthin isolated from brown algae *Sargassum angustifolium* has also been confirmed by Oliyaei et al. (2021), suggesting the potential application of fucoxanthin as a promising bioactive compound for natural diabetes therapy.

2.13 SKIN PROTECTIVE ACTIVITY

Fucoxanthin exhibits photoprotective potential, which protects skin against UV radiation, promotes skin restoration, and reduces wrinkle formation, thus suggesting the advantage of using it as an anti-photoaging agent in moisturizers or oral medication (Urikura et al., 2011). The skin protection effect of fucoxanthin could be attributed to the antioxidative and anti-inflammatory mechanisms, as demonstrated by several *in vitro* and in *vivo* studies (Mal et al., 2021; Natsume et al., 2020; Rodríguez-Luna et al., 2018; Tavares et al., 2020).

2.14 MARENNINE

It has been long accepted that *Haslea ostrearia* is the only species of diatom worldwide that synthesizes and excretes a water-soluble blue pigment, known as marennine-like pigments, and these thus often referred to as "blue diatoms" or "blue *Haslea*" (Gastineau et al., 2016). Marennine is the main factor responsible for greening oyster gills in western France (Gastineau et al., 2021). In recent years, some other new species of blue *Haslea,* such as *H. provincialis* (Gastineau et al., 2016), *H. silbo* (Gastineau et al., 2021), and *H. karadagensis* (Gastineau, Davidovich, et al., 2012; Gastineau et al., 2014) were also conveyed to produce marennine-like pigments which display structural and chemical similarities to marennine produced by *H. ostrearia*. Furthermore, this blue pigment exhibits several biological activities, such as anti-viral, antiproliferative, anti-bacterial, and antioxidant properties (Gastineau et al., 2014). Therefore, these unique pigments accumulate in certain diatom species, especially the *H. ostrearia* strain, and have great potential applications in the food, healthcare, cosmetics, and aquaculture industries (Ambati et al., 2019; Gastineau et al., 2014).

2.15 AQUACULTURE

Blue *Haslea* species have great potential to be used as probiotics in aquaculture, especially for "green-water" systems (Gastineau et al., 2012; Gastineau et al., 2012; Turcotte et al., 2016). In an *in vitro* study, marennine and marennine-like pigments isolated from *H. ostrearia* and *H. karadagensis*, respectively, showed anti-bacterial activities against three marine pathogenic bacteria (*Pseudoalteromonas elyakowii*, *Polaribacter irgensii*, and *Vibrio aestuarianus*). These bacterial pathogens induce diseases in aquaculture (Gastineau et al., 2012; Gastineau et al., 2012). Turcotte et al. (2016) investigated the potential application of marennine-containing culture supernatant of *H. ostrearia* in controlling the pathogenic marine bacteria *Vibrio splendidus* in bivalve hatcheries using the scallop *Placopecten magellanicus* and the blue mussel *Mytilus edulis* as model animals. As the marennine-treated larvae showed significantly higher survival with the improved physiological condition than controls (larvae not treated with marennine) marennine can be an exciting probiotic for pathogen control in hatcheries.

2.16 FOOD AND COSMETICS INDUSTRIES

Currently, Brilliant Blue (FD&C Blue No. 1) and Indigo Carmine (FD&C Blue No. 2) are the most commonly used blue coloring agents in the food and beverage industries (Gastineau et al., 2018). Nonetheless, both blue coloring agents are synthetic. Therefore, consumers avoid them (Sharma et al., 2011) as there are growing concerns over the safety of consuming foodstuffs containing artificial additives (Phong et al., 2022). In response to the increasing consumer demand for natural food ingredients, it is imperative to search for naturally derived blue colorants for product developers (Buchweitz, 2016; Phong et al., 2022).

Blue-colored compounds are widely found in natural sources such as animals, plants, fungi, and microbes. These blue compounds are grouped into different chemical classes, such as flavonoids and quinones (Newsome et al., 2014). However, developing a new blue colorant from natural resources is challenging as it is difficult to replicate blue hues in food. The developed new blue color must meet specific criteria such as color intensity and solubility. Furthermore, the pigment should remain stable in heat, acid, light, and oxygen, should not interact with other ingredients or contribute flavor while also meeting the relevant safety requirements (Gastineau et al., 2018; Newsome et al., 2014).

Currently, only a few natural blue colorants are used in the food and beverage industry (Gastineau et al., 2018; Newsome et al., 2014). Phycocyanin is a blue pigment found in cyanobacteria and the two eukaryote algal genera, *Cryptophyta* and *Rhodophyta* (Eriksen, 2008). However, its use as a food and cosmetics colorant remains limited. Due to its poor stability, phycocyanin cannot replace the synthetic Brilliant Blue colorant in most foods and beverages (Newsome et al., 2014).

Because of their water solubility and color intensity, the applications of marennine and marennine-like pigments as natural blue pigments are of great interest to the food and cosmetics industries (Gastineau et al., 2018). However, as the color of marennine changes with pH, its application as a coloring agent poses a challenge to the food and beverage industry. Hence, more studies are required to evaluate the pigment's possible variation and long-term stability before its use as a food additive (Gastineau et al., 2018).

2.17 CONCLUSIONS AND PERSPECTIVES

Diatoms are a promising candidate for the sustainable production of a wide range of industrial commodities due to their richness in various bioactive compounds. These bioactive compounds have demonstrated a high economic value and market demand in the food, cosmetic, health, agricultural, and nutraceuticals industries. Natural-derived diatoms' bioactive chemicals are preferred over chemical-based commercial products because of their accessibility and fewer side effects in humans. With the increased interest in the use of novel metabolites produced from marine microorganisms, these compounds can be included in regular human diets for advanced health benefits. Therefore, it is necessary to develop a sustainable and cost-effective process to produce natural food ingredients to replace synthetic ones. Nevertheless, the exploitation of bioactive compounds from diatoms for industrial applications remains in its early stages, mainly due to the low technology readiness level and lack of economy of scale in diatoms' upstream and downstream processing. The scaling up of the manufacturing process relies on various factors such as cultivation and harvesting techniques, gene modification, and omics information related to genomics, proteomics, and metabolomics. Currently, the cultivation of diatoms for commercial use is unsustainable and prohibitive. Therefore, further study of the cultivation of diatoms in a cost-effective manner for future applications is necessary. Biomass harvesting, dehydration, and disruption are known to be the main energy-consuming steps that contribute to the overall high production cost. These challenges could be overcome with advancements in effective bioreactors developed by applying ecological and engineering principles and improved genetic tools. In addition, new extraction methods need to be designed to replace conventional time-consuming and solvent-intensive methods. For example, the use of green solvents (ionic liquids, supercritical fluids, deep eutectic solvents) accompanied by microwaves, pulsed electric fields, or ultrasound would be a viable option. Integrated research is needed in all areas, including extraction methods, bioactivity, chemical composition, and biotechnology, to address all the practical limitations and technical hurdles to make the diatom-based industry a commercial success in the future. With efficient large-scale cultivation systems and cost-effective processing, diatoms can be a great source to satisfy the growing demand for bioactive compounds for applications in the food, nutraceuticals, pharmaceuticals, feed, and aquaculture sectors.

REFERENCES

Abeer, M. M., Trajkovic, S., & Brayden, D. J. (2021). Measuring the oral bioavailability of protein hydrolysates derived from food sources: A critical review of current bioassays. *Biomedicine & Pharmacotherapy* 144: 112275.

Almeida, T. P., Ferreira, J., Vettorazzi, A., Azqueta, A., Rocha, E., & Ramos, A. A. (2018). Cytotoxic activity of fucoxanthin, alone and in combination with the cancer drugs imatinib and doxorubicin, in CML cell lines. *Environmental Toxicology and Pharmacology* 59: 24–33.

Alzahrani, M. A. J., Perera, C. O., & Hemar, Y. (2018). Production of bioactive proteins and peptides from the diatom Nitzschia laevis and comparison of their in vitro antioxidant activities with those from Spirulina platensis and Chlorella vulgaris. *International Journal of Food Science & Technology* 53(3): 676–682.

Ambati, R. R., Gogisetty, D., Aswathanarayana, R. G., Ravi, S., Bikkina, P. N., Bo, L., & Yuepeng, S. (2019). Industrial potential of carotenoid pigments from microalgae: Current trends and future prospects. *Critical Reviews in Food Science and Nutrition* 59(12): 1880–1902.

Amin, S. A., Parker, M. S., & Armbrust, E. V. (2012). Interactions between diatoms and bacteria. *Microbiology and Molecular Biology Reviews* 76(3): 667–684.

Barkia, I., Al-Haj, L., Abdul Hamid, A., Zakaria, M., Saari, N., & Zadjali, F. (2019). Indigenous marine diatoms as novel sources of bioactive peptides with antihypertensive and antioxidant properties. *International Journal of Food Science & Technology* 54(5): 1514–1522.

Barkia, I., Saari, N., & Manning, S. R. (2019). Microalgae for high-value products towards human health and nutrition. *Marine Drugs* 17(5): 304.

Batista, A. P., Gouveia, L., Bandarra, N. M., Franco, J. M., & Raymundo, A. (2013). Comparison of microalgal biomass profiles as novel functional ingredient for food products. *Algal Research* 2(2): 164–173.

Bayu, A., Rachman, A., Noerdjito, D., Putra, M., & Widayatno, W. (2020). *High-Value Chemicals from Marine Diatoms: A Biorefinery Approach.* Paper presented at the IOP Conference Series: Earth and Environmental Science.

Beattie, A., Hirst, E., & Percival, E. (1961). Studies on the metabolism of the Chrysophyceae. Comparative structural investigations on leucosin (chrysolaminarin) separated from diatoms and laminarin from the brown algae. *Biochemical Journal* 79(3): 531.

Bhattacharjya, R., Marella, T. K., Tiwari, A., Saxena, A., Singh, P. K., & Mishra, B. (2020). Bioprospecting of marine diatoms Thalassiosira, Skeletonema and Chaetoceros for lipids and other value-added products. *Bioresource Technology* 318: 124073.

Brembu, T., Jørstad, M., Winge, P., Valle, K. C., & Bones, A. M. (2011). Genome-wide profiling of responses to cadmium in the diatom Phaeodactylum tricornutum. *Environmental Science & Technology* 45(18): 7640–7647.

Brown, M. R., & Jeffrey, S. (1995). The amino acid and gross composition of marine diatoms potentially useful for mariculture. *Journal of Applied Phycology* 7(6): 521–527.

Buchweitz, M. (2016). Natural solutions for blue colors in food. In: *Handbook on Natural Pigments in Food and Beverages*, pp. 355–384. Elsevier.

Caballero, M. A., Jallet, D., Shi, L., Rithner, C., Zhang, Y., & Peers, G. (2016). Quantification of chrysolaminarin from the model diatom Phaeodactylum tricornutum. *Algal Research* 20: 180–188.

Chakrabarti, S., Guha, S., & Majumder, K. (2018). Food-derived bioactive peptides in human health: Challenges and opportunities. *Nutrients 10*(11): 1738.

Chakrabarti, S., Jahandideh, F., & Wu, J. (2014). Food-derived bioactive peptides on inflammation and oxidative stress. *BioMed Research International* 2014.

Chalamaiah, M., Ulug, S. K., Hong, H., & Wu, J. (2019). Regulatory requirements of bioactive peptides (protein hydrolysates) from food proteins. *Journal of Functional Foods* 58: 123–129.

Chalamaiah, M., Yu, W., & Wu, J. (2018). Immunomodulatory and anticancer protein hydrolysates (peptides) from food proteins: A review. *Food Chemistry* 245: 205–222.

Chen, C.-S., Anaya, J. M., Zhang, S., Spurgin, J., Chuang, C.-Y., Xu, C., …. Jiang, Y. (2011). Effects of engineered nanoparticles on the assembly of exopolymeric substances from phytoplankton. *PLoS ONE* 6(7): e21865.

Cheung, R. C. F., Ng, T. B., Wong, J. H., & Chan, W. Y. (2015). Chitosan: an update on potential biomedical and pharmaceutical applications. *Marine Drugs* 13(8): 5156–5186.

Chiovitti, A., Molino, P., Crawford, S. A., Teng, R., Spurck, T., & Wetherbee, R. (2004). The glucans extracted with warm water from diatoms are mainly derived from intracellular chrysolaminaran and not extracellular polysaccharides. *European Journal of Phycology* 39(2): 117–128.

Cho, S., Lee, D. H., Won, C.-H., Kim, S. M., Lee, S., Lee, M.-J., & Chung, J. H. (2010). Differential effects of low-dose and high-dose beta-carotene supplementation on the signs of photoaging and type I procollagen gene expression in human skin in vivo. *Dermatology* 221(2): 160–171. https://doi.org/10.1159/000305548

Colgrave, M. L., Dominik, S., Tobin, A. B., Stockmann, R., Simon, C., Howitt, C. A.,…. Vanhercke, T. (2021). Perspectives on future protein production. *Journal of Agricultural and Food Chemistry* 69(50): 15076–15083.

Dufossé, L., Galaup, P., Yaron, A., Arad, S. M., Blanc, P., Murthy, K. N. C., & Ravishankar, G. A. (2005). Microorganisms and microalgae as sources of pigments for food use: a scientific oddity or an industrial reality? *Trends in Food Science & Technology* 16(9): 389–406.

Durkin, C. A., Mock, T., & Armbrust, E. V. (2009). Chitin in diatoms and its association with the cell wall. *Eukaryotic Cell* 8(7): 1038–1050.

Eilertsen, H. C., Eriksen, G. K., Bergum, J.-S., Strømholt, J., Elvevoll, E., Eilertsen, K.-E.,…. Svenning, J. B. (2022). Mass cultivation of microalgae: I. Experiences with vertical column airlift photobioreactors, diatoms and CO_2 sequestration. *Applied Sciences* 12(6): 3082.

Elieh-Ali-Komi, D., & Hamblin, M. R. (2016). Chitin and chitosan: production and application of versatile biomedical nanomaterials. *International Journal of Advanced Research* 4(3): 411.

Eriksen, N. T. (2008). Production of phycocyanin—a pigment with applications in biology, biotechnology, foods and medicine. *Applied Microbiology and Biotechnology* 80(1); 1–14.

Feng, C., Li, J., Wu, G. S., Mu, Y. Z., Kong, M., Jiang, C. Q.,…. Chen, X. G. (2016). Chitosan-coated diatom silica as hemostatic agent for hemorrhage control. *ACS Applied Materials & Interfaces* 8(50): 34234–34243.

Fleurence, J. (2021). Perspectives on the use of algae in agriculture and animal production. *Phycology* 1(2): 79–82.

Fu, Y., Chen, T., Chen, S. H. Y., Liu, B., Sun, P., Sun, H., & Chen, F. (2021). The potentials and challenges of using microalgae as an ingredient to produce meat analogues. *Trends in Food Science & Technology* 112: 188–200.

Gastineau, R., Davidovich, N. A., Bardeau, J.-F., Caruso, A., Leignel, V., Hardivillier, Y.,…. Gaudin, P. (2012). *Haslea karadagensis* (Bacillariophyta): a second blue diatom, recorded from the Black Sea and producing a novel blue pigment. *European Journal of Phycology* 47(4): 469–479.

Gastineau, R., Hansen, G., Davidovich, N. A., Davidovich, O., Bardeau, J.-F., Kaczmarska, I.,…. Jacquette, B. (2016). A new blue-pigmented hasleoid diatom, Haslea provincialis, from the Mediterranean Sea. *European Journal of Phycology* 51(2): 156–170.

Gastineau, R., Hansen, G., Poulin, M., Lemieux, C., Turmel, M., Bardeau, J.-F.,…. Fleurence, J. (2021). *Haslea silbo*, a novel cosmopolitan species of blue diatoms. *Biology* 10(4): 328.

Gastineau, R., Hardivillier, Y., Leignel, V., Tekaya, N., Morançais, M., Fleurence, J.,…. Hellio, C. (2012). Greening effect on oysters and biological activities of the blue pigments produced by the diatom *Haslea karadagensis* (Naviculaceae). *Aquaculture* 368: 61–67.

Gastineau, R., Pouvreau, J.-B., Hellio, C., Morançais, M., Fleurence, J. l., Gaudin, P.,.... Mouget, J.-L. (2012). Biological activities of purified marennine, the blue pigment responsible for the greening of oysters. *Journal of Agricultural and Food Chemistry* 60(14): 3599–3605.

Gastineau, R., Prasetiya, F. S., Falaise, C., Cognie, B., Decottignies, P., Morançais, M.,.... Tremblay, R. (2018). Marennine-like pigments: Blue diatom or green oyster cult. *Blue biotechnology: Production and Use of Marine Molecules* 1: 529–551.

Gastineau, R., Turcotte, F., Pouvreau, J.-B., Morançais, M., Fleurence, J., Windarto, E.,.... Babin, M. (2014). Marennine, promising blue pigments from a widespread Haslea diatom species complex. *Marine Drugs* 12(6): 3161–3189.

Gille, A., Stojnic, B., Derwenskus, F., Trautmann, A., Schmid-Staiger, U., Posten, C.,.... Ribot, J. (2019). A lipophilic fucoxanthin-rich *Phaeodactylum tricornutum* extract ameliorates effects of diet-induced obesity in C57BL/6J mice. *Nutrients* 11(4): 796.

Gómez-Loredo, A., Benavides, J., & Rito-Palomares, M. (2016). Growth kinetics and fucoxanthin production of *Phaeodactylum tricornutum* and *Isochrysis galbana* cultures at different light and agitation conditions. *Journal of Applied Phycology* 28(2): 849–860.

Grzesik, M., Romanowska-Duda, Z., Piotrowski, K., & Janas, R. (2015). Diatoms (Bacillariophyceae) as an effective base of a new generation of organic fertilizers. *Przemysl Chemiczny* 94(3): 391–396.

Gügi, B., Le Costaouec, T., Burel, C., Lerouge, P., Helbert, W., & Bardor, M. (2015). Diatom-specific oligosaccharide and polysaccharide structures help to unravel biosynthetic capabilities in diatoms. *Marine Drugs* 13(9): 5993–6018.

Gurmu, F., Hussein, S., & Laing, M. (2014). The potential of orange-fleshed sweet potato to prevent vitamin A deficiency in Africa. *International Journal of Vitamin and Nutrition Results* 84(1–2): 65–78.

Guruvayoorappan, C., & Kuttan, G. (2007). β-Carotene inhibits tumor-specific angiogenesis by altering the cytokine profile and inhibits the nuclear translocation of transcription factors in B16F-10 melanoma cells. *Integrated Cancer Therapy* 6(3): 258–270. doi:10.1177/1534735407305978

Guvatova, Z., Dalina, A., Marusich, E., Pudova, E., Snezhkina, A., Krasnov, G.,.... Moskalev, A. (2020). Protective effects of carotenoid fucoxanthin in fibroblasts cellular senescence. *Mechanisms of Ageing and Development* 189: 111260.

Guzmán-Murillo, M. A., López-Bolaños, C. C., Ledesma-Verdejo, T., Roldan-Libenson, G., Cadena-Roa, M. A., & Ascencio, F. (2007). Effects of fertilizer-based culture media on the production of exocellular polysaccharides and cellular superoxide dismutase by *Phaeodactylum tricornutum* (Bohlin). *Journal of Applied Phycology* 19(1): 33–41.

Haug, A., & Myklestad, S. (1976). Polysaccharides of marine diatoms with special reference to *Chaetoceros* species. *Marine Biology* 34(3): 217–222.

Hecky, R., Mopper, K., Kilham, P., & Degens, E. (1973). The amino acid and sugar composition of diatom cell-walls. *Marine Biology* 19(4): 323–331.

Hempel, F., & Maier, U. G. (2012). An engineered diatom acting like a plasma cell secreting human IgG antibodies with high efficiency. *Microbial Cell Factories* 11(1): 1–6.

Hosikian, A., Lim, S., Halim, R., & Danquah, M. K. (2010). Chlorophyll extraction from microalgae: a review on the process engineering aspects. *International Journal of Chemical Engineering* 228: 391362.

Hussein, H. A., & Abdullah, M. A. (2020). Anticancer compounds derived from marine diatoms. *Marine Drugs* 18(7): 356.

Ishika, T., Moheimani, N. R., Bahri, P. A., Laird, D. W., Blair, S., & Parlevliet, D. (2017). Halo-adapted microalgae for fucoxanthin production: effect of incremental increase in salinity. *Algal Research* 28: 66–73.

Jayakumar, R., Menon, D., Manzoor, K., Nair, S. V., & Tamura, H. (2010). Biomedical applications of chitin and chitosan based nanomaterials—A short review. *Carbohydrate Polymers* 82(2): 227–232.

Jayakumar, R., Tamura, H., Nair, S., & Furuike, T. (2010). Perspectives of chitin and chitosan nanofibrous scaffolds in tissue engineering. In *Tissue Engineering*. InTech. Publishers, London, UK. doi: 10.5772/8593

Kamalanathan, M., Chiu, M.-H., Bacosa, H., Schwehr, K., Tsai, S.-M., Doyle, S.,.... Bretherton, L. (2019). Role of polysaccharides in diatom Thalassiosira pseudonana and its associated bacteria in hydrocarbon presence. *Plant Physiology* 180(4): 1898–1911.

Kanazawa, K., Ozaki, Y., Hashimoto, T., Das, S. K., Matsushita, S., Hirano, M.,.... Nakatsuka, M. (2008). Commercial-scale preparation of biofunctional fucoxanthin from waste parts of brown sea algae Laminalia japonica. *Food Science and Technology Research* 14(6): 573–573.

Kang, K. H., Qian, Z. J., Ryu, B., Karadeniz, F., Kim, D., & Kim, S. K. (2012). Antioxidant peptides from protein hydrolysate of microalgae *Navicula incerta* and their protective effects in HepG2/CYP2E1 cells induced by ethanol. *Phytotherapy Research* 26(10): 1555–1563.

Karami, Z., & Akbari-Adergani, B. (2019). Bioactive food derived peptides: A review on correlation between structure of bioactive peptides and their functional properties. *Journal of Food Science and Technology* 56(2): 535–547.

Karkhane Yousefi, M., Seyed Hashtroudi, M., Mashinchian Moradi, A., & Ghasempour, A. (2018). In vitro investigating of anticancer activity of focuxanthin from marine brown seaweed species. *Global Journal of Environmental Science and Management* 4(1): 81–90.

Khatoon, H., Banerjee, S., Yusoff, F., & Shariff, M. (2009). Evaluation of indigenous marine periphytic Amphora, Navicula and Cymbella grown on substrate as feed supplement in Penaeus monodon postlarval hatchery system. *Aquaculture Nutrition* 15(2): 186–193.

Kilpert, F., & Podsiadlowski, L. (2006). The complete mitochondrial genome of the common sea slater, *Ligia oceanica* (Crustacea, Isopoda) bears a novel gene order and unusual control region features. *BMC Genomics* 7(1): 1–18.

Kim, D. E., Cho, S. H., Park, H. M., & Chang, Y. K. (2016). Relationship between bone mineral density and dietary intake of β-carotene, vitamin C, zinc and vegetables in postmenopausal Korean women: a cross-sectional study. *Journal of International Medical Research* 44(5): 1103–1114.

Kim, S. M., Jung, Y.-J., Kwon, O.-N., Cha, K. H., Um, B.-H., Chung, D., & Pan, C.-H. (2012). A potential commercial source of fucoxanthin extracted from the microalga *Phaeodactylum tricornutum. Applied Biochemistry and Biotechnology* 166(7): 1843–1855.

Kolmakova, A., & Kolmakov, V. (2019). Amino acid composition of green microalgae and diatoms, cyanobacteria, and zooplankton. *Inland Water Biology* 12(4): 452–461.

Kong, Z.-L., Sudirman, S., Hsu, Y.-C., Su, C.-Y., & Kuo, H.-P. (2019). Fucoxanthin-rich brown algae extract improves male reproductive function on streptozotocin-nicotinamide-induced diabetic rat model. *International Journal of Molecular Sciences* 20(18): 4485. Retrieved from www.mdpi.com/1422-0067/20/18/4485

Koo, S. Y., Hwang, J.-H., Yang, S.-H., Um, J.-I., Hong, K. W., Kang, K.,.... Kim, S. M. (2019). Anti-obesity effect of standardized extract of microalga *Phaeodactylum tricornutum* containing fucoxanthin. *Marine Drugs* 17(5): 311.

Kuczynska, P., Jemiola-Rzeminska, M., & Strzalka, K. (2015). Photosynthetic pigments in diatoms. *Marine Drugs* 13(9): 5847–5881.

Kumar, M. N. R. (2000). A review of chitin and chitosan applications. *Reactive and Functional Polymers* 46(1): 1–27.

Kusaikin, M., Ermakova, S., Shevchenko, N., Isakov, V., Gorshkov, A., Vereshchagin, A.,.... Zvyagintseva, T. (2010). Structural characteristics and antitumor activity of a new chrysolaminaran from the diatom alga *Synedra acus. Chemistry of Natural Compounds* 46(1): 1–4.

Lakatos, G. E., Ranglová, K., Manoel, J. C., Grivalský, T., Kopecký, J., & Masojídek, J. (2019). Bioethanol production from microalgae polysaccharides. *Folia Microbiologica* 64(5): 627–644.

Lebeau, T., & Robert, J.-M. (2003). Diatom cultivation and biotechnologically relevant products. Part II: Current and putative products. *Applied Microbiology and Biotechnology* 60(6): 624–632.

Lincoln, R. A., Strupinski, K., & Walker, J. M. (1990). Biologically active compounds from diatoms. *Diatom Research* 5(2): 337–349.

Liu, C.-L., Liang, A.-L., & Hu, M.-L. (2011). Protective effects of fucoxanthin against ferric nitrilotriacetate-induced oxidative stress in murine hepatic BNL CL.2 cells. *Toxicology in Vitro* 25(7): 1314–1319.

Lopes, F. G., Oliveira, K. A., Lopes, R. G., Poluceno, G. G., Simioni, C., Da Silva Pescador, G.,.... Garcez, R. C. (2020). Anti-cancer effects of fucoxanthin on human glioblastoma cell line. *Anticancer Research,* 40(12): 6799–6815.

Lorenz, R. T., & Cysewski, G. R. (2000). Commercial potential for *Haematococcus* microalgae as a natural source of astaxanthin. *Trends in Biotechnology* 18(4): 160–167.

Lourenço-Lopes, C., Fraga-Corral, M., Jimenez-Lopez, C., Carpena, M., Pereira, A., Garcia-Oliveira, P.,.... Simal-Gandara, J. (2021). Biological action mechanisms of fucoxanthin extracted from algae for application in food and cosmetic industries. *Trends in Food Science & Technology* 117: 163–181.

Ma, M., Li, Y., Chen, J., Wang, F., Yuan, L., Li, Y.,.... Jin, H. (2021). High-cell-density cultivation of the flagellate alga *Poterioochromonas malhamensis* for biomanufacturing the water-soluble β-1, 3-glucan with multiple biological activities. *Bioresource Technology* 337: 125447.

Mal, N., Srivastava, K., Sharma, Y., Singh, M., Rao, K. M., Enamala, M. K.,.... Chavali, M. (2021). Facets of diatom biology and their potential applications. *Biomass Conversion and Biorefinery* 12(5): 1959–1975.

Maoka, T. (2020). Carotenoids as natural functional pigments. *Journal of Natural Medicines* 74(1): 1–16.

Matsumoto, Y., Tousen, Y., & Ishimi, Y. (2018). β-Carotene prevents bone loss in hind limb unloading mice. *Journal of Clinical Biochemistry and Nutrition* 63(1): 42–49.

McClure, D. D., Luiz, A., Gerber, B., Barton, G. W., & Kavanagh, J. M. (2018). An investigation into the effect of culture conditions on fucoxanthin production using the marine microalgae *Phaeodactylum tricornutum*. *Algal Research* 29: 41–48.

McNeela, E. A., O'Connor, D., Jabbal-Gill, I., Illum, L., Davis, S. S., Pizza, M.,.... Mills, K. H. (2000). A mucosal vaccine against diphtheria: formulation of cross reacting material (CRM197) of diphtheria toxin with chitosan enhances local and systemic antibody and Th2 responses following nasal delivery. *Vaccine* 19(9–10): 1188–1198.

Méresse, S., Fodil, M., Fleury, F., & Chénais, B. (2020). Fucoxanthin, a marine-derived carotenoid from brown seaweeds and microalgae: A promising bioactive compound for cancer therapy. *International Journal of Molecular Sciences* 21(23): 9273.

Mezzomo, N., & Ferreira, S. R. (2016). Carotenoids functionality, sources, and processing by supercritical technology: a review. *Journal of Chemistry* 2016: 3164312.

Mills, K. H., Cosgrove, C., McNeela, E. A., Sexton, A., Giemza, R., Jabbal-Gill, I.,.... Podda, A. (2003). Protective levels of diphtheria-neutralizing antibody induced in healthy volunteers by unilateral priming-boosting intranasal immunization associated with restricted ipsilateral mucosal secretory immunoglobulin a. *Infection and Immunity* 71(2): 726–732.

Miyashita, K., Beppu, F., Hosokawa, M., Liu, X., & Wang, S. (2020). Bioactive significance of fucoxanthin and its effective extraction. *Biocatalysis and Agricultural Biotechnology* 26: 101639.

Mock, T., Otillar, R. P., Strauss, J., McMullan, M., Paajanen, P., Schmutz, J.,.... Ward, B. J. (2017). Evolutionary genomics of the cold-adapted diatom *Fragilariopsis cylindrus*. *Nature* 541(7638): 536–540.

Mock, T., Samanta, M. P., Iverson, V., Berthiaume, C., Robison, M., Holtermann, K.,.... Rodesch, M. (2008). Whole-genome expression profiling of the marine diatom Thalassiosira pseudonana identifies genes involved in silicon bioprocesses. *Proceedings of the National Academy of Sciences* 105(5): 1579–1584.

Morganti, P., Palombo, P., Palombo, M., Fabrizi, G., Cardillo, A., Svolacchia, F.,.... Mezzana, P. (2012). A phosphatidylcholine hyaluronic acid chitin–nanofibrils complex for a fast skin remodeling and a rejuvenating look. *Clinical, Cosmetic and Investigational Dermatology* 5: 213.

Natsume, C., Aoki, N., Aoyama, T., Senda, K., Matsui, M., Ikegami, A.,.... Fujita, T. (2020). Fucoxanthin ameliorates atopic dermatitis symptoms by regulating keratinocytes and regulatory innate lymphoid cells. *International Journal of Molecular Sciences* 21(6): 2180.

Nekvapil, F., Brezestean, I., Lazar, G., Firta, C., & Pinzaru, S. C. (2022). Resonance Raman and SERRS of fucoxanthin: Prospects for carotenoid quantification in live diatom cells. *Journal of Molecular Structure* 1250: 131608.

Neumann, U., Derwenskus, F., Flaiz Flister, V., Schmid-Staiger, U., Hirth, T., & Bischoff, S. C. (2019). Fucoxanthin, a carotenoid derived from *Phaeodactylum tricornutum* exerts antiproliferative and antioxidant activities in vitro. *Antioxidants* 8(6): 183.

Newsome, A. G., Culver, C. A., & Van Breemen, R. B. (2014). Nature's palette: the search for natural blue colorants. *Journal of Agricultural and Food Chemistry* 62(28): 6498–6511.

Nymark, M., Valle, K. C., Brembu, T., Hancke, K., Winge, P., Andresen, K.,.... Bones, A. M. (2009). An integrated analysis of molecular acclimation to high light in the marine diatom *Phaeodactylum tricornutum*. *PLoS One* 4(11): e7743.

Oliyaei, N., Moosavi-Nasab, M., Tamaddon, A. M., & Tanideh, N. (2021). Antidiabetic effect of fucoxanthin extracted from Sargassum angustifolium on streptozotocin-nicotinamide-induced type 2 diabetic mice. *Food Science & Nutrition* 9(7): 3521–3529.

Pangestuti, R., & Kim, S.-K. (2011). Biological activities and health benefit effects of natural pigments derived from marine algae. *Journal of Functional Foods* 3(4): 255–266.

Patel, A. K., Albarico, F. P. J. B., Perumal, P. K., Vadrale, A. P., Ntan, C. T., Chau, H. T. B.,.... Saini, R. (2022). Algae as an emerging source of bioactive pigments. *Bioresource Technology* 351: 126910.

Peng, J., Yuan, J.-P., Wu, C.-F., & Wang, J.-H. (2011). Fucoxanthin, a marine carotenoid present in brown seaweeds and diatoms: metabolism and bioactivities relevant to human health. *Marine Drugs* 9(10): 1806–1828.

Phong, W. N., Gibberd, M., Payne, A. D., Dykes, G. A., & Coorey, R. (2022). Methods used for extraction of plant volatiles have potential to preserve truffle aroma: a review. *Comprehensive Reviews in Food Science and Food Safety* 1–25. doi:10.1111/1541-4337.12927

Phong, W. N., Le, C. F., Show, P. L., Chang, J. S., & Ling, T. C. (2017). Extractive disruption process integration using ultrasonication and an aqueous two-phase system for protein recovery from *Chlorella sorokiniana*. *Engineering in Life Sciences* 17(4): 357–369. https://doi.org/10.1002/elsc.201600133

Phong, W. N., Show, P. L., Le, C. F., Tao, Y., Chang, J.-S., & Ling, T. C. (2018). Improving cell disruption efficiency to facilitate protein release from microalgae using chemical and mechanical integrated method. *Biochemical Engineering Journal* 135: 83–90.

Phong, W. N., Show, P. L., Ling, T. C., Juan, J. C., Ng, E.-P., & Chang, J.-S. (2018). Mild cell disruption methods for bio-functional proteins recovery from microalgae—Recent developments and future perspectives. *Algal Research* 31: 506–516. https://doi.org/10.1016/j.algal.2017.04.005

Pletikapić, G., Radić, T. M., Zimmermann, A. H., Svetličić, V., Pfannkuchen, M., Marić, D.,.... Žutić, V. (2011). AFM imaging of extracellular polymer release by marine diatom *Cylindrotheca closterium* (Ehrenberg) Reiman & JC Lewin. *Journal of Molecular Recognition* 24(3): 436–445.

Poojary, M. M., Barba, F. J., Aliakbarian, B., Donsì, F., Pataro, G., Dias, D. A., & Juliano, P. (2016). Innovative alternative technologies to extract carotenoids from microalgae and seaweeds. *Marine Drugs* 14(11): 214.

Pouvreau, J.-B., Taran, F. d. r., Rosa, P., Pin, S., Fleurence, J. l., & Pondaven, P. (2008). Antioxidant and free radical scavenging properties of marennine, a blue-green polyphenolic pigment from the diatom *Haslea ostrearia* (Gaillon/Bory) Simonsen responsible for the natural greening of cultured oysters. *Journal of Agricultural and Food Chemistry* 56(15): 6278–6286.

Reimann, B. E., Leivin, J. C., & Volcani, B. E. (1966). Studies on the biochemistry and fine structure of silica shell formation in diatoms. II. The structure of the cell wall of *Navicula pelliculosa* (Breb.) Hilse. *Journal of Phycology* 2(2): 74–84.

Rhodes, C., Bingham, A., Heard, A. M., Hewitt, J., Lynch, J., Waite, R., & Bell, M. D. (2017). Diatoms to human uses: linking nitrogen deposition, aquatic eutrophication, and ecosystem services. *Ecosphere 8*(7): e01858.

Ristivojević, P., Jovanović, V., Opsenica, D. M., Park, J., Rollinger, J. M., & Velicković, T. Ć. (2021). Rapid analytical approach for bioprofiling compounds with radical scavenging and antimicrobial activities from seaweeds. *Food Chemistry* 334: 127562.

Rodríguez-Luna, A., Ávila-Román, J., González-Rodríguez, M. L., Cózar, M. J., Rabasco, A. M., Motilva, V., & Talero, E. (2018). Fucoxanthin-containing cream prevents epidermal hyperplasia and UVB-induced skin erythema in mice. *Marine Drugs* 16(10): 378.

Sachindra, N. M., Sato, E., Maeda, H., Hosokawa, M., Niwano, Y., Kohno, M., & Miyashita, K. (2007). Radical scavenging and singlet oxygen quenching activity of marine carotenoid fucoxanthin and its metabolites. *Journal of Agricultural and Food Chemistry* 55(21): 8516–8522.

Saini, R. K., Nile, S. H., & Park, S. W. (2015). Carotenoids from fruits and vegetables: Chemistry, analysis, occurrence, bioavailability and biological activities. *Food Research International* 76: 735–750.

Sánchez, A., & Vázquez, A. (2017). Bioactive peptides: A review. *Food Quality and Safety* 1(1): 29–46.

Sathasivam, R., & Ki, J.-S. (2018). A review of the biological activities of microalgal carotenoids and their potential use in healthcare and cosmetic industries. *Marine Drugs* 16(1): 26.

Sathya, R., MubarakAli, D., MohamedSaalis, J., & Kim, J.-W. (2021). A systemic review on microalgal peptides: Bioprocess and sustainable applications. *Sustainability* 13(6): 3262.

Scala, S., & Bowler, C. (2001). Molecular insights into the novel aspects of diatom biology. *Cellular and Molecular Life Sciences CMLS* 58(11): 1666–1673.

Seth, K., Kumar, A., Rastogi, R. P., Meena, M., & Vinayak, V. (2021). Bioprospecting of fucoxanthin from diatoms—Challenges and perspectives. *Algal Research* 60: 102475.

Sethi, D., Butler, T. O., Shuhaili, F., & Vaidyanathan, S. (2020). Diatoms for carbon sequestration and bio-based manufacturing. *Biology* 9(8): 217.

Sharma, V., McKone, H. T., & Markow, P. G. (2011). A global perspective on the history, use, and identification of synthetic food dyes. *Journal of Chemical Education* 88(1): 24–28.

Smetacek, V. (1985). Role of sinking in diatom life-history cycles: ecological, evolutionary and geological significance. *Marine Biology* 84(3): 239–251.

Spolaore, P., Joannis-Cassan, C., Duran, E., & Isambert, A. (2006). Commercial applications of microalgae. *Journal of Bioscience and Bioengineering* 101(2): 87–96.

Staats, N., De Winder, B., Stal, L. J., & Mur, L. R. (1999). Isolation and characterization of extracellular polysaccharides from the epipelic diatoms *Cylindrotheca closterium* and *Navicula salinarum*. *European Journal of Phycology* 34(2): 161–169.

Steinrücken, P., Prestegard, S. K., de Vree, J. H., Storesund, J. E., Pree, B., Mjøs, S. A., & Erga, S. R. (2018). Comparing EPA production and fatty acid profiles of three *Phaeodactylum tricornutum* strains under western Norwegian climate conditions. *Algal Research* 30: 11–22.

Stonik, V., & Stonik, I. (2015). Low-molecular-weight metabolites from diatoms: structures, biological roles and biosynthesis. *Marine Drugs* 13(6): 3672–3709.

Suh, J.-K. F., & Matthew, H. W. (2000). Application of chitosan-based polysaccharide biomaterials in cartilage tissue engineering: a review. *Biomaterials* 21(24): 2589–2598.

Svenning, J. B., Dalheim, L., Vasskog, T., Matricon, L., Vang, B., & Olsen, R. L. (2020). Lipid yield from the diatom *Porosira glacialis* is determined by solvent choice and number of extractions, independent of cell disruption. *Scientific Reports* 10(1): 1–10.

Svetličić, V., Žutić, V., Pletikapić, G., & Radić, T. M. (2013). Marine polysaccharide networks and diatoms at the nanometric scale. *International Journal of Molecular Sciences* 14(10): 20064–20078.

Tang, X., & Bi, G. (2016). Complete mitochondrial genome of *Fistulifera solaris* (Bacillariophycidae). *Mitochondrial DNA Part A,* 27(6); 4405–4406.

Tavares, R. S. N., Kawakami, C. M., Pereira, K. d. C., do Amaral, G. T., Benevenuto, C. G., Maria-Engler, S. S.,…. Gaspar, L. R. (2020). Fucoxanthin for topical administration, a phototoxic vs. photoprotective potential in a tiered strategy assessed by in vitro methods. *Antioxidants* 9(4); 328.

Tesson, B., Genet, M. J., Fernandez, V., Degand, S., Rouxhet, P. G., & Martin-Jézéquel, V. (2009). Surface chemical composition of diatoms. *ChemBioChem* 10(12): 2011–2024.

Tesson, B., & Hildebrand, M. (2013). Characterization and localization of insoluble organic matrices associated with diatom cell walls: insight into their roles during cell wall formation. *PLoS ONE* 8(4); e61675.

Tiwari, A., & Marella, T. K. (2019). Potential and application of diatoms for industry-specific wastewater treatment. In *Application of Microalgae in Wastewater Treatment* (pp.321–339) Springer, Cham.

Tiwari, A., Melchor-Martínez, E. M., Saxena, A., Kapoor, N., Singh, K. J., Saldarriaga-Hernández, S.,…. Iqbal, H. M. (2021). Therapeutic attributes and applied aspects of biological macromolecules (polypeptides, fucoxanthin, sterols, fatty acids, polysaccharides, and polyphenols) from diatoms—a review. *International Journal of Biological Macromolecules* 171: 398–413.

Tokoro, A., Takewaki, N., Suzuki, K., Mikami, T., Suzuki, S., & Suzuki, M. (1988). Growth-inhibitory effect of hexa-N-acetylchitohexanse and chitohexaose against Meth-A solid tumor. *Chemical and Pharmaceutical Bulletin* 36(2): 784–790.

Torres-Monroy, I., & Ullrich, M. S. (2018). Identification of bacterial genes expressed during diatom-bacteria interactions using an in vivo expression technology approach. *Frontiers in Marine Science 5*: 200.

Turcotte, F., Mouget, J.-L., Genard, B., Lemarchand, K., Deschênes, J.-S., & Tremblay, R. (2016). Prophylactic effect of *Haslea ostrearia* culture supernatant containing the pigment marennine to stabilize bivalve hatchery production. *Aquatic Living Resources* 29(4): 401.

Udenigwe, C. C., & Aluko, R. E. (2012). Food protein-derived bioactive peptides: production, processing, and potential health benefits. *Journal of Food Science* 77(1): R11–R24.

Urikura, I., Sugawara, T., & Hirata, T. (2011). Protective effect of fucoxanthin against UVB-induced skin photoaging in hairless mice. *Bioscience, Biotechnology, and Biochemistry* 75(4): 757–760.

Van Etten, C., Hubbard, J., Mallan, J. M., Smith, A., & Blessin, C. (1959). Amino acids in soybeans, amino acid composition of soybean protein fractions. *Journal of Agricultural and Food Chemistry* 7(2): 129–131.

Venkatesan, J., Vinodhini, P. A., Sudha, P. N., & Kim, S.-K. (2014). Chitin and chitosan composites for bone tissue regeneration. *Advances in Food and Nutrition Research* 73: 59–81.

Viera, I., Pérez-Gálvez, A., & Roca, M. (2019). Green natural colorants. *Molecules* 24(1): 154.

Vigani, M., Parisi, C., Rodríguez-Cerezo, E., Barbosa, M. J., Sijtsma, L., Ploeg, M., & Enzing, C. (2015). Food and feed products from micro-algae: Market opportunities and challenges for the EU. *Trends in Food Science & Technology* 42(1): 81–92.

Wu, J., Gu, X., Yang, D., Xu, S., Wang, S., Chen, X., & Wang, Z. (2021). Bioactive substances and potentiality of marine microalgae. *Food Science & Nutrition* 9(9): 5279–5292.

Xia, S., Gao, B., Fu, J., Xiong, J., & Zhang, C. (2018). Production of fucoxanthin, chrysolaminarin, and eicosapentaenoic acid by *Odontella aurita* under different nitrogen supply regimes. *Journal of Bioscience and Bioengineering* 126(6): 723–729.

Xia, S., Gao, B., Li, A., Xiong, J., Ao, Z., & Zhang, C. (2014). Preliminary characterization, antioxidant properties and production of chrysolaminarin from marine diatom *Odontella aurita. Marine Drugs* 12(9): 4883–4897.

Xia, S., Wang, K., Wan, L., Li, A., Hu, Q., & Zhang, C. (2013). Production, characterization, and antioxidant activity of fucoxanthin from the marine diatom *Odontella aurita. Marine Drugs 11*(7): 2667–2681.

Xiao, R., & Zheng, Y. (2016). Overview of microalgal extracellular polymeric substances (EPS) and their applications. *Biotechnology Advances* 34(7): 1225–1244.

Yaklich, R. W. (2001). β-Conglycinin and glycinin in high-protein soybean seeds. *Journal of Agricultural and Food Chemistry* 49(2): 729–735.

Yu, R.-x., Yu, R.-t., & Liu, Z. (2018). Inhibition of two gastric cancer cell lines induced by fucoxanthin involves downregulation of Mcl-1 and STAT3. *Human Cell* 31(1): 50–63.

Yu, Y., Shen, M., Song, Q., & Xie, J. (2018). Biological activities and pharmaceutical applications of polysaccharide from natural resources: A review. *Carbohydrate Polymers* 183: 91–101.

3 Genetic Engineering of Diatoms

Innovations and Future Perspectives

Mohammed Rehmanji[1,2,3]*, Asha Arumugam Nesamma*[1]*,*
Nida Jamil Khan[2]*, Tasneem Fatma*[2]*, and Pannaga Pavan Jutur*[1]
[1]Omics of Algae Group and DBT-ICGEB Centre for Advanced Bioenergy Research, Industrial Biotechnology, International Centre for Genetic Engineering and Biotechnology, New Delhi, India
[2]Department of Biosciences, Jamia Millia Islamia, New Delhi, India
[3]Camborne School of Mines and Environment and Sustainability Institute, University of Exeter, Penryn, Cornwall, UK

3.1 INTRODUCTION

Diatoms are a group of eukaryotic phytoplankton, which emerged some 180 million years ago, occurring in great diversity and abundance, with 100,000 species comprised of intricately delicate membrane designs in different sizes and forms due to the silicified cell wall nature (Bozarth et al., 2009; Siaut et al., 2007). These unicellular stramenopiles have an ecologically important role in maintaining the biogeochemical cycles by regulating and facilitating carbon, phosphate, and silicon cycles (Bidle et al., 2002; Lopez et al., 2005). Diatoms are primary producers and responsible for over 40% of the production and account for one-fifth of the oxygen humans breathe (Falkowski et al., 1998). Novel bioactive compounds with medical, pharmaceutical, and nutraceutical applications, including vegetarian proteins, omega-3, and other vital fatty acids (Wen and Chen, 2001a; Wen and Chen, 2001b; Rehmanji et al., 2022), antioxidant pigments like fucoxanthin (Xia et al., 2013; Rehmanji et al., 2022), chrysolaminarin, etc. (Figure 3.1) make these diatoms suitable candidates for multifaceted microalgal biorefineries (Bhattacharyya et al., 2021).

A specialized molecular toolkit has recently been credited to diatoms, which is promising as a viable chassis. The introduction of the full nuclear genomes of ten diatom species is now available at the Joint Genome Institute (JGI), Genomics User Facility (www.phycocosm.jgi.doe.gov/phycocosm/home) along with 26 diatom genomes that have so far been uploaded to the NCBI database (www.ncbi.nlm.nih.gov/datasets/genomes). Functional genomics tools are available for *Thalassiosira pseudonana* (Poulsen et al., 2006), *Phaeodactylum tricornutum* (Apt et al., 1996; Falciatore et al., 1999), and a few other diatom species (Siaut et al., 2007), such as targeted genetic engineering via transcription activator-like effector nucleases (TALENs) and CRISPR-Cas9 techniques (George et al., 2020). The recently shown extrachromosomal technique is also included in next-generation engineering efforts and large episomes containing multiple DNA segments can be retained and expressed without the need for genomic incorporation (Karas et al., 2015). These technologies are important in diatom genetics because they enable multi-gene piling procedures

DOI: 10.1201/9781003436553-3

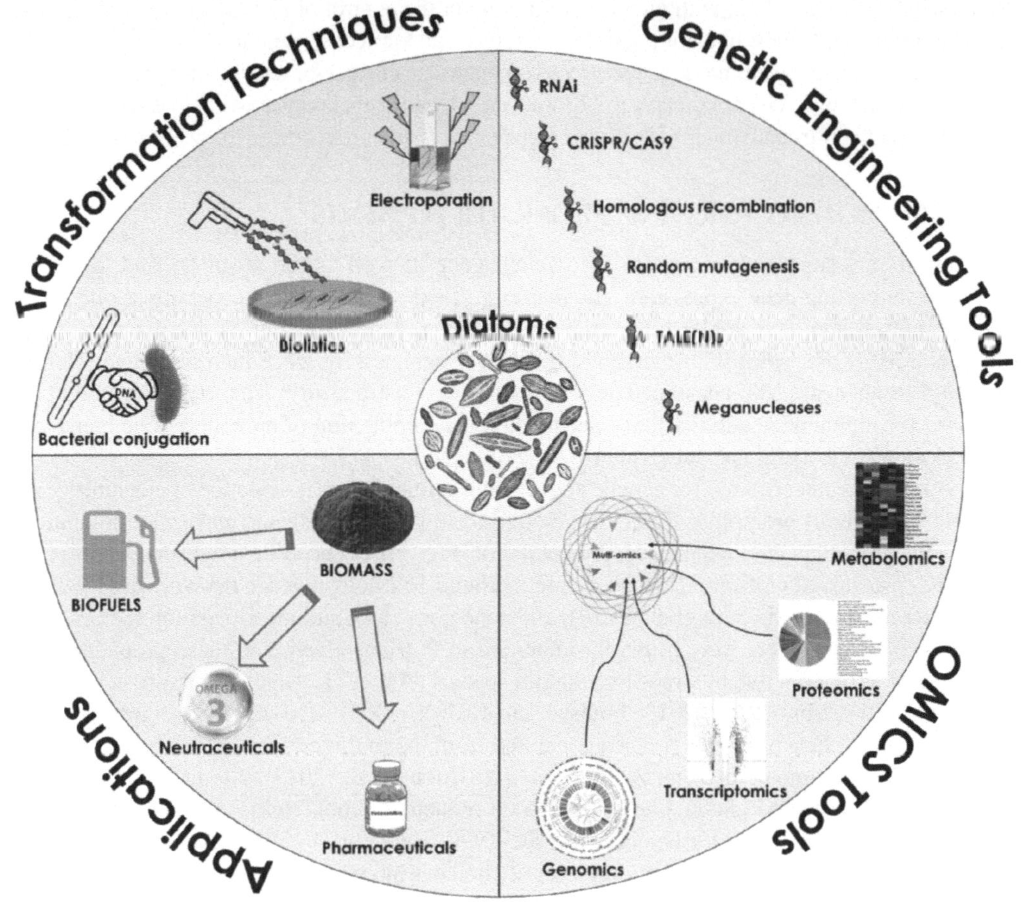

FIGURE 3.1 Schematic representation depicting different aspects of diatom genetic engineering showing various transformation techniques, gene-editing tools, omics tools, and applications of diatom biomass.

(Ainley et al., 2013; Goyal et al., 2009). Moreover, the recently created universal Loop (uLoop) assembly kit offers a set of important pieces for reconfigurable DNA assembly and high-throughput testing, and more complicated diatom gene-stacking designs (Pollak et al., 2019).

Currently a "100 Diatom Genomes Project (100DGP)" was sanctioned by the annual community science program (CSP) call, US Department of Energy–Joint Genome Institute (JGI) aiming to sequence 100 diatom species; this knowledge will be critical for advancing diatom-based biotechnology and synthetic biology platforms (www.jgi.doe.gov/csp-2021-100-diatom-genomes). A further 170 transcriptomes of diatoms were created (Sabatino et al., 2015). Altogether, the availability of these resources offers unparalleled potential of diatoms compared to other model algae. This chapter demonstrates the currently used genetic engineering techniques including advanced genome-editing technologies and their applications in fundamental and applied research aiding in improving our understanding of the cellular processes in diatoms.

3.2 DESIGN OF THE GENETIC TOOLBOX IN MODEL DIATOMS

Over the past few decades, there has been an advancement in the collection of genomic and transcriptome data, in cooperation with an improvement of screening technologies for the manufacture of particular molecules, along with vital knowledge about the control of diatom metabolism. Genetic manipulation methods have made it possible to engineer diatom cells' genetic heritage by changing, removing, and inserting genomic sequences. Such changes are critical for understanding gene function and boosting diatoms' metabolic capacity. Moreover, advanced techniques for delivering DNA into cells and further confirmatory steps for assuring gene expression are required for genetic engineering.

3.3 EXPRESSION VECTORS AND PROMOTER ELEMENTS

Gene carriers are supposed to contain the desired gene as well as the promoter and terminator sequences controlling gene expression. The majority of expression vectors can express one or two genes of interest and have applicability in overexpression, gene knockout, co-localization studies (Liu et al., 2016) and work with metabolic pathways regulated by more than one gene (Hempel et al., 2011; Siaut et al., 2007) designed destination vector constructs in *P. tricornutum* (pDEST) that were used for immunodetection, affinity purification [via production of an epitope-like hemagglutinin (HA)-tagged protein] and subcellular localization studies.

A second important criterion for genetic engineering studies includes selection of constitutive and inducible endogenous promoters (Table 3.1) employed to induce transgenic expression in diatoms. Endogenous promoters have been discovered in diatoms for genes encoding a light-harvesting complex (LHC) protein (Lhcf, formerly known as fcp; around 15 Lhcf genes are known, with Lhcf1–15 having been identified) (Lepetit et al., 2010), a histone gene (h4), and the elongation factor 2 (ef2) genes. The Lhcf promoters were primarily identified in *P. tricornutum*, and these promoters were demonstrated to be effective in a range of diatom species (Apt et al., 1996; Falciatore et al., 1999; Ifuku et al., 2015; Muto et al., 2013; Poulsen and Kröger, 2005). Red algal-like light-harvesting complex protein (Lhcr) promoter (Lhcr5) generated from the multigene family has been employed as a constitutive promoter in *Chaetoceros gracilis* (Ifuku et al., 2015). Along with promoters, terminators are limited to Lhcf1, Lhcf9, nr (nitrate reductase), rbcL (rubisco small subunit), and Lhcr14 genes (Apt et al., 1996; Falciatore et al., 1999; Karas et al., 2015; Poulsen and Kröger, 2005; Xie et al., 2014). One important promoter of choice is h4, which stimulates light-independent expression at reduced levels compared with Lhcf promoters, facilitating inducible trigger for gene expression (De Riso et al., 2009). Moreover, these inducible promoters may be utilized to regulate transgene expression reversibly in an effective way (Table 3.1).

The nitrate reductase (nr) promoter is often utilized as an inducible diatom promoter. This promoter is activated in the presence of ammonium ions along with the presence of nitrate in the medium (Poulsen and Kröger, 2005). The nr promoter was discovered and employed in *T. pseudonana*, *P. tricornutum*, and *Chaetoceros* after various studies carried out in *Cylindrotheca fusiformis* (Ifuku et al., 2015; Poulsen et al., 2006; Poulsen and Kröger, 2005; Schellenberger Costa et al., 2013). However, the promoter nr retains activity in the absence of nitrate, resulting in modest gene expression (Chu et al., 2016). Promoters activated by iron deficiency and derived from the iron starvation-induced protein 1 (Isi1), ferrichrome-binding protein 1 (fbp1), and flavodoxin (fld) genes (Yoshinaga et al., 2014) have been discovered, opening up new avenues for gene regulation in diatoms. *P. tricornutum*-derived Lhcf2 promoter utilized in *Fistulifera* species (Muto et al., 2013), and the *C. fusiformis*-derived Lhcf1 promoter used in *P. tricornutum* (Kadono et al., 2015), are two examples of promoters which induce transgenic expression without interfering with endogenous regulatory networks (Harada et al., 2005).

Furthermore, viral promoters like CaMV (cauliflower mosaic virus) promoter, the CMV (cytomegalovirus) promoter, and the RSV (Rous sarcoma virus) promoter are utilized for reporter gene expression in *P. tricornutum* (Sakaue et al., 2008). Similarly, diatom-infecting virus-derived ClP1 promoter may induce stable expression at higher levels as compared to endogenous and other viral promoters (Kadono et al., 2015). In the present scenario, the mentioned examples of promoters have been used to successfully drive transgenic expression in diatoms, however substantial advances in

TABLE 3.1
Endogenous Promoters Used in Marine Diatoms Species Derived from In-House Genes

Promoters	Category	Gene associated	References
fcpB promoter	Constitutive	fucoxanthin chlorophyll a/c-binding protein B gene	Apt et al. (1996)
fcpC promoter		fucoxanthin chlorophyll a/c-binding protein C gene	
fcpD promoter		fucoxanthin chlorophyll a/c-binding protein D gene	Slattery et al. (2018)
fcpE promoter		fucoxanthin chlorophyll a/c-binding protein E gene	Apt et al. (1996)
fcpF promoter		fucoxanthin chlorophyll a/c-binding protein F gene	Falciatore et al. (1999)
pRBCMT		ribulose-1,5-bisphosphate carboxylase/ oxygenase small subunit N-methyltransferase I gene	Slattery et al. (2018)
EF2 promoter		elongation factor 2 gene	Seo et al. (2015)
GLNA promoter		glutamine synthetase gene	Erdene-Ochir et al. (2016)
HASP1 promoter		highly abundant secreted protein 1 gene	Hildebrand et al. (2017)
p40SRPS8		40S ribosomal protein S8 gene	Slattery et al. (2018)
pAct2		actin-like gene	Adler-Agnon et al. (2018)
pEF-1α		elongation factor-1 alpha gene	Slattery et al. (2018)
pH4–1B		histone H4 gene	
pPUP		purine permease-like transporter gene	
pγ Tubulin		tubulin gamma chain gene	
Pt211		hypothetical protein (PHATRDRAFT_49211*) gene	Zou et al. (2018)
V-ATPaseC promoter		vacuolar ATP synthase 16-kDa proteolipid subunit gene	Watanabe et al. (2018)
ptca1 promoter	Inducible	beta-carbonic anhydrase 1 gene (-1292 to +61 from the transcription start site)	Harada et al. (2006)
nr promoter (pNR)		nitrate reductase gene	Chu et al. (2016)
pAMT		ammonium transporter gene	Adler-Agnon et al. (2018)
pDGAT1		acyl-CoA diacylglycerol acyltransferase gene	Shemesh et al. (2016)
PIsi1		iron starvation induced protein 1 (-1021 to +214 from the transcription start site)	Yoshinaga et al. (2014)
PFBP1		ferrichrome binding protein 1 (-1097 to +9 from the transcription start site)	
PFld		flavodoxin (-688 to +407 from the transcription start site)	
pPhAP1		alkaline phosphatase gene	Lin et al. (2017)

heterologous expression may lead to further discovery of novel promoters that powerfully drive the transcription of numerous transgenes and broaden the canvas of the microalgal genetic toolkit.

3.4 SELECTABLE MARKERS

To assist the separation of genetic transformants, selectable markers imparting antibiotic resistance have been designed for diatoms (Table 3.2). Nourseothricin (nat), neomycin (nptII), and phleomycin/zeocin (sh *ble*) are the most often employed selectable markers in diatoms, and the optimum doses vary across species. Zeocin, at a dosage of 50 μg mL^{-1}, would be a superior antibiotic for the selection of *P. tricornutum* (Apt et al., 1996). The protein resistant to zeocin is produced on the expression of sh *ble* gene, which is utilized for the screening of diatom transformants (Table 3.2). Moreover, reports mention the linkage of sh *ble* gene and resistance to the antibiotic phleomycin (Karas et al.,

TABLE 3.2
Summary of Transformation Techniques, Selectable Markers, and Target Organelles Involved in Genetic Engineering of Diatoms

Diatom species	Transformation method	Specifications	Selection markers	Antibiotic concentrations	Target organelle	References
Thalassiosira pseudonana	Electroporation		Nourseothricin	100 μg mL^{-1}	Nucleus	Samukawa et al. (2014)
	Bacteria-mediated conjugative delivery	Conjugation with *E. coli*	Phleomycin and nourseothricin			Karas et al. (2015)
	Microparticle bombardment	–	Nourseothricin			Poulsen et al. (2006)
Phaeodactylum tricornutum	Electroporation	The pulsed diatoms were maintained in non-selective medium before spreading into selection medium	Nitrate reductase			Niu et al. (2012)
	Microparticle bombardment		Zeocin			
		–			Mitochondria (translocation into mitochondria)	Seo et al. (2018)
		He pressure	DCMU, streptomycin, and zeocin		Chloroplast	Materna et al. (2009)
	Bacteria-mediated conjugative delivery	Conjugation with *E. coli.*	Phleomycin and nourseothricin	20–300 μg mL^{-1}	Nucleus	Karas et al. (2015)
Chaetoceros gracilis	Electroporation	The pulsed diatoms were incubated 16–20 h in non-selective medium before spreading into selection medium	Nourseothricin	400 μg mL^{-1}		Ifuku et al. (2015)
Cyclotella cryptica	Microparticle bombardment	He pressure	Geneticin	50–100 μg mL^{-1}		Dunahay et al. (1995)
Navicula saprophila						
Seminavis robusta			Nourseothricin	–		Kirupamurthy (2014)
Amphora coffeaeformis				–		Buhmann et al. (2014)
Cylindrotheca fusiformis			Zeocin	1 mg mL^{-1}		Fischer et al. (2002)
Fistulifera solaris			Genticin	–	Nucleus, chloroplast (translocation into chloroplast)	Sunaga et al. (2015)

2015) other than zeocin and phleomycin; G418 and nourseothricin encoded by the nptII and nat genes, respectively, are also utilized for diatom selection (Buhmann et al., 2014; Karas et al., 2015; Muto et al., 2013). The genetic reporter methods frequently employed to detect and characterize diatom promoters have also substantially aided investigations of gene expression and regulation, as well as protein spatial localization. The expression of several reporter genes (uidA–β-glucuronidase) and numerous kinds of GFP (green fluorescent proteins) were observed in diatoms (Velmurugan and Deka, 2018). Reporter genes, such as the bacterial β-glucoronidase gene (GUS, also known as uidA) (Huang and Daboussi, 2017), and the firefly luciferase (LUC) gene (Falciatore et al., 1999), can be evaluated based on enzymatic activity, which aids in measuring promoter activity. Green fluorescent protein (GFP) (Huang and Daboussi, 2017; Karas et al., 2015), cyan fluorescent protein (CFP) (Karas et al., 2015), and yellow fluorescent protein (YFP) (Huysman et al., 2015) have been utilized as reporters to detect protein localization *in vivo*. Under the direction of the fcpA promoter, the diatom *Amphora coffeaeformis* was demonstrated to produce yellow fluorescent protein (yfp) fusion with the tyrosine-rich adhesion protein from mussel footpads (Buhmann et al., 2014). The importance of codon use in the effective expression of reporter genes in diatoms reveals that a wide range of selectable markers and reporter genes may be expressed in the diatom species, increasing the organism's potential for investigating basic biology problems and industrial applications.

3.5 TRANSFORMATION TECHNIQUES IN DIATOMS

Transformation results in permanent changes in the phenotype due to genetic manipulation of the host system. Various methods are employed for delivering foreign DNA into the diatom system, which include electroporation, biolistic, and bacterial conjugation (Figure 3.1; Table 3.2). Electroporation is one of the easier and conventional genetic transformation techniques. It uses a high electrical pulse to stimulate pore development in the membrane of the cell and various protocols are utilized and optimized by changing the parameters (pulse length, electrical field, etc.) (Miyahara et al., 2013; Niu et al., 2012; Zhang and Hu, 2014). The first successful transformation using electroporation was established in *P. tricornutum* (Niu et al., 2012). The transformation efficiency using the electroporation technique is 10–100-fold higher than for the biolistic method (Miyahara et al., 2013; Zhang and Hu, 2014).

Another method, i.e., biolistic, uses fine particles of microgold or tungsten, for delivering coated DNA directly into diatom cells. This approach has the benefit of being applicable to a wide range of diatom species. Microparticle bombardment was used to produce around 90% of gene changes in diatoms. The transformation efficiency of diatom cells ranged from 10^7 to 10^8 cells μg^{-1} DNA (Daboussi et al., 2014; Falciatore et al., 1999; Kadono et al., 2015; Nymark et al., 2016) (Table 3.2). Finally, the episome delivery mechanism is based on bacterial conjugation; with a higher transformation efficiency (1000 times greater than biolistic), this method is useful for both DNA delivery and long-term protein synthesis (Karas et al., 2015). This technique avoids the primary issue of uncontrolled genetic alterations caused by plasmids' random incorporation into chromosomes. Transformation using a low-copy number episomal vector was performed in the diatoms *P. tricornutum* and *Thalassiosira pseudonana*. The yeast-based replication region was used to create the episomal vector and was transformed utilizing a bacterial-based conjugation process mediated by *Escherichia coli*. This technique has the potential to transfer complete metabolic components into diatoms, as well as for diatom screening and advanced research into diatoms.

3.6 PLASTID TRANSFORMATION

Genetic modification of the chloroplast genome has gained a lot of interest because of its relevance for photosynthesis and its ability to function as a central cellular energy system. This organelle contains the necessary features for efficient and stable gene expression, a high frequency of homologous recombination, operon system, and is devoid of epigenetic markers that ensures stability in gene expression (Purton et al., 2013).

Previously, *P. tricornutum* plastid transformation by homologous recombination has been described in two investigations (Materna et al., 2009; Xie et al., 2014). Alternate tactics have been widely utilized in the diatom community since plastid transformation mechanisms are not well established. For example, the gene of interest can be linked to a plastid-targeting signal sequence and integrated into the nuclear genome, allowing the protein to be generated and delivered into plastids. A signal peptide is followed by a transit peptide-like region in plasmid-targeting signal sequences. These bipartite sequences are enough for plastid import (Apt et al., 1996; Kilian and Kroth, 2005). Plastid import, on the other hand, necessitates the existence of the conserved "ASAFAP"-motif between the signal peptide and the transit peptide-like domains (Kilian and Kroth, 2005). Fusion to the N-terminus of GFP confirmed the functionality of plastid-targeting signal sequences (Apt et al., 1996; Gruber et al., 2007; Kilian and Kroth, 2005). The chloramphenicol resistance gene, chloramphenicol acetyltransferase (cat), was successfully inserted between the trnI and trnA genes in the *P. tricornutum* plastid genome's inverted repeat regions using an electroporation technique (Xie et al., 2014).

No matter the targeted result, these molecular and genetic tools offer potent methods to regulate and control the required gene expression. However, the creation of a comprehensive set of genetic tools in diatoms remains unfinished. The genetic diversity across species, which limits the use of genetic tools from one species to another, has been one obstacle in applying molecular tools in these tiny photosynthetic organisms. However, as the cost of genome sequencing decreases, more strains may be easily identified, and the tools customized to their genetic code. Furthermore, as compared to previous industrial microbe platforms, the present toolsets have substantially enhanced the speed and accuracy with which diatoms may be engineered to optimize production. More broadly, applicable genetic toolkits are being designed to enable many species of diatoms to function as industrial production systems.

3.7 GENETIC ENGINEERING APPROACHES EMPLOYED IN DIATOM SPECIES

The regulation of gene expression via overexpression and knockout generation of gene targets using genome-editing technologies via double-strand-break (DSB) mechanisms are regular approaches employed for gene manipulation. Forward genetic and reverse genetic engineering methodologies can be used to investigate gene function. The former method uses random insertional mutagenesis using a selectable marker gene introduced using the vector followed by screening for a particular phenotype (Zhang et al., 2014). Meanwhile, reverse genetic techniques entail the alteration of a single gene, and the purpose of this strategy is to explore the implications of variations in a specific gene and so deduce its function (George et al., 2020). The diatom community has a preference toward reverse genetics for relating physiological processes to gene activities, as well as the problems of gene inactivation (both gene alleles get inactivated by random insertion) (Daboussi et al., 2014). Further, in the upcoming section we discuss methods which were conventionally utilized for gene expression/silencing and their counterparts in terms of recent developments.

3.8 CONVENTIONAL METHODS

The recent advancement in improvements of genetic tool kits have made it feasible to modulate the expression of a given gene. Further, it aids in protein localization, reconstruction of metabolic pathways, and developing knockouts and functional genomics studies. For diatoms, methods based on gene expression knock-down have been developed, allowing genes to be linked to specific functions, for example, RNA interference (RNAi) (De Riso et al., 2009; Siaut et al., 2007; Trentacoste et al., 2013) and artificial miRNAs (Kaur and Spillane, 2015). RNAi-mediated methods are incredibly potent for modifying gene expression, having already been used to validate the activities of important protein targets in metabolic networks and to boost the synthesis of valuable chemicals (De Riso et al., 2009; Huysman et al., 2015; Siaut et al., 2007; Trentacoste et al., 2013). The GUS reporter gene was used to demonstrate the notion of RNAi (De Riso et al., 2009). Antisense (inverted repeat) constructs were introduced into *P. tricornutum* for inhibiting GUS production.

3.9 SITE-SPECIFIC GENE MANIPULATION

Gene-specific alterations using double-stranded DNA break (DSB) repair mechanisms have recently appeared as a game-changing tool. This targeted approach provides significant benefits over traditional methods in the sense that it aids in transgenic integration at a particular locus, which eliminates the difficulties of genomic inconsistency, and the lack of transgenic expression that might occur with traditional approaches. Secondarily, it may be used to create gene-specific knock-outs/mutants, which is a particularly tough process in diploid organisms but is necessary for certain results. Molecular scissors capable of inducing DSBs at a particular locus upgrade the lower frequency (less than 1026) of homologous recombination by at least three orders of magnitude in a microalgal system (Daboussi et al., 2014; Rouet et al., 1994). Three types of sequence-specific nucleases (meganucleases, TALENs, and CRISPR/Cas9) have been successfully employed to induce targeted alterations of diatom genomes. These are naturally derived gene-editing tools with fundamental activities for altering genomes.

Two factors impede diatom genome editing: first, diatoms are diploids, which means that a mutagenesis event needs to exist in both alleles to neutralize the target sequence; and second, the majority of DSBs are repaired successfully which results in false positives with a mixed population of cells (with and without mutations), as well as mosaicism. As a result, an additional sub-cloning step is required (Seligman et al., 2002).

The initial idea for targeted genome changes in diatoms was described previously; using two types of designer nucleases, meganucleases and TALENs, in *P. tricornutum* producing a high frequency of both targeted mutants and homologous recombinants (Daboussi et al., 2014; Seligman et al., 2002). Further metabolic engineering in diatoms was initiated by development of a strain generating huge amounts of lipids by deactivating a critical gene involved in starch biosynthesis. TALEN technology has since been employed in diatoms for genetic engineering (Fortunato et al., 2016; Weyman et al., 2014) (Table 3.3).

Further, the CRISPR/Cas9 method was employed in *P. tricornutum* and *T. pseudonana* by two groups (Hopes et al., 2016; Nymark et al., 2016), demonstrating genome editing by expressing codon-optimized Cas9 using Lhcf2 promoter, and placing the guide RNA under the control of the *P. tricornutum* U6 snRNA promoter (Nymark et al., 2016) (Table 3.3).

3.10 RECOMBINANT GENE-EDITING TECHNIQUES

Strain engineering to improve microalgal productivity by genetic or metabolic alterations or in vivo development of appropriate mutants using recombinant genetic techniques generates strains with exceptional potential (Park et al., 2019). For instance, the complement creation of the trp1 phenotype in *Saccharomyces cerevisiae* utilizing the *P. tricornutum* cDNA pool (Lippmeier et al., 2002). Recently, pPhaT-EF2-Luc vector with a replaced promoter sequence [elongation factor 2 (EF2) promoters instead of the fcpB promoter] was utilized for over-expression of luciferase activity in *P. tricornutum* and *Chlamydomonas reinhardtii* under varied light exposures (Seo et al., 2015). Moreover, a heterologous gene expression cassette incorporating CAT reporter gene via electroporation in diatoms was successful (Xie et al., 2014). Hence, these genetic engineering approaches resulted in multigene targeting models (Falciatore et al., 2020; Gruber and Kroth, 2017; Serif et al., 2018).

TABLE 3.3
Gene Targets and Proteins Expressed in Diatom Species via Advanced Genetic Engineering Technologies

Diatom species	Targets	Genome editing method	References
Phaeodactylum tricornutum (CCAP 1055/1)	• Expression of Acyl-ACP thioesterase, increased accumulation of shorter chain. • Malic enzyme for lipid production. • Over expression of G6PD, enhanced lipid productivity. • Heterologous biosynthesis of the MIAs by CrGES expression under phototrophic conditions. • Vanillin production. • PHBs for Bioplastics production. • Human IgGaHBsAg and IgG1/kappa Ab CL4mAb: antibody to hepatitis B virus surface protein against the nucleoprotein of Marburg virus. • Over expression of DXS increased fucoxanthin synthesis. • Generation of a lipid producing strain through the disruption of the UDP-glucose pyrophosphorylase gene.	Targeted mutagenesis methods: meganucleases, gene knockouts, TALENS, and CRISPR/Cas9	Radakovits et al. (2011) Trentacoste Emily et al. (2013) Wu et al. (2019) Erdene-Ochir et al. (2016), Slattery et al. (2018) Eilers et al. (2016), Hempel et al. (2011) Hempel et al. (2011) Daboussi et al. (2014) Poulsen and Kröger (2005)
Thalassiosira pseudonana (CCMP 1335)	• Localization of SiMat1-GFP. • Expression of the protective HsIbpA DR2 antigen for the production of a vaccine against bovine respiratory disease, • Expression of scFvTNT scFv and sdAbEA1 to detect Bacillus anthracis.		Davis et al. (2017) Ford et al. (2016) Kotzsch et al. (2016) Shrestha and Hildebrand (2014) Weyman et al. (2014) Muto et al. (2015)
Thalassiosira weissflogii (CCMP1030)	• Overexpression of the endogenous glycerol kinase improve lipid productivity.		
Fistulifera solaris (JPCC DA058)			

3.11 DIATOMS-OMICS

To uncover the metabolic processes related with diatom species, comprehensive techniques to discover, understand, and manage biological data were required. Omics intervention, namely the sequencing of diatom genomes, has changed diatom study worldwide. Furthermore, these studies have contributed transcriptome and proteome information to aid in the annotation of the whole genome. Some diatom strains, including *T. pseudonana* (Koester et al., 2018) and *P. tricornutum* (Rastogi et al., 2020), have recently been re-sequenced using microsatellite-based genotyping approaches, revealing essential polymorphisms along with evidence for changes in evolutionary pattern, genome structure, and environmental adaptability. For example, sequenced genome of *F. solaris* demonstrated the presence of allodiploidy and distinct evolutionary architecture along with genetic regulatory mechanisms (Tanaka et al., 2015). In the upcoming section, efforts are made to describe the role of various omics tools in diatom genetics study, and their application in understanding diatom physiology, metabolism, response to environmental conditions, and product formation mechanisms (Figure 3.1).

3.12 GENOMICS

Genome sequencing has proved to be a useful tool in understanding the diatom system, further aiding in rebuilding of metabolic pathways, elucidating evolutionary linkages, etc., and adding to the knowledge pool of systems biology. For example, the genome-derived methylome datasets of the oleaginous alga *Cyclotella cryptica* suggested an alternate system for pyruvate metabolism, by identifying key enzymes for triacylglycerol synthesis (Traller et al.. 2016) (Table 3.4). Further, the genome of *Seminavis robusta* recently published identified its phylogenetic linkage to phylum bacillariophyta, including the greatest number of protein-coding genes (36,254 and 438 RNA genes from organelle genomes). Moreover, its expression profiles show the effect of tandem duplication events that occurred previously resulting in about 70% of the genes being present as multicopy gene families, expanding the variety of gene activities (Osuna-Cruz et al., 2020). Annotation of *P. tricornutum* genome aided in the construction of a genome-scale metabolic model to predict metabolic flux distribution along with its biochemical composition, clarifying its potential for bioprospecting (Levering et al., 2016). Retrotransposons (RTs) and related elements have been identified in diatoms. These sequences aid in the prediction of genomic diversity and adaptive evolution. It has been observed that in the diatom genome there is a prevalence of three groups of copia elements (LTR), CoDi-1, CoDi-2, and CoDi6 (Aziz et al., 2010; Lescot et al., 2016).

TABLE 3.4
Depiction of Complete Sequenced Genome of Selected Diatoms and Related Parameters

Diatom species	Habitat	Genome size (Mbp)	Number of genes	Available genome database	References
Thalassiosira pseudonana	Marine	32.4	11,776	https://genome.jgi.doe.gov/Thaps3/Thaps3.home.html	Armbrust et al. (2004)
Thalassiosira oceania	oceanic	81.6	34,362	https://genome.jgi.doe.gov/Thaoce1/Thaoce1.info.html	Lommer et al. (2012)
Cyclotella cryptica	Marine	161.7	21,121	http://genomes.mcdb.ucla.edu/Cyclotella/download.html	Traller et al. (2016)
Skeletonema costatum	Coastal Marine	46.9	16,449	–	Ogura et al. (2018)
Phaeodactylum tricornutum	Coastal	27.4	10,402	http://protists.ensembl.org/Phaeodactylum_tricornutum/Info/Index	Bowler et al. (2008)
Fragilariopsis cylindrus	Cold adapted	61	21,006	https://genome.jgi.doe.gov/Fracy1/Fracy1.info.html	Mock et al. (2017)
Fistulifera solaris	Marine	49.7	20,455	https://trace.ddbj.nig.ac.jp/DRASearch/submission?acc5DRA002403	Tanaka et al. (2015)
Seminavis robusta	Benthic	117-151	36,254	https://genome.jgi.doe.gov/portal/Semrobnscriptome/Semrobnscriptome.info.html	Osuna-Cruz et al. (2020)
Fragilaria radians	Freshwater	98.4	11,184	-	Galachyants et al. (2019)
Pseudo-nitzschia multistriata	Marine	59	12,008	http://apollo.tgac.ac.uk/Pseudo-nitzschia_multistriata_V1_4_browser/sequences	Basu et al. (2017)

Various portals and websites containing diatom-related databases are listed as follows: Joint Genome Institute, BioCyc, Ensembl Protists, Pico-Plaza, Genome Portal for Algae, PlaNet, DiatomCyc databases, Ocean Gene Atlas, Marine Atlas of Tara Ocean Unigenes (MATOU), JGI IMG (Integrated Microbial Genomes and Microbiomes) (Chen et al., 2019; Villar et al., 2018); overall these databases provide a catalogue consisting of diatom-related information (Table 3.4).

3.13 TRANSCRIPTOMICS

The Gordon and Betty Moore Foundation proposed the Marine Microbial Eukaryotic Transcriptome Sequencing Project (MMETSP) in 2014 as a global initiative to re-annotate and curate metagenomic and metatranscriptomic datasets of various protists (Keeling et al., 2014). Metatranscriptomic investigations revealed novel insights into cellular processes and resource utilization of these cell factories (Alexander et al., 2015a; Alexander et al., 2015b; Cuvelier Marie et al., 2010; Delmont et al., 2015).

Transcriptome analysis of various diatoms revealed major changes in the cellular processes as a response to changing environmental conditions for better adaptability. Recently, meta transcriptomes of diatoms were mapped using Pfam protein domains resulting in the reconstruction of functional metabolic processes of the community in response to environmental variations (Elferink et al., 2020). Another application involves quantitative transcriptomic fingerprinting (QTF) capturing variations in transcripts markers in a particular condition in different diatom species. It was observed in *S. costatum* and *T. rotula* that various QTFs were identified in both species involving divergent genes responsible for various pathways (Alexander et al., 2015a; Alexander et al., 2015b). The studies based on transcriptomes highlighted functional variations leading to improved ecological tactics for the survival of diatoms (Caron et al., 2017). Comparative analysis of transcripts of *Pseudo-nitzschia arenysensis*, *Pseudo-nitzschia delicatissima*, and *Pseudo-nitzschia multistriata* feature variation patterns in oxylipin content in different species, further it predicted a novel homolog of nitric oxide synthase (NOS) in *Pseudo-nitzschia multistriata* (Di Dato et al., 2015). Functional annotation of these transcripts showed changes in isoprenoids and domoic acid (MVA and MEP pathways) biosynthetic pathways. To identify diatom-specific genes involved in cell wall construction, a comparative study was performed using the transcriptomes of three species: *Nitzschia palea*, *Achnanthes kuwaitensis*, and *Pseudoleyanella lunata*. The results emphasized endoplasmic reticulum-targeting proteins similar to silcanin-1 and SET domain members of the methyltransferase family.

Insights into lipid metabolism of *F. solaris* were gained by transcriptome analysis in nitrogen-depleted conditions. Parallel upregulation of transcripts involved in triacylglycerol biosynthesis and degradation (β-oxidation) along with constant ATP production was observed (Tanaka et al., 2015), these results concluded that in stress conditions *F. solaris* maintains growth as well as oil production. Further transcriptome-based enzyme identification was demonstrated for pharmaceutical applications. Enzymes responsible for biosynthesis and degradation of anti-inflammatory chemicals in *Cylindrotheca closterium* under different silicon concentrations have been recently explored (Elagoz et al., 2020). Overall, these studies demonstrate the prominence of transcriptome datasets and their significance in understanding diatoms' physiological and molecular functionality. Some of the transcriptome databases that contain marine omics data are AlgaePath, MMETSP, and Alganaut (Lauritano et al., 2019).

3.14 METABOLOMICS

The metabolomics technique aims to improve knowledge of complex biological networks and their activities by shedding light on metabolites and their interactions, particularly their involvement in cell signaling. *In silico* linkage of metabolites to upregulated microarray-detected transcripts, based on metabolic pathway mapping, have resulted as a potential metabolome-based tool. Bioactive pigments, long-chain polyunsaturated fatty acids (LC-PUFAs), polysaccharides, and vitamins are examples of secondary metabolites, which are high-value compounds generated by physiological and/or environmental stimuli (Rehmanji et al., 2022). Metabolic pathways along with metabolites are studied using high-throughput metabolite techniques such as nuclear magnetic resonance (NMR) and mass spectrometry (MS) (Lauritano et al., 2019). Gene expression and metabolic profiles of diatoms, while feeding (predator) on diatoms (prey) have been studied in *Skeletonema marinoi* cells. The results showed variations in lipid chain length as well as its accumulation, nitrogen metabolism, cell cycle, and frustule production in diatoms. Metabolic markers overexpressed were oxylipins, and transcript abundance shows upregulated expression of GPCRs, MAPK, MPAKK, and nitric oxide synthase (Amato et al., 2018).

Diatoms are important lipid producers, the lipidomics and metabolome profiles of various microalgae and different diatom lineages were compared under various growth conditions, reflecting changes induced by environmental parameters reflected in growth and metabolite accumulation providing metabolic species-specific footprints (Bromke et al., 2015; Kareya et al., 2021; Mariam et al., 2021; Paliwal and Jutur, 2021; Singh et al., 2022). For instance, an untargeted metabolomics study was used to measure changes in the metabolic profile occurring throughout the mating process in the diatom *Pseudo-nitzschia multistriata*. The results highlighted 70 metabolites in the control and mixed-cell populations undergoing sexual reproduction (Fiorini et al., 2020). Further, the effects of diatoms, *Nanofrustulum shiloi* and *Cylindrotheca closterium* on the reproductive success of sea urchins was studied via a metabolomics-based approach (Ruocco et al., 2018). Publicly accessible databases for metabolic studies are PubChem (Kim et al., 2019), The Kyoto Encyclopedia of Genes and Genomes (KEGG) (Kanehisa and Goto, 2000), MetaCyc, life (Caspi et al., 2018), and for analyzing metabolome datasets MetaboAnalyst5.0 and MetaboLights.

3.15 PROTEOMICS

Proteomics studies have been used to examine a variety of applications in diatoms, such as the effects of nutrition deprivation, light modulation, new protein mining, and domain and gene family analysis. These investigations offered hints for understanding the complex cellular systems that drive diatom physiology. Furthermore, the data collected have resulted in a greater knowledge of the metabolic mechanisms that add to their overall productivity. Various stressors cause substantial changes in diatom proteomes, with the goal of adjusting their cellular metabolism for survival in stress conditions.

Protein expression profiles allow researchers to obtain system-level knowledge regarding diatom responses. For example, a proteomic study of Phatr3 found a link between diverse protein ancestries using machine-learning techniques like DAMA (Domain Annotation by a Multi-objective Approach) and CLADE (Cluster learning-assisted directed evolution). This included prediction of the gene families and domain organization. The ASAFind and HECTAR tools were used to predict the subcellular location of the proteins (Rastogi et al., 2020). To unveil the nitrogen uptake mechanism a novel combination of marine meta-omics and machine learning techniques (Boosted Regression Tree method [BRT]) (Elith et al., 2008) was utilized which aids in understanding the expression of different nitrogen (N) transporters, belonging to the DiAMT1 and DiNRT2 families in diatoms (Busseni et al., 2019).

Further, a two-dimensional gel electrophoresis (2-DE)-based proteomic method demonstrated that the response to N deficiency in diatoms differs significantly from that of other eukaryotic photoautotrophs and is more similar to that of cyanobacteria (Hockin et al., 2012). Another proteomic investigation looked at the molecular change toward neutral lipid accumulation in *P. tricornutum* following nitrogen deficiency (Yang et al., 2014). The results showed that, in nitrogen deficiency, diatoms' metabolic flux shifts carbon storage toward lipid accumulation (Yang et al., 2014).

Recent developments in phosphopeptides enrichment, followed by liquid chromatography-tandem mass spectrometry (LC-MS/MS), have enabled system-level studies of proteins controlled by phosphorylation, and a substantial number of phosphorylation events have been found in numerous model species (Bodenmiller et al., 2010; Olsen et al., 2006; Pinkse et al., 2008;, Villén et al., 2007; Yang et al., 2013). The phospho-proteome study of *P. tricornutum* reveals fresh insights into how diatoms respond to and adapt to various stimuli in the marine environment. One of the important proteome-based applications was identification and quantification of redox-sensitive proteins (e.g., redoxome) highlighting redox-based processes that drive cell sensing and stress acclimatization (Leichert et al., 2008). Rosenwasser et al. (2014) demonstrated an elegant MS-based technique (OxICAT) for identifying the diatom redoxome and quantifying its degree of oxidation under oxidative stress conditions. OxICAT is a quantitative thiol-trapping approach that was recently developed and is used to analyze the ROS-sensitive proteome *in vivo* (Leichert et al., 2008). The prospective future use of diatom proteomics might aid in the knowledge of the physiological processes governing diatom cellular responses.

3.16 MULTI-OMICS

Recently, an integrated approach involving transcriptomics, metabolomics, and proteomics has been used in various studies to examine the growth and metabolism of diatoms, for example when cultivated on reduced carbon sources (Villanova et al., 2017), different salinity, etc. Mixotrophic treatment increased biomass productivity, TAG accumulation, and photosynthetic capability, implying novel media engineering strategies. To improve the understanding of organellar activity in diatoms, Schober et al. (2019) established a technique for isolating and purifying mitochondria and plastids from diatom cells. The thorough proteome analysis found 53 and 49 proteins from whole-cell lysates containing plastids and mitochondria, respectively. A relative proteome study through distinct taxa might improve the understanding of organelle-specific metabolite production and absorption along with transcriptome regulation (Schober et al., 2019). Likewise, Remmers et al. (2018) provided an integrated omics outline of *P. tricornutum*, highlighting differential regulation in terms of biomass and metabolite accumulation throughout a range of nitrogen concentration and various light intensities (Remmers et al., 2018). The extensive combination of bioinformatics and systems biology fields, and growing genome-sequencing studies, has enabled a synergistic understanding of diatom metabolism. Numerous international programs, such as the Marine Genomics Initiative in Europe, the European Marine Biological Resource Centre, and the Microbiology Initiative, have promoted genome-level research on diatoms across the world. Functional simulation of diatom models utilizing metaomics data has offered a reference for genomic and phylogenomic association in diatoms, allowing for a better understanding of the connections between ecology, genetics, and genomics (Blaby-Haas and Merchant, 2019).

3.17 CONCLUSIONS AND PERSPECTIVES

This chapter highlights recent advances in the genetic engineering of diatoms, ranging from gene-delivery methods to applications of genome tailoring, along with an in-depth description of omics investigations and their implications for fundamental and practical research in diatoms. In spite of the fact that diatoms are valuable assets, the current situation shows that relatively less knowledge of transcription factors, regulatory elements, reporter proteins, gene-knockout techniques, DNA delivery methods, etc. exists, and challenges such as controlled gene expression under inducible promoter in diatom system, omics of cellular organelles such as plastids, mitochondria and their transformation, and effective pre-treatment and post-transformation methodologies needs to be further developed. Moreover, a significant drawback of utilizing CRISPR/Cas9 in diatoms is that the Cas9 protein is continuously expressed, which has been shown to be hazardous to the cells (Jiang et al., 2014). As a result, improving the CRISPR/Cas9 system for diatoms is critical.

These challenges can be addressed by the collection of genomic and transcriptome data and linking them with the development of low-cost and simple genome-editing tools adding on to the rapid advancement of diatom biology. New subjects may now be investigated, such as the amazing ability of diatoms to adapt to changing environmental circumstances, studying biofilms, diatom morphological plasticity, and cell cycle studies in diatoms. As demonstrated by the production of high-value compounds, the discovery of these tools has proposed a roadmap toward metabolic pathway engineering for biotechnological applications. The use of these technologies in a growing number of diatom species will be extremely beneficial for both fundamental and practical research. Finally, via the development of advanced metabolic techniques and extensive engineering, functional investigations in diatoms may be enhanced and researched successfully. The modified designs can also be applied in a variety of biotechnological applications, particularly in oleaginous diatoms.

FUNDING

Support for this research was provided from the Department of Biotechnology (DBT)–BT/PB/Center/03/2011–Phase II.

REFERENCES

Adler-Agnon, Z., Leu, S., Zarka, A., Boussiba, S. & Khozin-Goldberg, I. (2018). Novel promoters for constitutive and inducible expression of transgenes in the diatom *Phaeodactylum tricornutum* under varied nitrate availability. *Journal of Applied Phycology* 30: 2763–2772.

Alexander, H., Jenkins Bethany, D., Rynearson Tatiana, A. & Dyhrman Sonya, T. (2015a). Metatranscriptome analyses indicate resource partitioning between diatoms in the field. *Proceedings of the National Academy of Sciences* 112: E2182–E2190.

Alexander, H., Rouco, M., Haley Sheean, T., Wilson Samuel, T., Karl David, M. & Dyhrman Sonya, T. (2015b). Functional group-specific traits drive phytoplankton dynamics in the oligotrophic ocean. *Proceedings of the National Academy of Sciences* 112: E5972–E5979.

Amato, A., Sabatino, V., Nylund, G.M., Bergkvist, J., Basu, S., Andersson, M.X., Sanges, R., Godhe, A., Kiørboe, T., Selander, E. & Ferrante, M.I. (2018). Grazer-induced transcriptomic and metabolomic response of the chain-forming diatom *Skeletonema marinoi*. *The ISME Journal* 12: 1594–1604.

Apt, K.E., Grossman, A.R. & Kroth-Pancic, P.G. (1996). Stable nuclear transformation of the diatom *Phaeodactylum tricornutum*. *Molecular and General Genetics* 252: 572–579.

Armbrust, E.V., Berges, J.A., Bowler, C., Green, B.R., Martinez, D., Putnam, N.H., Zhou, S., Allen, A.E., Apt, K.E., Bechner, M., Brzezinski, M.A., Chaal, B.K., Chiovitti, A., Davis, A.K., Demarest, M.S., Detter, J.C., Glavina, T., Goodstein, D., Hadi, M.Z., Hellsten, U., Hildebrand, M., Jenkins, B.D., Jurka, J., Kapitonov, V.V., Kröger, N., Lau, W.W., Lane, T.W., Larimer, F.W., Lippmeier, J.C., Lucas, S., Medina, M., Montsant, A., Obornik, M., Parker, M.S., Palenik, B., Pazour, G.J., Richardson, P.M., Rynearson, T.A., Saito, M.A., Schwartz, D.C., Thamatrakoln, K., Valentin, K., Vardi, A., Wilkerson, F.P. & Rokhsar, D.S. (2004). The genome of the diatom *Thalassiosira pseudonana*: ecology, evolution, and metabolism. *Science* 306: 79–86.

Aziz, R.K., Breitbart, M. & Edwards, R.A. (2010). Transposases are the most abundant, ubiquitous genes in nature. *Nucleic Acids Research* 38: 4207–4217.

Basu, S., Patil, S., Mapleson, D., Russo, M.T., Vitale, L., Fevola, C., Maumus, F., Casotti, R., Mock, T., Caccamo, M., Montresor, M., Sanges, R. & Ferrante, M.I. (2017). Finding a partner in the ocean: molecular and evolutionary bases of the response to sexual cues in a planktonic diatom. *New Phytologist* 215: 140–156.

Bhattacharjya, R., Tiwari, A., Marella, T.K., Bansal, H. & Srivastava, S. (2021). New paradigm in diatom omics and genetic manipulation. *Bioresource Technology* 325: 124708.

Bidle, K.D., Manganelli, M. & Azam, F. (2002). Regulation of oceanic silicon and carbon preservation by temperature control on bacteria. *Science* 298: 1980–1984.

Blaby-Haas, C.E. & Merchant, S.S. (2019). Comparative and functional algal genomics. *Annual Review of Plant Biology* 70: 605–638.

Bodenmiller, B., Wanka, S., Kraft, C., Urban, J., Campbell, D., Pedrioli, P.G., Gerrits, B., Picotti, P., Lam, H., Vitek, O., Brusniak, M.Y., Roschitzki, B., Zhang, C., Shokat, K.M., Schlapbach, R., Colman-Lerner, A., Nolan, G.P., Nesvizhskii, A.I., Peter, M., Loewith, R., Von Mering, C. & Aebersold, R. (2010). Phosphoproteomic analysis reveals interconnected system-wide responses to perturbations of kinases and phosphatases in yeast. *Science Signaling* 3: RS4–RS19.

Bowler, C., Allen, A.E., Badger, J.H., Grimwood, J., Jabbari, K., Kuo, A., Maheswari, U., Martens, C., Maumus, F., Otillar, R.P., Rayko, E., Salamov, A., Vandepoele, K., Beszteri, B., Gruber, A., Heijde, M., Katinka, M., Mock, T., Valentin, K., Verret, F., Berges, J.A., Brownlee, C., Cadoret, J.-P., Chiovitti, A., Choi, C.J., Coesel, S., De Martino, A., Detter, J.C., Durkin, C., Falciatore, A., Fournet, J., Haruta, M., Huysman, M.J.J., Jenkins, B.D., Jiroutova, K., Jorgensen, R.E., Joubert, Y., Kaplan, A., Kröger, N., Kroth, P.G., La Roche, J., Lindquist, E., Lommer, M., Martin–Jézéquel, V., Lopez, P.J., Lucas, S., Mangogna, M., Mcginnis, K., Medlin, L.K., Montsant, A., Secq, M.-P.O.L., Napoli, C., Obornik, M., Parker, M.S., Petit, J.-L., Porcel, B.M., Poulsen, N., Robison, M., Rychlewski, L., Rynearson, T.A., Schmutz, J., Shapiro, H., Siaut, M., Stanley, M., Sussman, M.R., Taylor, A.R., Vardi, A., Von Dassow, P., Vyverman, W., Willis, A., Wyrwicz, L.S., Rokhsar, D.S., Weissenbach, J., Armbrust, E.V., Green, B.R., Van De Peer, Y. & Grigoriev, I.V. (2008). The *Phaeodactylum* genome reveals the evolutionary history of diatom genomes. *Nature* 456: 239–244.

Bozarth, A., Maier, U.G. & Zauner, S. (2009). Diatoms in biotechnology: modern tools and applications. *Applied Microbiology and Biotechnology* 82: 195–201.

Bromke, M.A., Sabir, J.S., Alfassi, F.A., Hajarah, N.H., Kabli, S.A., Al-Malki, A.L., Ashworth, M.P., Méret, M., Jansen, R.K. & Willmitzer, L. (2015). Metabolomic profiling of 13 diatom cultures and their adaptation to nitrate-limited growth conditions. *PLOS ONE* 10: e0138965.

Buhmann, M.T., Poulsen, N., Klemm, J., Kennedy, M.R., Sherrill, C.D. & Kröger, N. (2014). A tyrosine-rich cell surface protein in the diatom *Amphora coffeaeformis* identified through transcriptome analysis and genetic transformation. *PLOS ONE* 9: e110369.

Busseni, G., Rocha Jimenez Vieira, F., Amato, A., Pelletier, E., Pierella Karlusich, J.J., Ferrante, M.I., Wincker, P., Rogato, A., Bowler, C., Sanges, R., Maiorano, L., Chiurazzi, M., Ribera D'alcalà, M., Caputi, L. & Iudicone, D. (2019). Meta-omics reveals genetic flexibility of diatom nitrogen transporters in response to environmental changes. *Molecular Biology and Evolution* 36: 2522–2535.

Caron, D.A., Alexander, H., Allen, A.E., Archibald, J.M., Armbrust, E.V., Bachy, C., Bell, C.J., Bharti, A., Dyhrman, S.T., Guida, S.M., Heidelberg, K.B., Kaye, J.Z., Metzner, J., Smith, S.R. & Worden, A.Z. (2017). Probing the evolution, ecology and physiology of marine protists using transcriptomics. *Nature Reviews Microbiology* 15: 6–20.

Caspi, R., Billington, R., Fulcher, C.A., Keseler, I.M., Kothari, A., Krummenacker, M., Latendresse, M., Midford, P.E., Ong, Q., Ong, W.K., Paley, S., Subhraveti, P. & Karp, P.D. (2018). The MetaCyc database of metabolic pathways and enzymes. *Nucleic Acids Research* 46: D633–D639.

Chen, I.M.A., Chu, K., Palaniappan, K., Pillay, M., Ratner, A., Huang, J., Huntemann, M., Varghese, N., White, J.R., Seshadri, R., Smirnova, T., Kirton, E., Jungbluth, S.P., Woyke, T., Eloe-Fadrosh, E.A., Ivanova, N.N. & Kyrpides, N.C. (2019). IMG/M v.5.0: an integrated data management and comparative analysis system for microbial genomes and microbiomes. *Nucleic Acids Research* 47: D666–D677.

Chu, L., Ewe, D., Río Bártulos, C., Kroth, P.G. & Gruber, A. (2016). Rapid induction of GFP expression by the nitrate reductase promoter in the diatom *Phaeodactylum tricornutum*. *PeerJ* 4: e2344.

Cuvelier Marie, L., Allen Andrew, E., Monier, A., Mccrow John, P., Messié, M., Tringe Susannah, G., Woyke, T., Welsh Rory, M., Ishoey, T., Lee, J.-H., Binder Brian, J., Dupont Chris, L., Latasa, M., Guigand, C., Buck Kurt, R., Hilton, J., Thiagarajan, M., Caler, E., Read, B., Lasken Roger, S., Chavez Francisco, P. & Worden Alexandra, Z. (2010). Targeted metagenomics and ecology of globally important uncultured eukaryotic phytoplankton. *Proceedings of the National Academy of Sciences* 107: 14679–14684.

Daboussi, F., Leduc, S., Maréchal, A., Dubois, G., Guyot, V., Perez-Michaut, C., Amato, A., Falciatore, A., Juillerat, A., Beurdeley, M., Voytas, D.F., Cavarec, L. & Duchateau, P. (2014). Genome engineering empowers the diatom *Phaeodactylum tricornutum* for biotechnology. *Nature Communications* 5: 3831–3838.

Davis, A., Crum, L.T., Corbeil, L.B. & Hildebrand, M. (2017). Expression of *Histophilus somni* IbpA DR2 protective antigen in the diatom *Thalassiosira pseudonana*. *Applied Microbiology and Biotechnology* 101: 5313–5324.

De Riso, V., Raniello, R., Maumus, F., Rogato, A., Bowler, C. & Falciatore, A. (2009). Gene silencing in the marine diatom *Phaeodactylum tricornutum. Nucleic Acids Research* 37: e96.

Delmont, T.O., Eren, A.M., Vineis, J.H. & Post, A.F. (2015). Genome reconstructions indicate the partitioning of ecological functions inside a phytoplankton bloom in the Amundsen Sea, Antarctica. *Frontiers in Microbiology* 6: 01090.

Di Dato, V., Musacchia, F., Petrosino, G., Patil, S., Montresor, M., Sanges, R. & Ferrante, M.I. (2015). Transcriptome sequencing of three *Pseudo-nitzschia* species reveals comparable gene sets and the presence of nitric oxide synthase genes in diatoms. *Scientific Reports* 5: 12329–12343.

Dunahay, T.G., Jarvis, E.E. & Roessler, P.G. (1995). Genetic transformation of the diatoms *Cyclotella cryptica* and *Navicula saprophila. Journal of Phycology* 31: 1004–1012.

Eilers, U., Bikoulis, A., Breitenbach, J., Büchel, C. & Sandmann, G. (2016). Limitations in the biosynthesis of fucoxanthin as targets for genetic engineering in *Phaeodactylum tricornutum. Journal of Applied Phycology* 28: 123–129.

Elagoz, A.M., Ambrosino, L. & Lauritano, C. (2020). *De novo* transcriptome of the diatom *Cylindrotheca closterium* identifies genes involved in the metabolism of anti-inflammatory compounds. *Scientific Reports* 10: 4138–4147.

Elferink, S., John, U., Neuhaus, S. & Wohlrab, S. (2020). Functional genomics differentiate inherent and environmentally influenced traits in dinoflagellate and diatom communities. *Microorganisms* 8: 567–589.

Elith, J., Leathwick, J.R. & Hastie, T. (2008). A working guide to boosted regression trees. *Journal of Animal Ecology* 77: 802–813.

Erdene-Ochir, E., Shin, B.-K., Huda, M.N., Kim, D.H., Lee, E.H., Song, D.-G., Kim, Y.-M., Kim, S.M. & Pan, C.-H. (2016). Cloning of a novel endogenous promoter for foreign gene expression in *Phaeodactylum tricornutum. Applied Biological Chemistry* 59: 861–867.

Falciatore, A., Casotti, R., Leblanc, C., Abrescia, C. & Bowler, C. (1999). Transformation of nonselectable reporter genes in marine diatoms. *Marine Biotechnology* 1: 239–251.

Falciatore, A., Jaubert, M., Bouly, J.-P., Bailleul, B. & Mock, T. (2020). Diatom molecular research comes of age: model species for sying phytoplankton biology and diversity. *The Plant Cell* 32: 547–572.

Falkowski, P.G., Barber, R.T. & Smetacek, V.V. (1998). Biogeochemical controls and feedbacks on ocean primary production. *Science* 281: 200–207.

Fiorini, F., Borgonuovo, C., Ferrante, M.I. & Brönstrup, M. (2020). A metabolomics exploration of the sexual phase in the marine diatom *Pseudo-nitzschia multistriata. Marine Drugs* 18: 313–330.

Fischer, H., Robl, I., Sumper, M. & Kröger, N. (2002). Targeting and covalent modification of cell wall and membrane proteins heterologously expressed in the diatom *Cylindrotheca fusiformis* (Bacillariophyceae). *Journal of Phycology* 35: 113–120.

Ford, N.R., Hecht, K.A., Hu, D., Orr, G., Xiong, Y., Squier, T.C., Rorrer, G.L. & Roesijadi, G. (2016). Antigen binding and site-directed labeling of biosilica-immobilized fusion proteins expressed in diatoms. *ACS Synthetic Biology* 5: 193–199.

Fortunato, A.E., Jaubert, M., Enomoto, G., Bouly, J.-P., Raniello, R., Thaler, M., Malviya, S., Bernardes, J.S., Rappaport, F., Gentili, B., Huysman, M.J.J., Carbone, A., Bowler, C., D'alcalà, M.R., Ikeuchi, M. & Falciatore, A. (2016). Diatom phytochromes reveal the existence of far-red-light-based sensing in the ocean. *The Plant Cell* 28: 616–628.

Galachyants, Y.P., Zakharova, Y.R., Volokitina, N.A., Morozov, A.A., Likhoshway, Y.V. & Grachev, M.A. (2019). *De novo* transcriptome assembly and analysis of the freshwater araphid diatom Fragilaria radians, Lake Baikal. *Scientific Data* 6: 183–194.

George, J., Kahlke, T., Abbriano, R.M., Kuzhiumparambil, U., Ralph, P.J. & Fabris, M. (2020). Metabolic engineering strategies in diatoms reveal unique phenotypes and genetic configurations with implications for algal genetics and synthetic biology. *Frontiers in Bioengineering and Biotechnology* 8: 00513.

Goyal, A., Bhowmik, P. K., Basu, S. K., (2009). Mini chromosomes: the second generation genetic engineering tool. *Plant Omics Journal* 2: 1–8.

Gruber, A. & Kroth, P.G. (2017). Intracellular metabolic pathway distribution in diatoms and tools for genome-enabled experimental diatom research. *Philosophical Transactions of the Royal Society of London Series B, Biological Sciences* 372: 20160402.

Gruber, A., Vugrinec, S., Hempel, F., Gould, S.B., Maier, U.G. & Kroth, P.G. (2007). Protein targeting into complex diatom plastids: functional characterisation of a specific targeting motif. *Plant Molecular Biology* 64: 519–530.

Harada, H., Nakajima, K., Sakaue, K. & Matsuda, Y. (2006). CO_2 sensing at ocean surface mediated by cAMP in a marine diatom. *Plant Physiology* 142: 1318–1328.

Harada, H., Nakatsuma, D., Ishida, M. & Matsuda, Y. (2005). Regulation of the expression of intracellular beta-carbonic anhydrase in response to CO_2 and light in the marine diatom *Phaeodactylum tricornutum*. *Plant Physiology* 139: 1041–1050.

Hempel, F., Bozarth, A.S., Lindenkamp, N., Klingl, A., Zauner, S., Linne, U., Steinbüchel, A. & Maier, U.G. (2011). Microalgae as bioreactors for bioplastic production. *Microbial Cell Factories* 10: 81–87.

Hempel, F., Lau, J., Klingl, A. & Maier, U.G. (2011). Algae as protein factories: Expression of a human antibody and the respective antigen in the diatom *Phaeodactylum tricornutum*. *PLOS ONE* 6: e28424.

Hildebrand, M., Manandhar-Shrestha, K. & Abbriano, R. (2017). Effects of chrysolaminarin synthase knockdown in the diatom *Thalassiosira pseudonana*: Implications of reduced carbohydrate storage relative to green algae. *Algal Research* 23: 66–77.

Hockin, N.L., Mock, T., Mulholland, F., Kopriva, S. & Malin, G. (2012). The response of diatom central carbon metabolism to nitrogen starvation is different from that of green algae and higher plants. *Plant Physiology* 158: 299–312.

Hopes, A., Nekrasov, V., Kamoun, S. & Mock, T. (2016). Editing of the urease gene by CRISPR-Cas in the diatom *Thalassiosira pseudonana*. *Plant Methods* 12: 49–61.

Huang, W. & Daboussi, F. (2017). Genetic and metabolic engineering in diatoms. *Philosophical Transactions of the Royal Society of London Series B, Biological Sciences* 372: 20160411.

Huysman, M.J.J., Tanaka, A., Bowler, C., Vyverman, W. & De Veylder, L. (2015). Functional characterization of the diatom cyclin-dependent kinase A2 as a mitotic regulator reveals plant-like properties in a non-green lineage. *BMC Plant Biology* 15: 86–97.

Ifuku, K., Yan, D., Miyahara, M., Inoue-Kashino, N., Yamamoto, Y.Y. & Kashino, Y. (2015). A stable and efficient nuclear transformation system for the diatom *Chaetoceros gracilis*. *Photosynthesis Research* 123: 203–211.

Jiang, W., Brueggeman, A.J., Horken, K.M., Plucinak, T.M., Weeks, D.P. (2014). Successful transient expression of Cas9 and single guide RNA genes in *Chlamydomonas reinhardtii*. *Eukaryotic Cell* 11: 1465–1469.

Kadono, T., Miyagawa-Yamaguchi, A., Kira, N., Tomaru, Y., Okami, T., Yoshimatsu, T., Hou, L., Ohama, T., Fukunaga, K., Okauchi, M., Yamaguchi, H., Ohnishi, K., Falciatore, A. & Adachi, M. (2015). Characterization of marine diatom-infecting virus promoters in the model diatom *Phaeodactylum tricornutum*. *Scientific Reports* 5: 18708.

Kanehisa, M. & Goto, S. (2000). KEGG: Kyoto Encyclopedia of Genes and Genomes. *Nucleic Acids Research* 28: 27–30.

Karas, B.J., Diner, R.E., Lefebvre, S.C., Mcquaid, J., Phillips, A.P.R., Noddings, C.M., Brunson, J.K., Valas, R.E., Deerinck, T.J., Jablanovic, J., Gillard, J.T.F., Beeri, K., Ellisman, M.H., Glass, J.I., Hutchison Iii, C.A., Smith, H.O., Venter, J.C., Allen, A.E., Dupont, C.L. & Weyman, P.D. (2015). Designer diatom episomes delivered by bacterial conjugation. *Nature Communications* 6: 6925–6937.

Karas, B.J., Suzuki, Y. & Weyman, P.D. (2015). Strategies for cloning and manipulating natural and synthetic chromosomes. *Chromosome Research* 23: 57–68.

Kareya, M.S., Mariam, I., Rajacharya, G.H., Nesamma, A.A. & Jutur, P.P. (2021). Valorization of carbon dioxide (CO_2) to enhance production of biomass, biofuels, and biorenewables (B3) in *Chlorella saccharophila* UTEX247: a circular bioeconomy perspective. *Biofuels, Bioproducts and Biorefining* 15: 1–16.

Kaur, S. & Spillane, C. (2015). Reduction in carotenoid levels in the marine diatom *Phaeodactylum tricornutum* by artificial microRNAs targeted against the endogenous phytoene synthase gene. *Marine Biotechnology* 17: 1–7.

Keeling, P.J., Burki, F., Wilcox, H.M., Allam, B., Allen, E.E., Amaral-Zettler, L.A., Armbrust, E.V., Archibald, J.M., Bharti, A.K., Bell, C.J., Beszteri, B., Bidle, K.D., Cameron, C.T., Campbell, L., Caron, D.A., Cattolico, R.A., Collier, J.L., Coyne, K., Davy, S.K., Deschamps, P., Dyhrman, S.T., Edvardsen, B., Gates, R.D., Gobler, C.J., Greenwood, S.J., Guida, S.M., Jacobi, J.L., Jakobsen, K.S., James, E.R., Jenkins, B., John, U., Johnson, M.D., Juhl, A.R., Kamp, A., Katz, L.A., Kiene, R., Kudryavtsev, A., Leander, B.S., Lin, S., Lovejoy, C., Lynn, D., Marchetti, A., Mcmanus, G., Nedelcu, A.M., Menden-Deuer, S., Miceli, C., Mock, T., Montresor, M., Moran, M.A., Murray, S., Nadathur, G., Nagai, S., Ngam,

P.B., Palenik, B., Pawlowski, J., Petroni, G., Piganeau, G., Posewitz, M.C., Rengefors, K., Romano, G., Rumpho, M.E., Rynearson, T., Schilling, K.B., Schroeder, D.C., Simpson, A.G.B., Slamovits, C.H., Smith, D.R., Smith, G.J., Smith, S.R., Sosik, H.M., Stief, P., Theriot, E., Twary, S.N., Umale, P.E., Vaulot, D., Wawrik, B., Wheeler, G.L., Wilson, W.H., Xu, Y., Zingone, A. & Worden, A.Z. (2014). The marine microbial eukaryote transcriptome sequencing project (MMETSP): Illuminating the functional diversity of eukaryotic life in the oceans through transcriptome sequencing. *PLOS Biology* 12: e1001889.

Kilian, O. & Kroth, P.G. (2005). Identification and characterization of a new conserved motif within the presequence of proteins targeted into complex diatom plastids. *The Plant Journal* 41: 175–183.

Kim, S., Chen, J., Cheng, T., Gindulyte, A., He, J., He, S., Li, Q., Shoemaker, B.A., Thiessen, P.A., Yu, B., Zaslavsky, L., Zhang, J. & Bolton, E.E. (2019). PubChem 2019 update: Improved access to chemical data. *Nucleic Acids Research* 47: D1102–D1109.

Kirupamurthy, D. (2014). Studies of light responses and the development of a transformation system for the benthic diatom *Seminavis robusta*. *Semantic Scholar* 82349212.

Koester, J.A., Berthiaume, C.T., Hiranuma, N., Parker, M.S., Iverson, V., Morales, R., Ruzzo, W.L. & Armbrust, E.V. (2018). Sexual ancestors generated an obligate asexual and globally dispersed clone within the model diatom species *Thalassiosira pseudonana*. *Scientific Reports* 8: 10492–10501.

Kotzsch, A., Pawolski, D., Milentyev, A., Shevchenko, A., Scheffel, A., Poulsen, N., Shevchenko, A. & Kröger, N. (2016). Biochemical composition and assembly of biosilica-associated insoluble organic matrices from the diatom *Thalassiosira pseudonana*. *The Journal of Biological Chemistry* 291: 4982–4997.

Lauritano, C., Ferrante, M.I. & Rogato, A. (2019). Marine natural products from microalgae: An–omics overview. *Marine Drugs* 17: 269–287.

Leichert, L.I., Gehrke, F., Gudiseva, H.V., Blackwell, T., Ilbert, M., Walker, A.K., Strahler, J.R., Andrews, P.C. & Jakob, U. (2008). Quantifying changes in the thiol redox proteome upon oxidative stress *in vivo*. *Proceedings of the National Academy of Sciences* 105: 8197–8202.

Lepetit, B., Volke, D., Gilbert, M., Wilhelm, C. & Goss, R. (2010). Evidence for the existence of one antenna-associated, lipid-dissolved and two protein-bound pools of diadinoxanthin cycle pigments in diatoms. *Plant Physiology* 154: 1905–1920.

Lescot, M., Hingamp, P., Kojima, K.K., Villar, E., Romac, S., Veluchamy, A., Boccara, M., Jaillon, O., Iudicone, D., Bowler, C., Wincker, P., Claverie, J.-M. & Ogata, H. (2016). Reverse transcriptase genes are highly abundant and transcriptionally active in marine plankton assemblages. *The ISME Journal* 10: 1134–1146.

Levering, J., Broddrick, J., Dupont, C.L., Peers, G., Beeri, K., Mayers, J., Gallina, A.A., Allen, A.E., Palsson, B.O. & Zengler, K. (2016). Genome-scale model reveals metabolic basis of biomass partitioning in a model diatom. *PLOS ONE* 11: e0155038.

Lin, H.-Y., Yen, S.-C., Kuo, P.-C., Chung, C.-Y., Yeh, K.-L., Huang, C.-H., Chang, J. & Lin, H.-J. (2017). Alkaline phosphatase promoter as an efficient driving element for exogenic recombinant in the marine diatom *Phaeodactylum tricornutum*. *Algal Research* 23: 58–65.

Lippmeier, J.C., Brown, A.M. & Apt, K.E. (2002). Isolation of algal genes by functional complementation of yeast. *Journal of Phycology* 38: 529–533.

Liu, X., Hempel, F., Stork, S., Bolte, K., Moog, D., Heimerl, T., Maier, U. & Zauner, S. (2016). Addressing various compartments of the diatom model organism *Phaeodactylum tricornutum* via sub-cellular marker proteins. *Algal Research* 20: 249–257.

Lommer, M., Specht, M., Roy, A.-S., Kraemer, L., Andreson, R., Gutowska, M.A., Wolf, J., Bergner, S.V., Schilhabel, M.B., Klostermeier, U.C., Beiko, R.G., Rosenstiel, P., Hippler, M. & Laroche, J. (2012). Genome and low-iron response of an oceanic diatom adapted to chronic iron limitation. *Genome Biology* 13: R66–R87.

Lopez, P.J., Desclés, J., Allen, A.E. & Bowler, C. (2005). Prospects in diatom research. *Current Opinion in Biotechnology* 16: 180–186.

Mariam, I., Kareya, M., Nesamma, A. & Jutur, P. (2021). Delineating metabolomic changes in native isolate *Aurantiochytrium* for production of docosahexaenoic acid in presence of varying carbon substrates. *Algal Research* 55: 102285.

Materna, A.C., Sturm, S., Kroth, P.G. & Lavaud, J. (2009). First induced plastid genome mutations in an alga with secondary plastids: psbA mutation in the diatom *Phaeodactylum tricornutum* (Bacillariophyceae) reveal consequences on the regulation of photosynthesis. *Journal of Phycology* 45: 838–846.

Miyahara, M., Aoi, M., Inoue-Kashino, N., Kashino, Y. & Ifuku, K. (2013). Highly efficient transformation of the diatom *Phaeodactylum tricornutum* by multi-pulse electroporation. *Bioscience, Biotechnology, and Biochemistry* 77: 874–876.

Mock, T., Otillar, R.P., Strauss, J., Mcmullan, M., Paajanen, P., Schmutz, J., Salamov, A., Sanges, R., Toseland, A., Ward, B.J., Allen, A.E., Dupont, C.L., Frickenhaus, S., Maumus, F., Veluchamy, A., Wu, T., Barry, K.W., Falciatore, A., Ferrante, M.I., Fortunato, A.E., Glöckner, G., Gruber, A., Hipkin, R., Janech, M.G., Kroth, P.G., Leese, F., Lindquist, E.A., Lyon, B.R., Martin, J., Mayer, C., Parker, M., Quesneville, H., Raymond, J.A., Uhlig, C., Valas, R.E., Valentin, K.U., Worden, A.Z., Armbrust, E.V., Clark, M.D., Bowler, C., Green, B.R., Moulton, V., Van Oosterhout, C. & Grigoriev, I.V. (2017). Evolutionary genomics of the cold-adapted diatom *Fragilariopsis cylindrus*. *Nature* 541: 536–540.

Muto, M., Fukuda, Y., Nemoto, M., Yoshino, T., Matsunaga, T. & Tanaka, T. (2013). Establishment of a genetic transformation system for the marine pennate diatom *Fistulifera* sp. strain JPCC DA0580—a high triglyceride producer. *Marine Biotechnology* 15: 48–55.

Muto, M., Tanaka, M., Liang, Y., Yoshino, T., Matsumoto, M. & Tanaka, T. (2015). Enhancement of glycerol metabolism in the oleaginous marine diatom *Fistulifera solaris* JPCC DA0580 to improve triacylglycerol productivity. *Biotechnology for Biofuels* 8: 4–11.

Niu, Y.F., Yang, Z.K., Zhang, M.H., Zhu, C.C., Yang, W.D., Liu, J.S. & Li, H.Y. (2012). Transformation of diatom *Phaeodactylum tricornutum* by electroporation and establishment of inducible selection marker. *BioTechniques* 52: 1–3.

Nymark, M., Sharma, A.K., Sparstad, T., Bones, A.M. & Winge, P. (2016). A CRISPR/Cas9 system adapted for gene editing in marine algae. *Scientific Reports* 6: 24951–24957.

Ogura, A., Akizuki, Y., Imoda, H., Mineta, K., Gojobori, T. & Nagai, S. (2018). Comparative genome and transcriptome analysis of diatom, *Skeletonema costatum*, reveals evolution of genes for harmful algal bloom. *BMC Genomics* 19: 765–777.

Olsen, J.V., Blagoev, B., Gnad, F., Macek, B., Kumar, C., Mortensen, P. & Mann, M. (2006). Global, in vivo, and site-specific phosphorylation dynamics in signaling networks. *Cell* 127: 635–648.

Osuna-Cruz, C.M., Bilcke, G., Vancaester, E., De Decker, S., Bones, A.M., Winge, P., Poulsen, N., Bulankova, P., Verhelst, B., Audoor, S., Belisova, D., Pargana, A., Russo, M., Stock, F., Cirri, E., Brembu, T., Pohnert, G., Piganeau, G., Ferrante, M.I., Mock, T., Sterck, L., Sabbe, K., De Veylder, L., Vyverman, W. & Vandepoele, K. (2020). The *Seminavis robusta* genome provides insights into the evolutionary adaptations of benthic diatoms. *Nature Communications* 11: 3320–3333.

Paliwal, C. & Jutur, P.P. (2021). Dynamic allocation of carbon flux triggered by task-specific chemicals is an effective non-gene disruptive strategy for sustainable and cost-effective algal biorefineries. *Chemical Engineering Journal* 418: 129413–129425.

Park, S., Nguyen, T.H.T. & Jin, E. (2019). Improving lipid production by strain development in microalgae: Strategies, challenges and perspectives. *Bioresource Technology* 292: 121953.

Pinkse, M.W., Mohammed, S., Gouw, J.W., Van Breukelen, B., Vos, H.R. & Heck, A.J. (2008). Highly robust, automated, and sensitive online TiO2-based phosphoproteomics applied to study endogenous phosphorylation in *Drosophila melanogaster*. *Journal of Proteome Research* 7: 687–697.

Pollak, B., Matute, T., Nuñez, I., Cerda, A., Lopez, C., Vargas, V., Kan, A., Bielinski, V., von Dassow, P., Dupont, CL., Federici, F. (2020). Universal loop assembly: open, efficient and cross-kingdom DNA fabrication. *Synthetic Biology* 5: 21–30.

Poulsen, N., Chesley, P.M. & Kröger, N. (2006). Molecular genetic manipulation of the diatom *Thalassiosira pseudonana* (Bacillariophyceae). *Journal of Phycology* 42: 1059–1065.

Poulsen, N. & Kröger, N. (2005). A new molecular tool for transgenic diatoms: control of mRNA and protein biosynthesis by an inducible promoter-terminator cassette. *The FEBS Journal* 272: 3413–3423.

Purton, S., Szaub-Newton, J., Brocklehurst, T., Young, R. & Economou, C. (2013). Genetic engineering of algal chloroplasts: Progress and prospects. *Russian Journal of Plant Physiology* 60: 491–499.

Radakovits, R., Eduafo, P.M. & Posewitz, M.C. (2011). Genetic engineering of fatty acid chain length in *Phaeodactylum tricornutum*. *Metabolic Engineering* 13: 89–95.

Rastogi, A., Vieira, F.R.J., Deton-Cabanillas, A.-F., Veluchamy, A., Cantrel, C., Wang, G., Vanormelingen, P., Bowler, C., Piganeau, G., Hu, H. & Tirichine, L. (2020). A genomics approach reveals the global genetic polymorphism, structure, and functional diversity of ten accessions of the marine model diatom *Phaeodactylum tricornutum*. *The ISME Journal* 14: 347–363.

Rehmanji, M., Nesamma, A. A., Khan, N. J., Fatma, T., & Jutur, P. P. (2022). Media engineering in marine diatom *Phaeodactylum tricornutum* employing cost-effective substrates for sustainable production of high-value renewables. *Biotechnology Journal* 3: 11.

Rehmanji, M., Singh, R., Nesamma, A .A., Khan, N. J., Fatma, T., Narula, A. & Jutur, P.P. (2022). Multifaceted applications of microalgal biomass valorization to enriched biorenewables, a review of futuristic biorefinery paradigm. *Bioresource Technology Reports* 17: 100972.Remmers, I.M., D'adamo, S., Martens, D.E., De Vos, R.C.H., Mumm, R., America, A.H.P., Cordewener, J.H.G., Bakker, L.V., Peters, S.A., Wijffels, R.H. & Lamers, P.P. (2018). Orchestration of transcriptome, proteome and metabolome in the diatom *Phaeodactylum tricornutum* during nitrogen limitation. *Algal Research* 35: 33–49.

Rosenwasser, S., Graff Van Creveld, S., Schatz, D., Malitsky, S., Tzfadia, O., Aharoni, A., Levin, Y., Gabashvili, A., Feldmesser, E. & Vardi, A. (2014). Mapping the diatom redox-sensitive proteome provides insight into response to nitrogen stress in the marine environment. *Proceedings of the National Academy of Sciences* 111: 2740–2745.

Rouet, P., Smih, F. & Jasin, M. (1994). Expression of a site-specific endonuclease stimulates homologous recombination in mammalian cells. *Proceedings of the National Academy of Sciences* 91: 6064–6068.

Ruocco, N., Costantini, S., Zupo, V., Lauritano, C., Caramiello, D., Ianora, A., Budillon, A., Romano, G., Nuzzo, G., D'ippolito, G., Fontana, A. & Costantini, M. (2018). Toxigenic effects of two benthic diatoms upon grazing activity of the sea urchin: morphological, metabolomic and de novo transcriptomic analysis. *Scientific Reports* 8: 5622–5635.

Sabatino, V., Russo, M., Patil, S., D'ippolito, G., Fontana, A. & Ferrante, M. (2015). Establishment of genetic transformation in the sexually reproducing diatoms *Pseudo-nitzschia multistriata* and *Pseudo-nitzschia arenysensis* and inheritance of the transgene. *Marine Biotechnology* 17: 452–462.

Sakaue, K., Harada, H. & Matsuda, Y. (2008). Development of gene expression system in a marine diatom using viral promoters of a wide variety of origin. *Physiologia Plantarum* 133: 59–67.

Samukawa, M., Shen, C., Hopkinson, B.M. & Matsuda, Y. (2014). Localization of putative carbonic anhydrases in the marine diatom, *Thalassiosira pseudonana. Photosynthesis Research* 121: 235–249.

Schellenberger Costa, B., Sachse, M., Jungandreas, A., Bartulos, C.R., Gruber, A., Jakob, T., Kroth, P.G. & Wilhelm, C. (2013). Aureochrome 1a is involved in the photoacclimation of the diatom *Phaeodactylum tricornutum. PLOS ONE* 8: e74451.

Schober, A.F., Río Bártulos, C., Bischoff, A., Lepetit, B., Gruber, A. & Kroth, P.G. (2019). Organelle studies and proteome analyses of mitochondria and plastids fractions from the diatom *Thalassiosira pseudonana. Plant and Cell Physiology* 60: 1811–1828.

Seligman, L.M., Chisholm, K.M., Chevalier, B.S., Chadsey, M.S., Edwards, S.T., Savage, J.H. & Veillet, A.L. (2002). Mutations altering the cleavage specificity of a homing endonuclease. *Nucleic Acids Research* 30: 3870–3879.

Seo, S., Jeon, H., Chang, K.S. & Jin, E. (2018). Enhanced biomass production by *Phaeodactylum tricornutum* overexpressing phosphoenolpyruvate carboxylase. *Algal Research* 31: 489–496.

Seo, S., Jeon, H., Hwang, S., Jin, E. & Chang, K.S. (2015). Development of a new constitutive expression system for the transformation of the diatom *Phaeodactylum tricornutum. Algal Research* 11: 50–54.

Serif, M., Dubois, G., Finoux, A.-L., Teste, M.-A., Jallet, D. & Daboussi, F. (2018). One-step generation of multiple gene knock-outs in the diatom *Phaeodactylum tricornutum* by DNA-free genome editing. *Nature Communications* 9: 3924–3934.

Shemesh, Z., Leu, S., Khozin-Goldberg, I., Didi-Cohen, S., Zarka, A. & Boussiba, S. (2016). Inducible expression of *Haematococcus* oil globule protein in the diatom *Phaeodactylum tricornutum*: Association with lipid droplets and enhancement of TAG accumulation under nitrogen starvation. *Algal Research* 18: 321–331.

Shrestha, R. & Hildebrand, M. (2014). Evidence for a regulatory role of diatom silicon transporters in cellular silicon responses. *Eukaryotic Cell* 14: 29–40.

Siaut, M., Heijde, M., Mangogna, M., Montsant, A., Coesel, S., Allen, A., Manfredonia, A., Falciatore, A. & Bowler, C. (2007). Molecular toolbox for studying diatom biology in *Phaeodactylum tricornutum. Gene* 406: 23–35.

Singh, R., Nesamma, A.A., Narula, A. & Jutur, P.P. (2022). Multi-fold enhancement of tocopherol yields employing high CO_2 supplementation and nitrate limitation in native isolate *Monoraphidium* sp. *Cells* 11: 1315–1334.

Slattery, S.S., Diamond, A., Wang, H., Therrien, J.A., Lant, J.T., Jazey, T., Lee, K., Klassen, Z., Desgagné-Penix, I., Karas, B.J. & Edgell, D.R. (2018). An expanded plasmid-based genetic toolbox enables Cas9 genome editing and stable maintenance of synthetic pathways in *Phaeodactylum tricornutum*. *ACS Synthetic Biology* 7: 328–338.

Sunaga, Y., Maeda, Y., Yabuuchi, T., Muto, M., Yoshino, T. & Tanaka, T. (2015). Chloroplast-targeting protein expression in the oleaginous diatom *Fistulifera solaris* JPCC DA0580 toward metabolic engineering. *Journal of Bioscience and Bioengineering* 119: 28–34.

Tanaka, T., Maeda, Y., Veluchamy, A., Tanaka, M., Abida, H., Maréchal, E., Bowler, C., Muto, M., Sunaga, Y., Tanaka, M., Yoshino, T., Taniguchi, T., Fukuda, Y., Nemoto, M., Matsumoto, M., Wong, P.S., Aburatani, S. & Fujibuchi, W. (2015). Oil accumulation by the oleaginous diatom *Fistulifera solaris* as revealed by the genome and transcriptome. *The Plant Cell* 27: 162–176.

Traller, J.C., Cokus, S.J., Lopez, D.A., Gaidarenko, O., Smith, S.R., Mccrow, J.P., Gallaher, S.D., Podell, S., Thompson, M., Cook, O., Morselli, M., Jaroszewicz, A., Allen, E.E., Allen, A.E., Merchant, S.S., Pellegrini, M. & Hildebrand, M. (2016). Genome and methylome of the oleaginous diatom *Cyclotella cryptica* reveal genetic flexibility toward a high lipid phenotype. *Biotechnology for Biofuels* 9: 258–278.

Trentacoste Emily, M., Shrestha Roshan, P., Smith Sarah, R., Glé, C., Hartmann Aaron, C., Hildebrand, M. & Gerwick William, H. (2013). Metabolic engineering of lipid catabolism increases microalgal lipid accumulation without compromising growth. *Proceedings of the National Academy of Sciences* 110: 19748–19753.

Velmurugan, N. & Deka, D. (2018). Transformation techniques for metabolic engineering of diatoms and haptophytes: current state and prospects. *Applied Microbiology and Biotechnology* 102: 4255–4267.

Villanova, V., Fortunato, A., Singh, D., Dal Bo, D., Conte, M., Obata, T., Jouhet, J., Fernie, A., Marechal, E., Falciatore, A., Pagliardini, J., Monnier, A., Poolman, M., Curien, G., Petroutsos, D. & Finazzi, G. (2017). Investigating mixotrophic metabolism in the model diatom *Phaeodactylum tricornutum*. *Philosophical Transactions of the Royal Society of London Series B, Biological Sciences* 372: 20160404.

Villar, E., Vannier, T., Vernette, C., Lescot, M., Cuenca, M., Alexandre, A., Bachelerie, P., Rosnet, T., Pelletier, E., Sunagawa, S. & Hingamp, P. (2018). The Ocean Gene Atlas: exploring the biogeography of plankton genes online. *Nucleic Acids Research* 46: W289–W295.

Villén, J., Beausoleil Sean, A., Gerber Scott, A. & Gygi Steven, P. (2007). Large-scale phosphorylation analysis of mouse liver. *Proceedings of the National Academy of Sciences* 104: 1488–1493.

Watanabe, Y., Kadono, T., Kira, N., Suzuki, K., Iwata, O., Ohnishi, K., Yamaguchi, H. & Adachi, M. (2018). Development of endogenous promoters that drive high-level expression of introduced genes in the model diatom *Phaeodactylum tricornutum*. *Marine Genomics* 42: 41–48.

Wen, Z.Y. & Chen, F. (2001a). Application of statistically-based experimental designs for the optimization of eicosapentaenoic acid production by the diatom *Nitzschia laevis*. *Biotechnology and Bioengineering* 75: 159–169.

Wen, Z.Y. & Chen, F. (2001b). A perfusion-cell bleeding culture strategy for enhancing the productivity of eicosapentaenoic acid by *Nitzschia laevis*. *Applied Microbiology and Biotechnology* 57: 316–322.

Weyman, P., Beeri, K., Lefebvre, S., Rivera, J., Mccarthy, J., Heuberger, A., Peers, G., Allen, A. & Dupont, C. (2014). Inactivation of *Phaeodactylum tricornutum* urease gene using transcription activator-like effector nuclease-based targeted mutagenesis. *Plant Biotechnology Journal* 13: 460–470.

Wu, S., Gu, W., Huang, A., Li, Y., Kumar, M., Lim, P.E., Huan, L., Gao, S. & Wang, G. (2019). Elevated CO_2 improves both lipid accumulation and growth rate in the glucose-6-phosphate dehydrogenase engineered *Phaeodactylum tricornutum*. *Microbial Cell Factories* 18: 161–177.

Xia, S., Wang, K., Wan, L., Li, A., Hu, Q. & Zhang, C. (2013). Production, characterization, and antioxidant activity of fucoxanthin from the marine diatom *Odontella aurita*. *Marine Drugs* 11: 2667–2681.

Xie, W.H., Zhu, C.C., Zhang, N.S., Li, D.W., Yang, W.D., Liu, J.S., Sathishkumar, R. & Li, H.Y. (2014). Construction of novel chloroplast expression vector and development of an efficient transformation system for the diatom *Phaeodactylum tricornutum*. *Marine Biotechnology* 16: 538–546.

Yang, M.K., Qiao, Z.X., Zhang, W.Y., Xiong, Q., Zhang, J., Li, T., Ge, F. & Zhao, J.D. (2013). Global phosphoproteomic analysis reveals diverse functions of serine/threonine/tyrosine phosphorylation in the model cyanobacterium *Synechococcus* sp. strain PCC 7002. *Journal of Proteome Research* 12: 1909–1923.

Yang, Z.K., Ma, Y.H., Zheng, J.W., Yang, W.D., Liu, J.S. & Li, H.Y. (2014). Proteomics to reveal metabolic network shifts towards lipid accumulation following nitrogen deprivation in the diatom *Phaeodactylum tricornutum*. *Journal of Applied Phycology* 26: 73–82.

Yoshinaga, R., Niwa-Kubota, M., Matsui, H. & Matsuda, Y. (2014). Characterization of iron-responsive promoters in the marine diatom *Phaeodactylum tricornutum*. *Marine Genomics* 16: 55–62.

Zhang, C. & Hu, H. (2014). High-efficiency nuclear transformation of the diatom *Phaeodactylum tricornutum* by electroporation. *Marine Genomics* 16: 63–66.

Zhang, R., Patena, W., Armbruster, U., Gang, S.S., Blum, S.R. & Jonikas, M.C. (2014). High-throughput genotyping of green algal mutants reveals random distribution of mutagenic insertion sites and endonucleolytic cleavage of transforming DNA. *The Plant Cell* 26: 1398–1409.

Zou, L.-G., Chen, J.-W., Zheng, D.-L., Balamurugan, S., Li, D.-W., Yang, W.-D., Liu, J.-S. & Li, H.-Y. (2018). High-efficiency promoter-driven coordinated regulation of multiple metabolic nodes elevates lipid accumulation in the model microalga *Phaeodactylum tricornutum*. *Microbial Cell Factories* 17: 54–62.

4 Lipids from Diatoms

Ee Von Goh
Future Food Malaysia & School of Biosciences, University of Nottingham Malaysia, Selangor, Malaysia

4.1 INTRODUCTION

Lipids are a diverse group of compounds that are predominantly non-polar in nature. This is due to the fact that they are largely made up of hydrocarbons, with mostly non-polar carbon–carbon or carbon–hydrogen bonds (Alberts et al., 2002). Non-polar molecules are naturally hydrophobic, which means they are "water hearing" and are insoluble in water but are highly soluble in the non-polar or weakly polar organic solvents (Alberts et al., 2002). Lipids are the basic building blocks of all cells and play a wide range of important and diverse roles. They are essential components of the plasma membrane as well as other cellular compartments such as the nuclear membrane, endoplasmic reticulum, Golgi apparatus, and transport vesicles such as endosomes and lysosomes (Muro et al., 2014). The lipid composition of various organelles, cell types, and tissues can vary significantly, implying that different lipids are required for different functions (Muro et al., 2014).

While lipids include fats, oils, waxes, phospholipids, and steroids, the scope of this chapter will focus on fats and oils. Fat molecules are made up of two major components: glycerol and fatty acids (FAs). Glycerol is a three-carbon, five-hydrogen, and three-hydroxyl (OH) group organic compound (alcohol). FAs are composed of a long chain of hydrocarbons to which a carboxyl group is attached, thus the name "fatty acid." The number of carbons in a fatty acid can range from 4 to 36, with 12–18 carbons being the most common. The FAs in a fat molecule are ester-bonded to each of the three carbons of the glycerol molecule via an oxygen atom. FAs are essential molecules for the maintenance of physiological functions of many organisms (National Research Council, 1989).

Their nomenclature is represented by X:YωZ, where X represents the number of carbon atoms in the chain, Y represents the number of double bonds, and Z indicates the position of the first double bond counting from the methyl group (National Research Council, 1989). The two major groups of FAs are saturated fatty acids (SFAs) and unsaturated fatty acids (UFAs), differing by the existence of double bonds in UFAs. The position of the double bonds is determined by the desaturase enzyme activity, which performs in different positions on the carbon chain depending on the type of enzyme present in different organisms (e.g., Δ3 in animals, Δ6 in animals and plants, and Δ15 in algae and plants with chlorophyll) (Brett & Müller-Navarra, 1997).

Saturated fatty acids are primarily metabolized for energy and lipid storage. The number of double bonds distinguishes UFAs: monounsaturated fatty acids (MUFAs) have one double bond and can be synthesized de novo by nearly all organisms (National Research Council, 1989). Polyunsaturated fatty acids (PUFAs) are fatty acids with two or more double bonds that play an important role in brain development (Liu et al., 2015) as well as many physiological functions such as inflammation regulation (Calder, 2005) and cellular signaling (Stillwell & Wassall, 2003).

 DOI: 10.1201/9781003436553-4

Essential fatty acids (EFAs) are PUFAs that play critical roles in physiological and biochemical processes. They are termed "essential" due to the fact that they must be obtained by consumers from external sources through dietary intake. Due to a lack of the desaturase enzyme, most animals are unable to synthesize them de novo (Brett & Müller-Navarra, 1997). While some animals can synthesize EFAs from linolenic precursors such as linolenic acid (C18:3n3) which is found almost exclusively in plants (Brett & Müller-Navarra, 1997), the rate at which this conversion occurs is insufficient to supply the necessary amount of EFA for optimal growth and development. In fact, in a systematic review conducted by Lane et al. (2014), it was concluded that ALA from common plant sources, e.g., nut and seed oils, was minimally converted to DHA if at all.

Most high-value fatty acids belong to long-chain PUFAs like omega-3 fatty acids [e.g., eicosapentaenoic acid (EPA 20:5n3) and docosahexaenoic acid (DHA 22:6n3)] because the physiological activities of EFAs are primarily represented by EPA and DHA in animals. EPA and DHA are essential for brain development and the preservation of brain structure and function (Innis, 2007). Many physiological processes are influenced by EPA intake, including reproduction, immunity efficiency, and osmoregulation (Brett & Müller-Navarra, 1997). DHA is essential for the health and development of neurons, as well as neurotransmission, and has a strong influence on cognition and behavior function (Innis, 2007). It has also been shown to be important in the protection against oxidative stress function (Innis, 2007). Due to the various benefits of EPA and DHA and the fact that they cannot be synthesized by humans and animals, the market is growing, with high demand for these PUFA supplementations, particularly fish oil (Adarme-Vega et al., 2014).

4.1.1 Fatty Acids in the Ecosystems and Roles of Diatoms

Lipid energy is transferred from the primary producer algae to vertebrates to higher trophic level species via zooplankton: EFAs connect primary producers and secondary consumers as they move through the food web (Arts et al., 2001). EPA and DHA are the two important EFAs that phytoplankton synthesize de novo and animals bioaccumulate. In the aquatic food web for instance, EFAs must be obtained through external sources, e.g., direct feeding on phytoplanktonic, plant, or bacteria species (Veloza et al., 2006), or by ingesting EFA-rich lipid emulsions (Coutteau & Sorgeloos, 1997). Long-chain PUFAs are bioactive molecules that are essential for the maintenance of ecosystem processes and for the reproduction and growth of the aquatic marine food web. Even with adequate energy intake, consumers in the food web who lack these fatty acids may experience poor reproduction and growth (Arts et al., 2001). While awareness of the benefits of PUFAs has led to the exploitation of fish oil in recent years, fish are not true PUFA producers; rather, they bioaccumulate PUFAs through the consumption of PUFA-rich phytoplankton.

Diatoms, the most common phytoplankton, are a major primary producer of EPA and DHA. Many diatoms are oleaginous species capable of storing large amounts of lipids. Diatoms are considered oleaginous if they can generate at least 20% of total lipids from their dry biomass (Zulu et al., 2018). Diatoms have a remarkably high total lipid content that includes a significant proportion of very long-chain EPA and DHA PUFAs (Table 4.1). Terrestrial oil crops, on the other hand, are not a typical source of EPA and DHA. Although they contain varying amounts of the omega-3 fatty acid alpha-linolenic acid (ALA), the conversion of ALA to the essential fatty acids DHA and EPA is extremely inefficient (Lane et al., 2014). Diatoms, particularly those of marine origin, have received increased attention as potential sources of long-chain PUFAs. For example, *Phaeodactylum tricornutum* has been identified as one of the most promising industrial species for supplying omega-3 fatty acids in a sustainable manner (Gu et al., 2022).

Oleaginous diatoms have also been identified as promising candidates for biofuel production (d'Ippolito et al., 2015). Biofuels, as a low-carbon alternative to fossil fuels, may help to reduce greenhouse gas (GHG) emissions and the associated climate change impact from transportation.

TABLE 4.1
Lipid Content of Selected Diatom Species in Comparison to Other Sources [Oil Content (% of Dry Mass) and EPA/DHA Content (% of Total Fatty Acids)]

Sources	Total lipids	EPA	DHA	References
		Diatom		
Amphora exigua	33–45	5–13	ND	Chen, 2012[a]
Chaetoceros muelleri	33.6	12.8	0.8	Rodolfi et al., 2009[a]; Zhukova & Aizdaicher, 1995[a]
Fistulifera solaris	48–60	5–20	—	Matsumoto et al., 2010[a]; Satoh et al., 2013[a]
Melosira nummuloides	33	6.5–12	3.9–11	Chen, 2012[a]
Navicula lyra	37–42	5.5–9.3	< 1	Chen, 2012[a]
Nitzschia laevis	25–69	19.1	—	Wen & Chen, 2003[a]
Phaeodactylum tricornutum	20–30	34.5	< 1	Yongmanitchai & Ward, 1991[a]; Chisti, 2007[a]
Seminavis gracilenta	35–43	11–16	< 1	Darko et al., 2014[a]
Skeletonema costatum	21.1	29.2	3.1	Rodolfi, 2009[a]; Zhukova & Aizdaicher, 1995[a]
Thalassiosira pseudonana	21–31	12.2	3.4	Tonon et al., 2002[a]; Bromke et al., 2013[a]
Cylindrotheca closterium	42	5	ND	Demirel, 2016
Amphora capitellata	24	14	ND	Demirel, 2016
Nitzschia communis	32	18.3	ND	Demirel, 2016
Nitzschia thermalis	32	10	2.0	Demirel, 2016
		Terrestrial oil crops		
Soybean seed	18–20	—	—	Abdelghany et al., 2020
Canola seed	30.6–48.3	—	—	Matthaus et al., 2016
Olive fruit	10–30	—	—	El Riachy et al., 2019
Palm fruit	56	—	—	Poku, 2002
Maize germ	35–56	—	—	Ni et al., 2016
Sunflower seed	35–42	—	—	Guo et al., 2017
		Aquatic animals		
Krill (*Euphausia superba*)	6–43	16.8–19.4	7.2–11.2	Virtue, 1995
Pilchard (*Sardina pilchardus*)	12.8–28	9.9–11.4	24.1–26.7	Sibert, 2001; Bouderoua et al., 2011
Sprat (*Sprattus sprattus*)	11.5–18	1.32	6.8	Hardie & Mackie, 1969; Stancheva et al., 2010
Pacific herring (*Clupea pallasii*)	31.8–41.7	6.5–7.4	5.1–7.0	Murzina et al., 2016
Atlantic salmon (*Salmo salar*)	21.5	2.1	3.4	Aas et al., 2019

[a]As cited in Sayanova et al. (2017).

Currently, biofuel is produced from various vegetable oils such as palm oil, sunflower oil, soybean oil, and canola using various types of catalysts, with soybean oil serving as the primary feedstock for biofuel production (AOCS, 2022). The use of these terrestrial oil crops for biofuel, however, is not without controversy (this will be discussed in a later part of this chapter). As a result, the focus has shifted to diatoms as a better source of lipids for biofuels.

Unlike for food and dietary purposes, the ability to produce a large amount of PUFAs is not necessarily a desired trait for biofuel production. High PUFAs would result in biofuel with lower oxidation stability, but with better cold flow properties, while high levels of SFAs would inherently lead to biodiesel with poor cold flow properties (Nascimento et al., 2013). Ideal FA profiles for a good-quality biofuel should have higher levels of SFAs and MUFAs to strike a balance between cold flow and other properties (e.g., enhanced energy yields, higher oxidative stability) (Hoekman et al.,

2012). FAs such as C14:0, C16:0, C18:0, C18:1n9, and C18:3 are thought to be suitable for high-quality biofuel (Song et al., 2013). Studies have suggested a number of species, e.g., *Navicula* sp., *Thalassiosira* sp., *Chaetoceros calcitrans*, and *Skeletonema* sp., all contained significant amounts and good ratios of those FAs (Joseph et al., 2017).

4.2 LIPID SYNTHESIS IN DIATOMS

4.2.1 Proposed Pathway of FA and PUFA Synthesis in Diatoms

There is limited knowledge about how diatoms carry out and regulate lipid biosynthesis and metabolism, and current models of diatom lipid synthesis pathways suggest that it occurs in a manner similar to FA synthesis in flowering plants, with acetyl-CoA as a precursor (Zulu et al., 2018). The biosynthesis pathways are summarized in Figure 4.1.

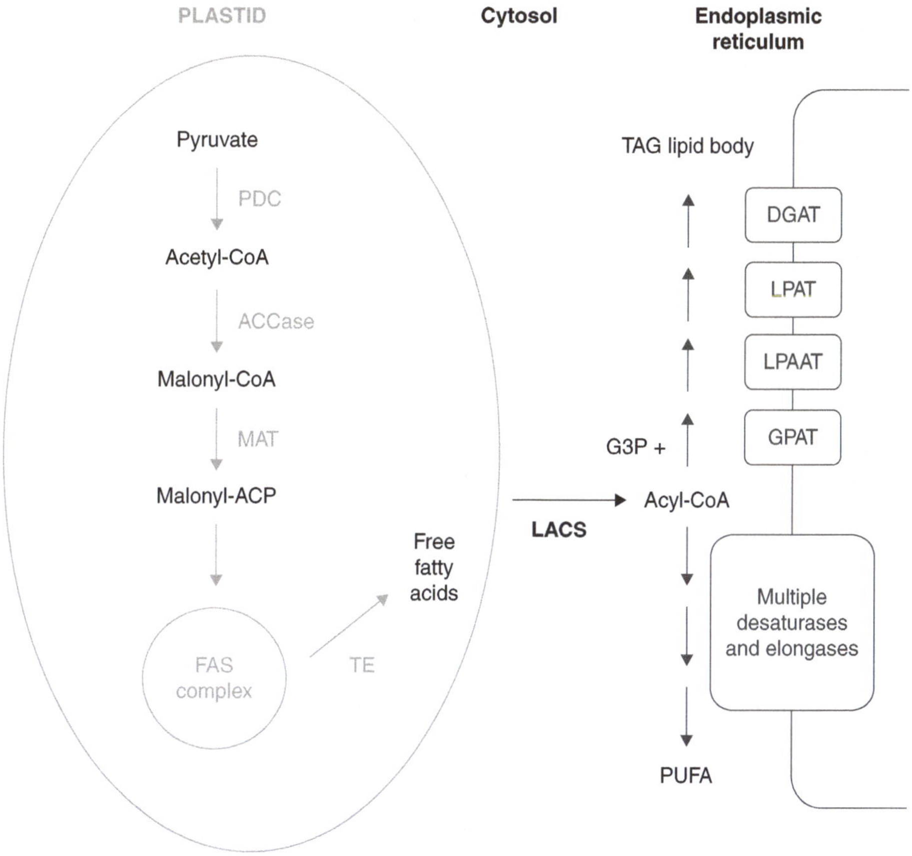

FIGURE 4.1 Simplified overview of the major pathways of lipid biosynthesis in a diatom cell, adapted from Radakovits et al. (2010) and Gimpel et al. (2015). Enzyme names are gray-scaled according to their location. Light gray: plastid; black solid: cytosol; dark gray: endoplasmic reticulum. PDC, pyruvate dehydrogenase complex; ACCase, acetyl-CoA carboxylase; MAT, malonyl-CoA/ACP transacylase; ACP, acyl-carrier protein; LACS, long-chain acyl-CoA synthetase; FAS, FA synthase; TE, fatty acyl-ACP thioesterase; GPAT, glycerol-3-phosphate acyltransferase; LPAAT, lyso-phosphatidic acid acyltransferase; LPAT, lyso-phosphatidylcholine acyltransferase; DGAT, diacylglycerol acyltransferase; CoA, coenzyme A; G3P, glycerate-3-phosphate; TAG, triacylglycerol; PUFA, polyunsaturated fatty acids.

Plastids are thought to be the primary sites of FA biosynthesis in diatom cells. The acetyl-CoA formed within the plastid would be the precursor for the biosynthesis of lipids. It would serve as a building block for the acyl chain's elongation by two carbons per elongation step. The malonyl-acyl carrier proteins (ACP) would catalyze the acyl chain elongation processes. Malonyl-ACPs are formed from acetyl-CoA via carboxylation and the transfer of the malonyl group to ACP, which are catalyzed by the enzymes ACCase and MAT, respectively. The enzyme thioesterases release fatty acids (e.g., 16:0, 18:0, 18:1, etc.) from acyl carrier proteins and convert them to acyl-CoAs. After a series of desaturation and elongation reactions, EPA and DHA would be synthesized from MUFA (18:1), primarily via the major route depicted in Figure 4.1. For TAG synthesis, the acyl-CoAs bind to glycerol-3-phosphate, which is esterified to lysophosphatidic acid, phosphatidic acid, and diacylglycerol (DAG). DGAT catalyzes the final step in TAG synthesis by acylating DAG (Figure 4.1).

4.2.2 Conditions that Influence Lipid Accumulation in Diatoms

Growth conditions of diatoms pose a significant influence on the lipid accumulation and their ratios, as well as the partitioning of FAs into various lipids in diatoms (Zulu et al., 2018). In theory, growth conditions can be modulated to increase lipid accumulation and alter the FA composition of diatoms according to the desired ratios.

Among the factors to be manipulated are carbon types and their availability. Because lipids are organic compounds, carbon is required for their production (Fields et al., 2014). While diatoms are capable of growing under different modes, namely photoautotrophic, heterotrophic, and mixotrophic, carbon source, light intensity, and wavelength have been reported to change the lipid metabolism in microalgae by enhancing the biomass production and lipid content (Zulu et al., 2018). Mixotrophic conditions are found to be far more effective than photoautotrophic conditions in increasing lipid production (Wang et al., 2012; Villanova et al., 2017). The type of carbon source used is also important: glucose, starch, and acetate, for example, have been shown to increase lipid production by 4.6-, 2-, and 1.7-fold, respectively, when compared to photoautotrophically grown *P. tricornutum* (Wang et al., 2012). Some research has also found that increasing CO_2 concentrations would boost biomass and lipid production in other strains (Wu et al., 2015).

Introduction of physiological stresses such as nutrient limitation can alter the ratio of accumulated lipids. Diatoms use the storage polysaccharide, chrysolaminarin, as their carbon sink under normal growth conditions (Kroth et al., 2008). However, diatoms shift the carbon flux toward lipid synthesis when exposed to stress (Zulu et al., 2018). The syntheses of both storage polysaccharide and lipids require the same carbon precursors, hence the inhibition of storage polysaccharide synthesis increases lipid accumulation (Li et al., 2010). However, this will inadvertently result in lower biomass production and is generally overcome by a two-stage cultivation strategy in which diatoms are first cultivated under optimal growth conditions to desired levels of biomass prior to stress exposure in the second stage (Zulu et al., 2018).

Other nutrient stresses commonly applied to suppress growth for enhanced lipid productivity are nitrogen limitation and phosphorus limitation (Abida et al., 2015; Popko et al., 2016; Yodsuwan et al., 2017). Both nitrogen and phosphorus play significant roles in photosynthesis and growth, with nitrogen limitation resulting in more drastic results in lipid productivity (Abida et al., 2015; Zulu et al., 2018). Because silicon is an essential component of diatoms, its limitation is frequently exploited to boost lipid production in diatoms (Adams & Bugbee, 2014). Manipulation of light (Jayakumar et al., 2021) and temperature (Hemme et al., 2014) conditions are yet other approaches used to improve lipid metabolism in diatoms.

Diatom lipid components are extremely sensitive to environmental changes, making them an effective assessment tool for monitoring of toxicological effects on marine biota and as bioindicators of ecosystem health (Gonçalves et al., 2017). Environmental stressors can induce structural changes in diatom biochemical composition, which are then reflected in their FA profiles (Filimonova et al., 2016; Mohy El-Din, 2018). Climate change and habitat modification can both dampen the switch from carbohydrate to fatty acid synthesis in diatoms; for instance, higher ocean temperatures and lower pH resulting from climate change have been reported to reduce the ability of diatoms to make long-chain PUFAs, and habitat degradation reduces the numbers of diatoms in the area, thereby reducing the absolute amounts of long-chain PUFAs produced (Baird, 2022).

4.3 PRODUCTION OF LIPIDS FROM DIATOMS

4.3.1 Cultivation and Harvesting

Diatoms are initially cultivated in closed, batch systems before seeding them into open ponds (Ratledge & Cohen, 2008). However, large-scale production of closed, batch systems faces numerous challenges, with the most significant ones being the initial capital costs, the requirement to remove heat, and the overwhelming difficulty and high cost of maintenance activities (Wang & Seibert, 2017). Another option is to cultivate diatoms in open water. However, for lipid production per se, oleaginous diatoms may face the problem of bioaccumulation of environmental pollutants in open water (Wang & Wang, 2005). Lipid content, rather than cell size, is an important factor in determining the uptake of organic pollutants by the diatom cells due to the lipophilicity of organic pollutants (e.g., pesticides) (Debenest et al., 2010). These residues can then be passed along the trophic levels by way of ingestion of "contaminated food." Nonetheless, the relationship between accumulation of pollutants in diatoms remains understudied as ecotoxicity data are only available for single pollutants, and the collective toxicity of many different pollutants in open water bodies has not been assessed (Debenest et al., 2010). Similar to pesticides, heavy metals can be transferred to the lipids in diatoms due to their lipophilicity and their toxicity is also understudied (Marella et al., 2020a).

During harvesting, separating diatoms from growth medium can be done via traditional methods such as centrifugation, filtration, or flocculation. These methods are either energy-intensive, difficult to operate, or necessitate the use of chemicals. Continuous or more frequent harvesting will help maximize diatom yield, but the ability to harvest diatoms efficiently and cheaply is critical for overall economic viability (Wang & Seibert, 2017).

4.3.2 Extraction and Downstream Processing

Downstream processing of diatom oil includes the process of cell wall breaking and oil extraction. Diatom cells are small and have silicified defensive cell walls. Hence, diatoms are not good candidates for the conventional mechanical pressing methods that are widely applied in vegetable oil extraction. Commonly used diatom cell crushing methods are comprised of mechanical (e.g., ball mill, homogeneous, and cavitation) and non-mechanical methods (e.g., chemical and biological methods) (Xue et al., 2020). Each method has its pros and cons in terms of energy consumption and oil extraction effect. Mechanical methods typically incur high energy consumption and may cause lipid oxidation, thereby affecting product quality. Chemical and biological non-mechanical methods on the other hand consume lower amounts of energy but pose a significant risk of pollution. A combination of several methods provides complementary advantages and often delivers an improved oil extraction effect (Xue et al., 2020).

After cell fragmentation, the choice of solvent used for oil extraction is considered. Selection of solvent is usually determined by its low boiling point for better recovery of the product, cost, availability, flammability, disposal procedures, and toxicity. Conventional extraction techniques are based on hydrophobicity and involve the use of organic solvents like hexane, petroleum ether, methanol, and chloroform (Alfio et al., 2021). Based on the "similarity compatibility principle", non-polar solvents dissolve and destroy non-polar lipids on diatom cell membranes to extract oil. A number of monophasic solvents (mixtures of one or more polar/non-polar solvents and water miscible, e.g., hexane/ethanol) and biphasic solvents (mixtures of two or more insoluble substances of different polarity, e.g., hexane in water) are effective for lipid extraction (Xue et al., 2020).

Despite its high extraction efficiency, organic solvents are currently restricted for use in laboratories only owing to the fact that organic solvents are toxic, volatile, and difficult to recycle. Green solvents such as supercritical fluids are generally used in environmentally conscious and sustainable processing facilities. A supercritical fluid (SCF) is one that has a higher temperature and pressure than its critical point. The most commonly used supercritical fluid is CO_2. Because CO_2 is in gas phase under atmospheric conditions, this non-polar solvent can be fully removed to form a solvent-free extract (Molino et al., 2020). While it has many benefits (e.g., non-toxic, easy to recycle, operates at low temperatures, etc.), the high cost of equipment and production has limited the application of supercritical fluid extraction mainly in the production of diatom biofuel (Xue et al., 2020). In terms of the production of edible diatom oil, there is still a shortage of cost-effective and feasible extraction strategies.

The hydroenzymatic method is yet another relatively new type of "mild", relatively energy-efficient, and highly specific method that does not produce harmful or volatile organic compounds but is capable of co-producing a large number of bioactive substances, which has piqued the interest of researchers of late (Xue et al., 2020). However, many challenges remain. First, it requires pre-treatment methods (e.g., ultrasound-assisted) beforehand to increase its oil extraction efficiency (Hou et al., 2018). On top of that, an emulsifying layer will be formed on top of the process. As a result, this method necessitates additional investment of resources into studying the composition and properties of the emulsifying layer in order to select an appropriate demulsification method (Yusoff et al., 2015). While there are other "mild" oil extraction methods developed in the recent years, it is worth highlighting that, to date, no efficient, reliable, and reasonably cost-effective downstream processing method has yet been identified. In reality, the downstream operations cost is estimated to represent 50–80% of the total processing cost (Vinayak et al., 2015), which is high enough to deter the industrialization and commercialization of diatom oil.

4.3.3 Edible vs. Industrial Oil

The development of edible oil from diatoms differs from biofuel or oil for other industrial uses in that the edible safety of the oils must be considered (Xue et al., 2020). The edible oil from diatoms must undergo multiple studies including the chemical and physical properties, toxicological studies to check for potential toxicity and the possibility of risk levels of heavy metals and pathogenic microorganisms, and the safety of the oil production process (Draaisma et al., 2013). While it is encouraging that several DHA-rich algal species, such as *Schizochytrium* sp., *Chlorella protothecoides*, *Ulkenia* sp. SAM2179 etc., have been accepted as "Generally Recognized as Safe" (GRAS) by the Food and Drug Administration (FDA) (Szabo et al., 2014), more evidence of fitness for consumption is required before diatom oil can be widely used as food (Xue et al., 2020).

When diatoms are cultured in an open-system or with wastewater, refining processes, similar to that of fish oil processing, are required to remove or reduce most of the undesirable components to an acceptable level fit for human consumption. Metals and persistent organic pollutants in diatom oil can be removed by neutralization, deodorization, and bleaching, while halogenated compounds can be removed by molecular distillation (Alfio et al., 2021). These techniques, somehow, could potentially change the quality of the oil by promoting its oxidation (Alfio et al., 2021). Under milder conditions, activated carbon may be employed to absorb dioxins and PCBs without affecting the DHA and EPA contents (Alfio et al., 2021).

As with other edible oils, diatom oil faces similar challenges in terms of oxidative stability and shelf life. This is especially crucial for diatom oil due to its higher PUFA content (thus having less antioxidant capacity) than other sources of oils. The most common antioxidant method is to add natural or synthetic antioxidants to the oil, and because combined oxidants have a greater effect than single oxidants, more research should be conducted to optimize the formulation and mixture ratio of antioxidant combinations for use in diatom oil extracts (Xue et al., 2020). Some new methods, e.g., nanotechnology to microencapsulate oil or make microemulsions, have great potential in preventing oil oxidation while improving the physical and nutritional properties of oil (Linke et al., 2020), but their safety evaluation and many other aspects are currently lacking.

4.4 CURRENT TRENDS AND THE SUSTAINABILITY MOVEMENT

4.4.1 Sustainable Oil Production

The ever-expanding global population and an increasing number of applications for vegetable oils are driving up demand for these oils, bringing about major threats to the environment. The production of vegetable oil is recognized as a major source of greenhouse gas emissions by means of lost carbon storage opportunity and agriculture and processing emissions (Alcock et al., 2022). The demand for vegetable oil has been steadily increasing over the last 60 years, causing an expansion of the cultivation area and intensifying production practices (Alcock et al., 2022). Strikingly, only these four crops produce more than 85% of the world's vegetable oil: oil palm, soybean, rapeseed, and sunflower (FAO, 2022). Although sharing similar basic photosynthetic mechanisms, diatoms are capable of producing up to 30 times the volume of oil per unit land area compared to the aforementioned commercial oilseed crops because diatoms, being single-celled organisms, do not need to support an array of non-photosynthetic cells as in plants (Wang & Seibert, 2017). Oil from diatoms may also serve as a good alternative to cooking oil due to its excellent qualities for human consumption (Wang & Seibert, 2017; Alcock et al., 2022).

Many nations are promoting the use of biofuels made from food crops in order to reduce reliance on imports and slow global warming caused by fossil fuel emissions, and such effort of converting food grain into automobile fuel has proven controversial (Gan et al., 2019). Unlike terrestrial oilseed crops like soybean, which when used for biofuel competes with food production, diatom can be grown without impacting food production (Vinayak et al., 2015). On top of that, diatoms can be cultivated on non-arable land or non-potable water while yielding a high volume of oil per unit land area, all of which are not possible in oilseed crop cultivation (Vinayak et al., 2015; Wang & Seibert, 2017; Upadhyay et al., 2021).

It is generally recognized that the main sources of PUFAs on the market are marine fish, since terrestrial plants are not good sources of EPA and DHA (Martins et al., 2013). Although the content of DHA and EPA in fish oil can exceed 20% (Table 4.1), there remain many issues to be resolved. The conundrum here is that an ever-increasing human population and fishing efforts have collectively reduced the fish stock of some of the most desirable (in terms of FA profile) fish species of the highest trophic level. Fish farming is rapidly developing to meet consumer demand in the face of fish stock depletion. Commercial fish feeds are currently made up of fishmeal (ground fish) and fish oil. The production of these feeds is both resource intensive and unsustainable. Because diatoms are natural PUFA producers, using them to produce PUFAs has more advantages than using fatty fish.

4.4.2 Sustainable Approach of Resource Reutilization, Waste Management, and Lipid Production

In line with the waste-to-wealth concept, diatoms have been widely recognized as promising tools for phycoremediation of wastewater and carbon fixation for obtaining value, such as high-value lipids, from the biomass (Upadhyay et al., 2021). Diatoms are important primary producers, accounting for up to one-fifth of global carbon fixation (Ohno et al., 2012). Their great ability for carbon dioxide fixation has important implications: diatoms absorb 10–20 billion tonnes of carbon dioxide every year, which is equivalent to the amount of carbon captured annually by all of the world's rainforests (Ohno et al., 2012). The unique photosynthetic, cellular, and metabolic characteristics of diatoms allow them to use pollutants such as nitrate, iron, phosphate, molybdenum, silica, and heavy metals such as copper, cadmium, chromium, and lead, making diatoms a good agent for wastewater treatment (Marella et al., 2020b). The abundant wastewater and CO_2 generated in urban and industrial localities daily provide the nourishment required by diatoms for oil generation. Ramachandra et al. (2013) found that algae grown in wastewater systems collected from sewage-fed urban lakes and sewage treatment plants of Bangalore and Mysore have remarkable lipid content, profile, and yield potential for biofuel purposes.

On the other hand, while the aquaculture industry has been actively promoted to overcome the supply issues, fish farming has resulted in serious concerns regarding environmental pollution related to the aquaculture wastewater and over-exploitation of the wild populations used as feed ingredients. Cultivation of diatoms with wastewater in fish farms serves as an affordable and productive source for valuable EFAs in aquaculture feed (Tossavainen et al., 2019). The diatom-supplemented feed eliminates the need for expensive feed and fish oil, leading to improved growth of the aquaculture species. This would generate higher profits for the aquaculture sector while simultaneously allowing waste management, thereby promoting a circular economy model.

4.5 POTENTIAL PROSPECTS AND CHALLENGES

Having discussed the different aspects of diatoms in the preceding sections, it can be summarized that diatoms have great potential as a sustainable source of valuable fatty acids as food for human consumption, animal feed, and biofuel, while at the same time serving as a sustainable solution to wastewater management and carbon sequestration. However, it is worth noting that many of these ideas are still at the conceptualization or lab-scale stage. The key concerns and challenges of each prospect from different perspectives will be discussed in the following.

4.5.1 Consumers' Perspectives and Marketing

Due to the scarcity of fish oil and the increased demand from the aquaculture feed and nutraceutical industries, researchers have been looking for alternative sources of long-chain omega-3 PUFAs. While diatoms can assist in meeting the high demand, the lipid class composition of oils from lower trophic levels, including diatoms, frequently differs from conventional fish oil, which may affect the human bioavailability of the fatty acids in the oils. The allergenic potential of diatom-derived edible oil is also unknown.

As discussed earlier, alternative terrestrial plant sources cannot be converted to EPA and DHA efficiently (Lane et al., 2014). Hence, diatom oils have competitive advantages over fish oils for being vegan or vegetarian friendly because they do not contain any animal products. In addition, the environmentally friendly feature of diatoms makes diatom oils a strong substitute for fish oils that would appeal to eco-conscious consumers. Nevertheless, diatoms are technically an enigma, i.e., neither plant nor animal, as they share biochemical features of both. It is generally considered vegan as it is non-sentient.

Although the concept of waste to wealth is very compelling, the incorporation of diatoms grown in wastewater into the human diet, whether as whole cells or oil extracts, may be hampered by social stigma, cultural practices, and widely held misconceptions. Other similar concepts, such as the incorporation of oil or textured protein harvested from black soldier fly larvae into the human diet, have already encountered such challenges (Wong et al., 2021). At present, consumer perceptions and acceptance of diatom oil are not fully known. Nonetheless, many foods have acquired tastes, and taste preferences are strongly shaped by popular culture. For example, while it has for centuries formed part of the diet of populations in China, Japan, and the Republic of Korea, Western countries have just recently become acquainted with edible seaweeds (e.g., nori and dulse), which can be described as slightly fishy and may not be to everyone's liking the first time they try it. The nutritional value of diatoms may persuade people to look past the taste and appreciate them.

The most successful microalgae exemplars thus far are chlorella and spirulina, which have found their place in health and wellness food market as whole-cell products. Diatomaceous earth, which is made from the fossilized remains of diatoms, is currently marketed by health food companies as a source of silica allegedly for treating a myriad of ailments. The FDA has approved the use of silica in food and agriculture, and the International Agency for Research on Cancer has classified it as non-carcinogenic. This could be seen as an encouragement to promote the consumption of diatoms as food, whether as whole cell or oil extract products.

4.5.2 Industrial Perspectives

The sustainable marketing strategy focuses on increasing human consumption of renewable products while also helping to reduce carbon emissions. When sustainability is used as a marketing or selling point to entice consumers to buy, it is critical to develop more sustainable and energy-efficient downstream processing. Scaling up the process for large-scale manufacturing, which is dependent on many micro and macro factors such as cultivation, harvesting, drying, and genetic modification, is one of the major challenges in diatom-based industries.

Apart from facility investments, the major costs of algae production are fertilizers, carbon dioxide, and electricity. These major costs alone would make it nearly impossible to produce competitively priced diatom biocrude (Wang & Seibert, 2017). For the production to be economically viable, diatom producers should strive to reduce or eliminate input costs by consuming pollutants such as industrial heavy metal wastewater, municipal wastewater, and carbon dioxide. Renewable energy, such as solar or wind power, could be used to reduce electricity costs where appropriate. An open diatom cultivation system is ideal in such instances. An open pond system can be designed to be embedded within the area of a renewable energy farm and/or waste discharge point to harness the ability to complement each other in energy use, space allocation, and the cost of land preparation (Wang & Seibert, 2017).

As with current petrochemical refining facilities, the economic viability of a diatom biorefinery facility will necessitate the production of as many co-products as possible. Depending on the source of the water used to produce the diatoms, many potential co-products are possible. This includes specific organics such as food-grade beta-carotene, pharmaceuticals, and pigments, as well as compounds such as polysaccharides, carbohydrates, surfactants, and other polymers. Making full use of diatom resources to meet energy, chemical, food, medicine, cosmetics, and other product needs in a variety of industries would increase the overall economic viability and profitability of the processes (Wang & Seibert, 2017).

4.5.3 Research Perspectives

It is known that organic waste is generated during the extraction and purification operations, which also consumes a lot of energy. Alternative downstream procedures must be developed urgently. Milking is one of the options, and it implies that the extraction should not destroy the diatom cells (Vinayak et al., 2015). As a result, there is no need to re-grow the diatoms. Biocompatible organic solvents pulsed electric field (PEF), spontaneous oozing, and membrane-bound protein pumps are some of the extraction methods that could be used to milk diatoms (Vinayak et al., 2015) and warrant further investigations.

Furthermore, research efforts should be made to achieve the simultaneous production of a group of valuable and metabolite compounds, as diatoms are capable of generating a wide range of such products. Comprehensive economic and environmental studies, including detailed sustainability and life cycle assessments, cost–benefit and trade-offs analyses, on the production of high-value compounds from diatoms are urgently required to develop a more economically feasible biorefining route in order to reduce microalgae processing costs and achieve integrated microalgae product production.

One of the most important requirements for efficient lipid production is the selection of potential lipid-accumulating microalgal strains through bioprospecting. Research has shown that species richness and complementary association of a microalgal consortia would yield better results for the simultaneous waste reutilization, phycoremediation, and lipid production at scale (Upadhyay et al., 2021). However, there is currently a scarcity of genomic, transcriptomic, and metabolic data.

Diatoms are a robust model for genome editing and cell transformation; in fact, a few companies have reportedly used genetically modified microalgae for biofuel production (Sharma et al., 2021). Further research may focus on improving lipid productivity in diatoms via targeted gene modifications of lipid metabolism in diatoms. Advanced metabolic engineering tools, such as the CRISPR/Cas9 system, can be used to generate strains that accumulate specific fatty acids like EPA and DHA (Yi et al., 2017). Understanding the biosynthetic pathways of lipids in model and other diatom species could lead to the development of synthetic biology tools for high-throughput strain development. In various cultivation scenarios, engineering strategies can be optimized. Global research efforts are currently being stepped up with the goal of increasing and modifying lipid accumulation in diatoms through genetic engineering. The availability of advanced sequencing technologies, e.g., the Synthetic Diatoms Project, heralds a revolution in genetic engineering research by greatly facilitating access to genome sequences.

Last but not least, the realization of the goal of sustainable diatom oil production requires multi-sectoral collaboration, bringing in multiple stakeholders including academic researchers, bio-entrepreneurs, and public authorities. This multi-sectoral collaboration would speed up the process through to wider acceptance and consumption by consumers.

REFERENCES

Aas, T. S., Ytrestøyl, T., & Åsgård, T. (2019). Utilization of feed resources in the production of Atlantic salmon (Salmo salar) in Norway: An update for 2016. *Aquaculture Reports* 15, 100216.

Abdelghany, A. M., Zhang, S., Azam, M., Shaibu, A. S., Feng, Y., Li, Y., … & Sun, J. (2020). Profiling of seed fatty acid composition in 1025 Chinese soybean accessions from diverse ecoregions. *The Crop Journal* 8(4), 635–644.

Abida, H., Dolch, L. J., Meï, C., Villanova, V., Conte, M., Block, M. A., … & Maréchal, E. (2015). Membrane glycerolipid remodeling triggered by nitrogen and phosphorus starvation in Phaeodactylum tricornutum. *Plant physiology* 167(1), 118–136.

Adams, C., & Bugbee, B. (2014). Enhancing lipid production of the marine diatom Chaetoceros gracilis: synergistic interactions of sodium chloride and silicon. *Journal of Applied Phycology* 26(3), 1351–1357.

Adarme-Vega, T. C., Thomas-Hall, S. R., & Schenk, P. M. (2014). Towards sustainable sources for omega-3 fatty acids production. *Current Opinion in Biotechnology* 26, 14–18.

Alberts, B., Johnson, A., Lewis, J., Raff, M., Roberts, K., & Walter, P. (2002). The Lipid Bilayer. In: *Molecular Biology of the Cell*. 4th edition. Garland Science.

Alcock, T. D., Salt, D. E., Wilson, P., & Ramsden, S. J. (2022). More sustainable vegetable oil: Balancing productivity with carbon storage opportunities. *Science of the Total Environment* 829, 154539.

Alfio, V. G., Manzo, C., & Micillo, R. (2021). From fish waste to value: an overview of the sustainable recovery of omega-3 for food supplements. *Molecules* 26(4), 1002.

AOCS (2022). *The continued struggle over biofuel feedstocks–March 2022*. Available at www.aocs.org/stay-informed/inform-magazine/featured-articles/the-continued-struggle-over-biofuel-feedstocks-march-2022

Arts, M. T., Ackman, R. G., & Holub, B. J. (2001). "Essential fatty acids" in aquatic ecosystems: a crucial link between diet and human health and evolution. *Canadian Journal of Fisheries and Aquatic Sciences* 58(1), 122–137.

Baird, P. (2022). Diatoms and fatty acid production in Arctic and estuarine ecosystems a reassessment of marine food webs, with a focus on the timing of shorebird migration. *Marine Ecology Progress Series* 688, 173–196.

Bouderoua, K., Mourot, J., Benmehdi-Tabet-Aoull, F., & Selselet-Attou, G. (2011). The effects of season and site of catch on morphometric characteristics, mineral content, and fatty acids of sardines (*Sardina pilchardus*) caught on the Algerian coast. *Journal of Aquatic Food Product Technology* 20(4), 412–420.

Brett, M., & Müller-Navarra, D. (1997). The role of highly unsaturated fatty acids in aquatic foodweb processes. *Freshwater Biology* 38(3), 483–499.

Bromke, M. A., Giavalisco, P., Willmitzer, L., & Hesse, H. (2013). Metabolic analysis of adaptation to short-term changes in culture conditions of the marine diatom Thalassiosira pseudonana. *PLoS One* 8(6), e67340.

Calder, P. C. (2005). Polyunsaturated fatty acids and inflammation. *Biochemical Society Transactions* 33(2), 423–427.

Chen, Y. C. (2012). The biomass and total lipid content and composition of twelve species of marine diatoms cultured under various environments. *Food Chemistry* 131(1), 211–219.

Chisti, Y. (2007). Biodiesel from microalgae. *Biotechnology Advances* 25(3), 294–306.

Coutteau, P., & Sorgeloos, P. (1997). Manipulation of dietary lipids, fatty acids and vitamins in zooplankton cultures. *Freshwater Biology* 38(3), 501–512.

d'Ippolito, G., Sardo, A., Paris, D., Vella, F. M., Adelfi, M. G., Botte, P., … & Fontana, A. (2015). Potential of lipid metabolism in marine diatoms for biofuel production. *Biotechnology for Biofuels* 8(1), 1–10.

Darko, E., Heydarizadeh, P., Schoefs, B., & Sabzalian, M. R. (2014). Photosynthesis under artificial light: the shift in primary and secondary metabolism. *Philosophical Transactions of the Royal Society B: Biological Sciences* 369(1640), 20130243.

Debenest, T., Silvestre, J., Coste, M., & Pinelli, E. (2010). Effects of pesticides on freshwater diatoms. *Reviews of Environmental Contamination and Toxicology* 203, 87–103.

Demirel, Z. (2016). Identification and fatty acid composition of coccolithophore and diatom species isolated from Aegean Sea. *Romanian Biotechnological Letters* 21(4), 11747–11753.

Draaisma, B. R., Wijffels, H. R., Slegers, P. M. E., Brentner, L. B., Roy, A., and Barbosa, M. J. (2013). Food commodities from microalgae. *Current Opinionn in Biotechnology* 24, 169–177.

El Riachy, M., Hamade, A., Ayoub, R., Dandachi, F., & Chalak, L. (2019). Oil content, fatty acid and phenolic profiles of some olive varieties growing in Lebanon. *Frontiers in Nutrition* 6, 94.

FAO (Food and Agriculture Organization of the United Nations). *FAOSTAT Statistical Database* (2022). www.fao.org/faostat/en/#data

Fields, M. W., Hise, A., Lohman, E. J., Bell, T., Gardner, R. D., Corredor, L., … & Gerlach, R. (2014). Sources and resources: importance of nutrients, resource allocation, and ecology in microalgal cultivation for lipid accumulation. *Applied Microbiology and Biotechnology* 98(11), 4805–4816.

Filimonova, V., Goncalves, F., Marques, J. C., De Troch, M., & Goncalves, A. M. (2016). Fatty acid profiling as bioindicator of chemical stress in marine organisms: a review. *Ecological Indicators* 67, 657–672.

Gan, J., Stupak, I., & Smith, C. T. (2019). Integrating policy, market, and technology for sustainability governance of agriculture-based biofuel and bioeconomic development in the US. *Energy, Sustainability and Society* 9(1), 1–25.

Gimpel, J. A., Henríquez, V., & Mayfield, S. P. (2015). In metabolic engineering of eukaryotic microalgae: potential and challenges come with great diversity. *Frontiers in Microbiology* 6, 1376.

Gonçalves, A. M., Marques, J. C., & Gonçalves, F. (2017). Fatty acids' profiles of aquatic organisms: Revealing the impacts of environmental and anthropogenic stressors. In *Fatty Acids*, Catala, A., Ed. (pp. 89–118). IntechOpen: London, UK, doi:10.5772/intechopen.68544.

Gu, W., Kavanagh, J. M., & McClure, D. D. (2022). Towards a sustainable supply of omega-3 fatty acids: Screening microalgae for scalable production of eicosapentaenoic acid (EPA). *Algal Research* 61, 102564.

Guo, S., Ge, Y., & Na Jom, K. (2017). A review of phytochemistry, metabolite changes, and medicinal uses of the common sunflower seed and sprouts (Helianthus annuus L.). *Chemistry Central Journal* 11(1), 1–10.

Hardy, R., & Mackie, P. (1969). Seasonal variation in some of the lipid components of sprats (Sprattus sprattus). *Journal of the Science of Food and Agriculture* 20(4), 193–198.

Hemme, D., Veyel, D., Mühlhaus, T., Sommer, F., Jüppner, J., Unger, A. K., … & Schroda, M. (2014). Systems-wide analysis of acclimation responses to long-term heat stress and recovery in the photosynthetic model organism Chlamydomonas reinhardtii. *The Plant Cell* 26(11), 4270–4297.

Hoekman, S. K., Broch, A., Robbins, C., Ceniceros, E., & Natarajan, M. (2012). Review of biodiesel composition, properties, and specifications. *Renewable and Sustainable Energy Reviews* 16(1), 143–169.

Hou, K., Yang, X., Bao, M., Chen, F., Tian, H., and Yang, L. (2018). Composition, characteristics and antioxidant activities of fruit oils from *Idesia polycarpa* using homogenate-circulating ultrasound-assisted aqueous enzymatic extraction. *Industrial Crops and Production* 117, 205–215.

Innis, S. M. (2007). Dietary (n-3) fatty acids and brain development. *The Journal of Nutrition* 137(4), 855–859.

Jayakumar, S., Bhuyar, P., Pugazhendhi, A., Rahim, M. H. A., Maniam, G. P., & Govindan, N. (2021). Effects of light intensity and nutrients on the lipid content of marine microalga (diatom) *Amphiprora* sp. for promising biodiesel production. *Science of the Total Environment* 768, 145471.

Joseph, M. M., Renjith, K. R., John, G., Nair, S. M., & Chandramohanakumar, N. (2017). Biodiesel prospective of five diatom strains using growth parameters and fatty acid profiles. *Biofuels* 8(1), 81–89.

Kroth, P. G., Chiovitti, A., Gruber, A., Martin-Jezequel, V., Mock, T., Parker, M. S., … & Bowler, C. (2008). A model for carbohydrate metabolism in the diatom *Phaeodactylum tricornutum* deduced from comparative whole genome analysis. *PloS One*, 3(1), e1426.

Lane, K., Derbyshire, E., Li, W., & Brennan, C. (2014). Bioavailability and potential uses of vegetarian sources of omega-3 fatty acids: a review of the literature. *Critical Reviews in Food Science and Nutrition* 54(5), 572–579.

Li, Y., Han, D., Hu, G., Sommerfeld, M., & Hu, Q. (2010). Inhibition of starch synthesis results in overproduction of lipids in *Chlamydomonas reinhardtii. Biotechnology and Bioengineering* 107(2), 258–268.

Linke, A., Weiss, J., and Kohlus, R. (2020). Oxidation rate of the non-encapsulated–and encapsulated oil and their contribution to the overall oxidation of microencapsulated fish oil particles. *Food Research International* 127, 108705.

Liu, J. J., Green, P., Mann, J. J., Rapoport, S. I., & Sublette, M. E. (2015). Pathways of polyunsaturated fatty acid utilization: implications for brain function in neuropsychiatric health and disease. *Brain Research* 1597, 220–246.

Marella, T. K., Saxena, A., & Tiwari, A. (2020a). Diatom mediated heavy metal remediation: A review. *Bioresource Technology* 305, 123068.

Marella, T. K., López-Pacheco, I. Y., Parra-Saldívar, R., Dixit, S., & Tiwari, A. (2020b). Wealth from waste: Diatoms as tools for phycoremediation of wastewater and for obtaining value from the biomass. *Science of the Total Environment* 724, 137960.

Martins, D. A., Custódio, L., Barreira, L., Pereira, H., Ben-Hamadou, R., Varela, J., & Abu-Salah, K. M. (2013). Alternative sources of n-3 long-chain polyunsaturated fatty acids in marine microalgae. *Marine Drugs* 11(7), 2259–2281.

Matsumoto, M., Sugiyama, H., Maeda, Y., Sato, R., Tanaka, T., & Matsunaga, T. (2010). Marine diatom, *Navicula* sp. strain JPCC DA0580 and marine green alga, *Chlorella* sp. strain NKG400014 as potential sources for biodiesel production. *Applied Biochemistry and Biotechnology* 161(1), 483–490.

Matthaus, B., Özcan, M. M., & Al Juhaimi, F. (2016). Some rape/canola seed oils: Fatty acid composition and tocopherols. *Zeitschrift für Naturforschung C*, 71(3–4), 73–77.

Mohy El-Din, S. M. (2018). Fatty acid profiling as bioindicator of chemical stress in marine *Pterocladia capillacea*, *Sargassum hornschuchii* and *Ulva lactuca*. *International Journal of Environmental Science and Technology* 15(4), 791–800.

Molino, A., Mehariya, S., Sanzo, D. G., Larocca, V., Martino, M., Leone, G. P., et al. (2020). Recent developments in supercritical fluid extraction of bioactive compounds from microalgae: role of key parameters, technological achievements and challenges. *Journal of CO_2 Utility* 36, 196–209.

Muro, E., Atilla-Gokcumen, G. E., & Eggert, U. S. (2014). Lipids in cell biology: how can we understand them better? *Molecular Biology of the Cell* 25(12), 1819–1823.

Murzina, S. A., Nefedova, Z. A., Pekkoeva, S. N., Ruokolainen, T. R., Ripatti, P. O., Semushin, A. V., & Nemova, N. N. (2016). Lipids and fatty acids of the White Sea herring *Clupea pallasi marisalbi* Berg (Clupeiformes, Clupeidae) from different habitats of the White Sea. *Fishes* 1(1), 65–76.

Nascimento, I. A., Marques, S. S. I., Cabanelas, I. T. D., Pereira, S. A., Druzian, J. I., de Souza, C. O., … & Nascimento, M. A. (2013). Screening microalgae strains for biodiesel production: lipid productivity and estimation of fuel quality based on fatty acids profiles as selective criteria. *Bioenergy Research* 6(1), 1–13.

National Research Council (US) Committee on Diet and Health. Diet and Health: *Implications for Reducing Chronic Disease Risk.* Washington (DC): National Academies Press (US); 1989. 7, Fats and Other Lipids. Available from: www.ncbi.nlm.nih.gov/books/NBK218759/

Ni, S., Zhao, W., Zhang, Y., Gasmalla, M. A., & Yang, R. (2016). Efficient and eco-friendly extraction of corn germ oil using aqueous ethanol solution assisted by steam explosion. *Journal of Food Science and Technology* 53(4), 2108–2116.

Ohno, N., Inoue, T., Yamashiki, R., Nakajima, K., Kitahara, Y., Ishibashi, M., & Matsuda, Y. (2012). CO2-cAMP-responsive cis-elements targeted by a transcription factor with CREB/ATF-like basic zipper domain in the marine diatom Phaeodactylum tricornutum. *Plant Physiology* 158(1), 499–513.

Poku, K. (2002). *Small-scale palm oil processing in Africa* (Vol. 148). Food & Agriculture Org.

Popko, J., Herrfurth, C., Feussner, K., Ischebeck, T., Iven, T., Haslam, R., … & Feussner, I. (2016). Metabolome analysis reveals betaine lipids as major source for triglyceride formation, and the accumulation of sedoheptulose during nitrogen-starvation of Phaeodactylum tricornutum. *PLoS One* 11(10), e0164673.

Radakovits, R., Jinkerson, R. E., Darzins, A., & Posewitz, M. C. (2010). Genetic engineering of algae for enhanced biofuel production. *Eukaryotic Cell* 9(4), 486–501.

Ramachandra, T. V., Madhab, M. D., Shilpi, S., & Joshi, N. V. (2013). Algal biofuel from urban wastewater in India: Scope and challenges. *Renewable and Sustainable Energy Reviews* 21, 767–777.

Ratledge, C., & Cohen, Z. (2008). Microbial and algal oils: do they have a future for biodiesel or as commodity oils? *Lipid Technology* 20(7), 155–160.

Rodolfi, L., Chini Zittelli, G., Bassi, N., Padovani, G., Biondi, N., Bonini, G., & Tredici, M. R. (2009). Microalgae for oil: Strain selection, induction of lipid synthesis and outdoor mass cultivation in a low-cost photobioreactor. *Biotechnology and Bioengineering* 102(1), 100–112.

Satoh, A., Ichii, K., Matsumoto, M., Kubota, C., Nemoto, M., Tanaka, M., … & Tanaka, T. (2013). A process design and productivity evaluation for oil production by indoor mass cultivation of a marine diatom, Fistulifera sp. JPCC DA0580. *Bioresource Technology* 137, 132–138.

Sayanova, O., Mimouni, V., Ulmann, L., Morant-Manceau, A., Pasquet, V., Schoefs, B., & Napier, J. A. (2017). Modulation of lipid biosynthesis by stress in diatoms. *Philosophical Transactions of the Royal Society B: Biological Sciences* 372(1728), 20160407.

Sharma, N., Simon, D. P., Diaz-Garza, A. M., Fantino, E., Messaabi, A., Meddeb-Mouelhi, F., … & Desgagné-Penix, I. (2021). Diatoms biotechnology: various industrial applications for a greener tomorrow. *Frontiers in Marine Science* 8, 106.

Sibert, J. (2001). Electronic tagging and tracking in marine fisheries. In: *Electronic Tagging and Tracking in Marine Fisheries* (pp. 1–6). Springer, Dordrecht.

Song, M., Pei, H., Hu, W., & Ma, G. (2013). Evaluation of the potential of 10 microalgal strains for biodiesel production. *Bioresource Technology* 141, 245–251.

Stancheva, M., Merdzhanova, A., Dobreva, D. A., & Makedonski, L. (2010). Fatty acid composition and fat-soluble vitamins content of sprat (*Sprattus sprattus*) and goby (*Neogobius rattan*) from Bulgarian Black Sea. *Ovidius University Annals of Chemistry* 21(1), 23–28.

Stillwell, W., & Wassall, S. R. (2003). Docosahexaenoic acid: membrane properties of a unique fatty acid. *Chemistry and Physics of Lipids* 126(1), 1–27.

Szabo, N. J., Matulka, R. A., Marone, P. A., Bauter, M. R., Chan, T., Franklin, S., et al. (2014). Safety evaluation of oleic-rich triglyceride oil produced by a heterotrophic microalgal fermentation process. *Food and Chemical Toxicology* 65, 301–311.

Tossavainen, M., Lahti, K., Edelmann, M., Eskola, R., Lampi, A. M., Piironen, V., … & Romantschuk, M. (2019). Integrated utilization of microalgae cultured in aquaculture wastewater: wastewater treatment and production of valuable fatty acids and tocopherols. *Journal of Applied Phycology* 31(3), 1753–1763.

Tonon, T., Harvey, D., Larson, T. R., & Graham, I. A. (2002). Long chain polyunsaturated fatty acid production and partitioning to triacylglycerols in four microalgae. *Phytochemistry* 61(1), 15–24.

Upadhyay, A. K., Singh, R., Singh, D. V., Singh, L., & Singh, D. P. (2021). Microalgal consortia technology: A novel and sustainable approach of resource reutilization, waste management and lipid production. *Environmental Technology & Innovation* 23, 101600.

Veloza, A. J., Chu, F. L. E., & Tang, K. W. (2006). Trophic modification of essential fatty acids by heterotrophic protists and its effects on the fatty acid composition of the copepod *Acartia tonsa. Marine Biology* 148(4), 779–788.

Villanova, V., Fortunato, A. E., Singh, D., Bo, D. D., Conte, M., Obata, T., ... & Finazzi, G. (2017). Investigating mixotrophic metabolism in the model diatom *Phaeodactylum tricornutum. Philosophical Transactions of the Royal Society B: Biological Sciences* 372(1728), 20160404.

Vinayak, V., Manoylov, K. M., Gateau, H., Blanckaert, V., Hérault, J., Pencreac'h, G., ... & Schoefs, B. (2015). Diatom milking: a review and new approaches. *Marine Drugs* 13(5), 2629–2665.

Virtue, P. (1995). *Lipids in Euphausia superba.* Doctoral dissertation, University of Tasmania.

Wang, H., Fu, R., & Pei, G. (2012). A study on lipid production of the mixotrophic microalgae *Phaeodactylum tricornutum* on various carbon sources. *African Journal of Microbiology Research* 6(5), 1041–1047.

Wang, J. K., & Seibert, M. (2017). Prospects for commercial production of diatoms. *Biotechnology for Biofuels* 10(1), 1–13.

Wang, X., & Wang, W. X. (2005). Uptake, absorption efficiency and elimination of DDT in marine phytoplankton, copepods and fish. *Environmental Pollution* 136(3), 453–464.

Wen, Z. Y., & Chen, F. (2003). Heterotrophic production of eicosapentaenoic acid by microalgae. *Biotechnology Advances* 21(4), 273–294.

Wong, C. Y., Kiatkittipong, K., Kiatkittipong, W., Ntwampe, S. K., Lam, M. K., Goh, P. S., ... & Lim, J. W. (2021). Black soldier fly larval valorization benefitting from ex-situ fungal fermentation in reducing coconut endosperm waste. *Processes* 9(2), 275.

Wu, S., Huang, A., Zhang, B., Huan, L., Zhao, P., Lin, A., & Wang, G. (2015). Enzyme activity highlights the importance of the oxidative pentose phosphate pathway in lipid accumulation and growth of Phaeodactylum tricornutum under CO_2 concentration. *Biotechnology for Biofuels* 8(1), 1–11.

Xue, Z., Yu, Y., Yu, W., Gao, X., Zhang, Y., & Kou, X. (2020). Development prospect and preparation technology of edible oil from microalgae. *Frontiers in Marine Science* 7, 402.

Yi, Z., Xu, M., Di, X., Brynjolfsson, S., & Fu, W. (2017). Exploring valuable lipids in diatoms. *Frontiers in Marine Science* 4, 17.

Yodsuwan, N., Sawayama, S., & Sirisansaneeyakul, S. (2017). Effect of nitrogen concentration on growth, lipid production and fatty acid profiles of the marine diatom *Phaeodactylum tricornutum. Agriculture and Natural Resources* 51(3), 190–197.

Yongmanitchai, W. A. R. D., & Ward, O. P. (1991). Growth of and omega-3 fatty acid production by *Phaeodactylum tricornutum* under different culture conditions. *Applied and Environmental Microbiology* 57(2), 419–425.

Yusoff, M. M., Gordon, H. M., and Niranjan, K. (2015). Aqueous enzyme assisted oil extraction from oilseeds and emulsion de-emulsifying methods: a review. *Trends in Food Science Technology* 41, 60–82.

Zhukova, N. V., & Aizdaicher, N. A. (1995). Fatty acid composition of 15 species of marine microalgae. *Phytochemistry* 39(2), 351–356.

Zulu, N. N., Zienkiewicz, K., Vollheyde, K., & Feussner, I. (2018). Current trends to comprehend lipid metabolism in diatoms. *Progress in Lipid Research* 70, 1–16.

5 Photonic Materials from Diatoms

Gou Rean Wong, Fong Fong Liew, and Der Jiun Ooi
Department of Oral Biology & Biomedical Sciences, Faculty of Dentistry, MAHSA University, Saujana Putra Campus, Selangor, Malaysia

5.1 INTRODUCTION

Diatoms of various varieties may be found in almost every aquatic ecosystem, from the oceans to lakes to rivers and streams. Often present in light green, red, or brown colors, diatoms are among nature's most complex, fascinating, and beautiful organisms. Being the photosynthetic organisms that carry out photosynthesis, diatoms play an essential part in the cycling of nutrients and the aquatic ecosystem. Other than producing about half of all the oxygen in the atmosphere, diatoms serve as a major food source for many organisms, including fish, crayfish, and other invertebrates (Marella et al., 2020).

Despite their small size, these organisms have sparked research interest in their potential to operate as photonic nanoparticles. Having a distinct pigment composition in comparison to other plants enables diatoms to concomitantly capture solar energy for photosynthesis and actively afford photoprotective functionality, thereby facilitating the efficient generation of biomass and oxygen (Kuczynska et al., 2015).

Apart from the photosynthetic pigments, the uniqueness of diatoms is further bestowed by the presence of frustules, the rigid nanostructured cell walls made of amorphous hydrated silica. Only a few diatom species complement the morphology with silica, whereas most multicellular life has a calcium-based skeleton. In particular, biosilica or amorphous hydrated biogenic silica (SiO_2) is exclusively observed as the basic constituents that form skeleton in the protists and some higher plants (Lowenstam & Weiner, 1989; Sardo et al., 2021).

The silica arranged in layers forms the thick outer wall, thin inner wall, and central cavity of frustules. Each species of diatom has its own distinct biosilica cell wall, which may be identified by the presence of regularly ordered slits or holes ranging in size from 10 to 1,000 nm. As part of the organism responsible for the functions of diatoms, the outer wall or periplasm of frustule is composed of two layers of overlapping plates, while the inner wall or the pith provides nutrients to the cell cavity (Sardo et al., 2021).

The highly ordered silica nanostructures of frustules are synthesized *in vivo* through a series of complicated biomineralization processes. The silica deposition vesicle, a specialized membrane-bound chamber in the diatom cell, is where biosilica morphogenesis, also known as biosilicification, takes place. It has been proposed that the silica deposition vesicle comprises an organic macromolecule matrix that not only drives silica synthesis but also acts as a template for biosilica nanopatterning. This has since led to the recognition of diatoms as the microscopic biofactories of nanostructured biosilica (Bedoshvili & Likhoshway, 2019).

The specific characteristics of diatom frustules' structure render diatoms the properties to filter and remove contaminants such as suspended particles and small organic molecules from water.

DOI: 10.1201/9781003436553-5

This makes them an excellent choice for water filtration systems, even in small quantities. The remarkable porosity and biocompatibility of frustules have also been used for biomedicine and drug-delivery applications (Leonardo et al., 2016; Sharma et al., 2021).

Additionally, diatom frustules have been used as models to understand some aspects of photonic crystal behavior such as light trapping and sorting. As the microcrystalline silica composition of frustules can emit light when struck by sunlight, it could potentially serve as versatile light sources. The photonic crystals' characteristics can also be employed to create efficient and transparent optical devices (Mcheik et al., 2018). The contemporary use of diatoms in sensing is also being discussed.

5.2 OPTICAL PROPERTIES OF DIATOMS

The recorded responses of a substance to electromagnetic radiations, particularly visible light, are defined by optical qualities such as its index of refraction and color. The optical properties are determined by the arrangement of atoms and molecules inside the material. Both physics and chemistry will be determined based on the structure of matter, which includes the color, density, crystalline structure, and electrical characteristics of the substance (Flory et al., 2011).

The optical properties of diatoms originated from the photosynthetic pigments present in diatoms, as well as their well-organized nano-scaled silica frustules' structures. Chemically, diatoms are comprised of both light-harvesting chlorophyll pigments and photoprotective carotenoids (Kuczynska et al., 2015). The regular arrangement of pores arranged through a highly ordered superimposition of biosilica layers, on the other hand, allows for specialized characteristics such as a hierarchically periodic pattern of high and low refractive index, as well as varying absorption coefficient in the visible range. This makes them fascinating for optical investigations and opens up the possibility of frustules being used as natural micro-lenses, optical filters, polarizers, couplers, and laser cavities (Wang et al., 2013).

Coscinodiscus frustules have two valves and girdle bands that link the two facing valves. During cell division, two new valves develop inside the parental valves, but at least one girdle band occurs in conjunction with each daughter valve, and more girdle bands may arise throughout time depending on the species (Kröger & Poulsen, 2008; Romann et al., 2015) as depicted in Figure 5.1. Valves and girdle bands have distinct optical characteristics because valves attenuate shorter wavelengths of visible light, while girdle bands attenuate longer wavelengths. Girdle bands feature an iridescent hue that shifts with the direction of light (Goessling et al., 2019).

While most pennates have long and bilaterally symmetrical valves, *Coscinodiscus* species has big and flat valves with well-organized holes on various levels (Figure 5.2). This has attracted a lot of attention in previous research. Several overlapping nanoporous patterns were seen on cleaned frustule valves. A single valve may be compared to a three-layer superposition. The layer on the

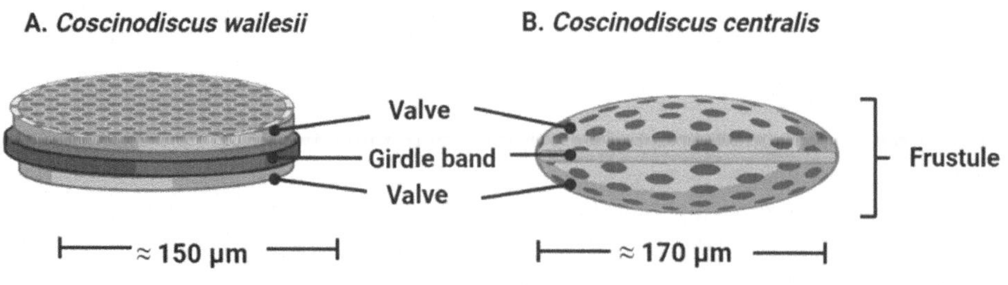

FIGURE 5.1 Frustule of (A) *Coscinodiscus wailesii* and (B) *Coscinodiscus centralis* diatoms. The diatom frustule contains photonic crystal-like properties, which is important in modifying diatom photosynthesis through light focusing, wave guiding, or spectral filtering of photosynthetically productive radiation (Goessling et al., 2019; Romann et al., 2015; Romann et al., 2015; Yamanaka et al., 2008).

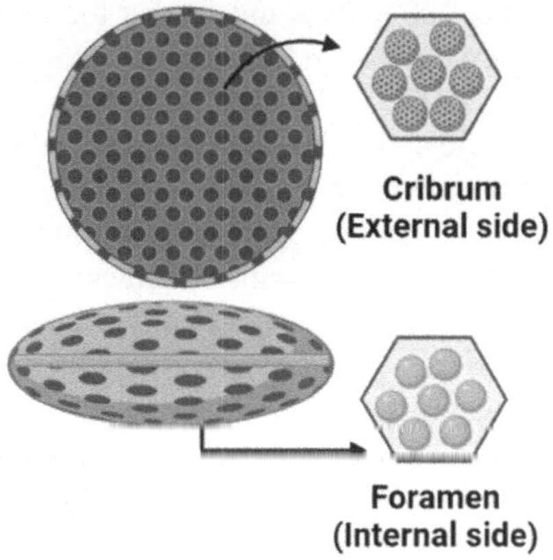

FIGURE 5.2 General illustration of *Coscinodiscus centralis* bio-silica valve cribrum side and foramen side (Goessling et al., 2019; Romann et al., 2015; Yamanaka et al., 2008).

exterior side of the valve (cribrum) has a multi-scale structure with flower-like microporosity motifs with hexagonal symmetry. The cribrum layer is about 200 nm thick, with thin silica branches splitting each flower-like pattern pore into several smaller pores (cribellum). The areola is a honeycomb-like structure that runs parallel to the cribrum. The flower-like and porous shape of the cribrum revolves around a single cavity that is characterized by the honeycomb structure of the areola. The foramen, or third layer, forms the inner side of the valve and features a hexagonal array of spherical pores. The porous structure of the foramen is similar to that of the other layers, with each pore of the foramen including an areola hexagonal chamber as well as a cribrum flower-like porous pattern (Chen et al., 2015; Romann et al., 2015).

5.3 LIGHT HARVESTING AND PHOTOSYNTHESIS

The presence of silica in cell walls is considered to have a substantial impact on diatom ecological success. As diatoms cannot move, they can only compete by changing the surface of their cells. In contrast to other more spherical species, such as most eukaryotic phytoplankton, diatom silica frustules allow for considerable surface area to volume ratios, which improves nutrient uptake (Mitchell et al., 2013). Furthermore, it affords mechanical protection (Friedrichs et al., 2013) by increasing the shell thickness and reducing their development rate in response to diffusible copepods, which cues as a defense against predators (Grønning & Kiørboe, 2020). The silica frustule also aids in adapting to variations in spectrum light. Although diatoms use photosynthesis in a comparable way to plants, their chloroplasts were obtained in a different manner (Hoek et al., 1995). One of the most widely studied diatoms, along with others, is the *Coscinodiscus* species. At least 100–200 μm wide, they have an extremely well-defined cell structure that can be seen in both two-dimensional (2D) and three-dimensional (3D) views. In particular, studies revealed that *Coscinodiscus granii* can withstand high levels of light, but does not use chloroplast high-light avoidance motions for photoprotection (Goessling et al., 2016).

Diatoms carry light-absorbing molecules that absorb solar energy and convert it into chemical energy through photosynthesis. Carbon dioxide (CO_2) is removed from the atmosphere by diatoms. Diatoms often live in fluctuating environments, requiring adaptation to drastic variations in water temperature, food availability, and light intensity. These changes have had a substantial influence on the physiology and survival of these tiny organisms. Diatoms often present at the blue light only (450–500 nm) region below the ocean's surface as water preferentially absorbs light of longer wavelengths (M. Ragni & d'Alcalà, 2007). This means that the light intensity diatoms get varies throughout the day. Diatoms employ photoreceptors like aureochromes to perceive the continually varying blue light (Takahashi et al., 2007).

Harvesting sunlight is critical to life on Earth. In order to maintain the long-term viability of food and energy production, this process must be carried out effectively. To do this, all potential light-harvesting power must be employed, which necessitates the improvement of both biological and artificial systems. The study of photosynthetic organisms shows that the efficiency of light gathering is highly reliant on the environment. The process constantly establishes a balance between photon collecting and photodamage.

The optical characteristics of diatom frustules, on the other hand, have sparked interest in the possibility of photo-biological uses. Nano-scale lattice structures in the frustules have been used to modulate light harvesting and to reduce hazardous radiation exposure, mostly through diffraction or forward-scattering waves through its nano-scale lattices. Fucoxanthin chlorophyll a/c-binding proteins (FCPs) are a large family of light-harvesting antennae found in diatoms that encircle the cores of both photosystem (PS) I and II to form PS–FCP supercomplexes, where FCPs harvest and transport solar energy to the PS cores. Chlorophyll (Chl) c and fucoxanthin (Fx), which are linked with FCPs, have distinct light absorption features in comparison to Chl b, β-carotene (Bcr), and lutein, which are present in green lineage photosynthetic animals. Chl c lacks the phytol tail seen in Chl a/b, but it does have a conjugated system at its protoporphyrin ring that extends to its propionic acid tail, which considerably improves its absorption in the blue-green light range (460–510 nm). Fx, on the other hand, has a shorter conjugated system than luteins and Bcr, which gives FCPs the capacity to collect green light (510–560 nm) (Croce & Van Amerongen, 2014; Demmig-Adams et al., 2014; Gundermann et al., 2014; Zapata et al., 2006).

Since more blue-green photons are available under water, these pigments are essential for diatoms and other related algae to capture light. Furthermore, the FCP antennae contain a diatoxanthin–diadinoxanthin cycle that aids in the quenching of excess energy under strong light conditions. The extraordinary capacity of FCPs to absorb blue-green light and release surplus energy might be one of the reasons diatoms thrive in aquatic environments. FCPs must have a stronger mechanism to protect diatoms from photodamage due to the visible continuous flow and mixing of water layers, which is another important element of FCPs. Aside from the above, diatoms have enzymes that are more sensitive to changes in pH in the lumen than the enzymes involved in the zeaxanthin–violaxanthin cycle of green algae and higher plants. This increases the diatoms' ability for non-photochemical quenching (Croce & Van Amerongen, 2014; Falkowski et al., 2003; Goss & Lepetit, 2015; Yu et al., 2018).

Silica deposition vesicles (SDVs) are specialized internal chambers bound by a lipid bilayer in diatoms. SDVs are silica synthesis sites that may be classified into two types. One produces plate- or dome-shaped silica structures known as valves, while the other produces biosilica rings known as girdle bands. Each daughter cell generates one valve during cell division, whereas girdle bands form during interphase. The silica structure is exocytosed once it is formed at the site of silica synthesis and forms the cell wall, which fully encases and protects the cell from the environment (Heintze et al., 2020).

It has been shown in the past that Fxs interact with Chls in a manner that is analogous to how they do so with FCP units, which occupy a large percentage of the available space in the FCP. It helps diatoms catch blue-green light and successfully transmit it to the Chls in low-light settings, whereas in high-light conditions they take energy from neighboring Chls to dissipate excess energy. Diatoms are able to endure the aquatic environment, which is characterized by a dynamic shift in the quality of available light due to the constant movement and mixing of the different layers of water. This is made possible by the combination of a complex network of Chl and Fxs, which aids diatoms in adapting to adverse environments (Wang et al., 2020).

Chl helps to capture photons of solar light. Briefly, among the four main forms of chlorophyll, a and b are the most commonly present. These Chl structures vary, and the variations impact the

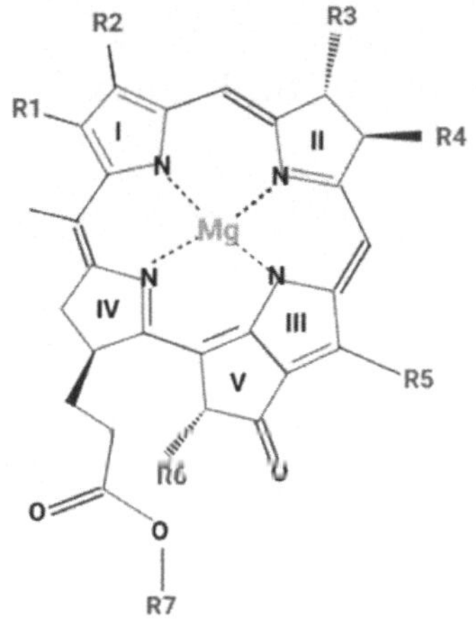

FIGURE 5.3 Structure of chlorophyll, porphyrin with a closed tetrapyrrole structure (Fiedor et al., 2019).

exact wavelength of light that each can absorb, which subsequently permits a distinct range of microorganisms to gather the whole spectrum of light, each absorbing a different range of wavelengths (Croce & Van Amerongen, 2014; Green et al., 2003). The unique tetrapyrrole ring with a Mg^{2+} core, as shown in Figure 5.3, has a lengthy 20-carbon phytol tail that helps tie them to the photosynthetic membrane. There are differences in the substitutions that take place all around the ring, in addition to differences in the length of the phytol tail and the replacements that take place there. A significant number of conjugated double bonds are capable of light absorption in the solar spectrum's visible and near-infrared regions (Aravindu et al., 2013; Reddy et al., 2013; Schliep et al., 2013).

Chl a, which appears in all oxygenic photosynthesis reaction centers, is found in almost all antenna complexes. It possesses a high dipole strength transition to the lowest excited state that grows linearly with the absorption coefficient for efficient excitation energy transfer. As a result, it is a good option for light harvesting and gathering. One of the most basic and significant chemical processes in molecular systems is excitation energy transfer. Excitation energy is a pair consisting of one electron in the excited state of the molecule and the "hole" it has left behind in the ground state. In addition to Chl a, there are a number of Chls that capture light in various parts of the solar spectrum, giving the adaptability required for adaptation to multiple environmental circumstances with changing light quality. Plant and green algal light-harvesting complexes (LHCs) constitute a vast LHC family with many similar proteins (Ballottari et al., 2012; Stomp et al., 2007).

Chl c is important and, with the capability to absorb wavelength at 450 nm mostly, Chl c depicts usefulness in underwater circumstances. Typically, Chl b and c transmit their excitation energy to Chl a, while electron energy transport progresses via Chl a (Croce & van Amerongen, 2011; Songaila et al., 2013). Chl c2 and c3 have an additional double bond which is part of the conjugated system. Chl c3 generally predominates in haptophytes and dinoflagellates, although diatoms often contain Chl c1 and some Chl c2 (Fawley, 1989) (Table 5.1).

More often than not, creatures that undertake oxygenic photosynthesis are constrained by the limitations rendered by PS I and PS II. There are, however, exceptions for a smaller variety of organisms with different photosynthetic reaction centers that do not split water and utilize the much longer wavelength light. Even if they might use a variety of light-gathering techniques, the chlorosomes of green bacteria stand out as the most important ones. These chlorosomes are the biological intervention that occurs in deep water, where very little light is able to reach the surface. These contain hundreds of thousands of Bacterio Chl (BChl) c, BChl d, and/or BChl e molecules, which may self-organize into multilayer structures. These structures are a crucial source of inspiration for the building of artificial antennae (Ganapathy et al., 2009).

TABLE 5.1
Chlorophyll that Has Been Characterized and Is Crucial for Light Harvesting as well as Electron Transfer Function

Pigment	Absorption	Organism	R1	R2	R3	R4	R5	R6	R7
Chlorophyll a	Red light (around 680 nm)	Algae, cyanobacteria	CH_3	$CH=CH_2$	CH_3	C_2H_5	CH_3	$COOCH_3$	Phytol
Chlorophyll b	Red light (660 nm)	Higher plant, prochlorophytes	CH_3	$CH=CH_2$	CHO	C_2H_5	CH_3	$COOCH_3$	Phytol
Chlorophyll c1	Red light (450–640 nm)	Eukaryotic microbes, marine and freshwater algae	CH_3	$CH=CH_2$	CH_3	C_2H_5	CH_3	$COOCH_3$	various
Chlorophyll c2	Red light (450–640 nm)	Eukaryotic microbes, marine and freshwater algae	CH_3	$CH=CH_2$	CH_3	$CH=CH_2$	CH_3	$COOCH_3$	various
Chlorophyll c3	Red light (450–640 nm)	Eukaryotic microbes, marine and freshwater algae	CH_3	$CH=CH_2$	$COOCH_3$	$CH=CH_2$	CH_3	$COOCH_3$	various
Chlorophyll d	Red infrared light (700–730 nm)	Underneath corals and algae, cyanobacterium	CH_3	CHO	CH_3	C_2H_5	CH_3	$COOCH_3$	Phytol
Chlorophyll f	Red infrared light (700–750 nm)	Cyanobacterium	CHO	C_2H_5	CH_3	C_2H_5	CH_3	$COOCH_3$	Phytol

Source: Croce & Van Amerongen (2014); Fiedor et al. (2019).

The confinement of optical radiation transmitted by a single valve and its wavelength dependency have since been explored beginning with incoherent sources. It is necessary to investigate the behaviors of light impinging out of the plane of the valve, passing through it, and reaching the interior of the cell (Joannopoulos et al., 2008). In normal settings, light strikes the diatom cell from the outside of the frustule (the convex side), but part of the light may be absorbed by the (concave) interior of the opposite valve as it goes through the four cells. In order to investigate the influence of the frustule nano-pattern on incoming light under experimental settings, the frustules are routinely cleansed of organic material and disassembled into their component components—the two valves and the girdle bands (Romann et al., 2015). Typically, measurements or simulations on a single valve could be conducted and measured. Nevertheless, the direction from which the light strikes the valve is not always obvious and the issue revolves around the impact of valve orientation in affecting the readings. It remains questionable if the photonic characteristics of the valve may alter depending on whether the incident light strikes from the outside or the interior of the cell. Another concern for measurements on a layer of frustules lies on the doubts if there is a specific pattern of how a single valve may generally fall when immersed in a drop of water or another liquid matrix. De Stefano et al. were the first to demonstrate the ability of a single valve of *Coscinodiscus wailesii*, with a diameter of around 150 μm, to confine red, coherent light in a tiny 10 μm region at a distance of 100 μm along the optical axis. This effect, which was not induced by valve disk refraction, was thought to be caused by the coherent superposition of diffraction contributions from single pores (De Stefano et al., 2007). The diffractive nature of the effect by using an Euler cradle to analyze the intensity and orientation of light transmitted by single valves of *Coscinodiscus wailesii* diatoms after irradiation with blue (472 nm), green (543 nm), and red (633 nm) laser light has been reported (Noyes et al., 2008).

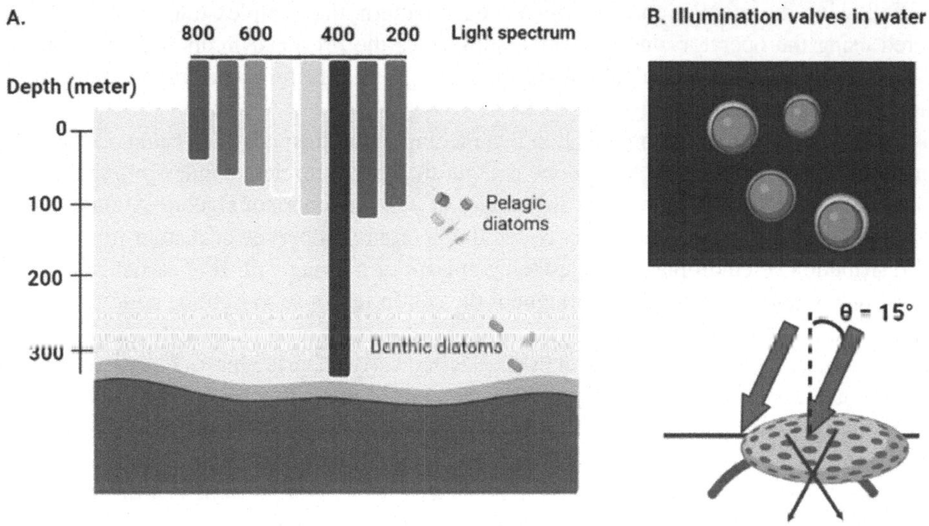

FIGURE 5.4 (A) Light spectrum composition varies with water depth. (B) Scattering characteristics of cleaned valves and living cells, illustration of the valve interaction with white light at a 15° incidence angle, illustrating probable interaction and focus of blue radiation. (Depauw et al., 2012; Goessling et al., 2018).

When pondering about possible photonic uses for diatom frustules, it is important to think about the light environment that diatoms are used to in their natural environments. At certain wavelengths, water partially polarizes light and absorbs it more strongly than air. Red, orange, and yellow wavelengths are quickly absorbed by water and can only reach a few meters beneath the surface, as depicted in Figure 5.4. Importantly, blue light is not strongly absorbed by water and, depending on the amount of dissolved matter and particles, can go much deeper (Depauw et al., 2012).

At the surface, only about 3% of the total light intensity is in the UV range, which is below 400 nm and goes down a few meters. Since diatoms live in places with a lot of UV and blue light, the frustule may be very important for protecting them from UV light. However, it is important to highlight that the organic material contained in diatom frustules is often overlooked in measurements or simulations since most studies are done on frustules that have been previously washed or rinsed (Johnsen & Sosik, 2004). Research with air and water revealed significant differences in the wavelength-dependent scattering properties of the cleaned valves. When cleaned valves were examined in water, they showed a significant amount of blue light forward scattering when illuminated by light entering at an angle of 15°. Living cells displayed forward scattering of green-yellow light that was not dependent on the angle of incidence. On the other hand, a bluish tint was only seen in cells that were larger than 200 m when they were subjected to light that was entering at an angle of incidence of 15° (Goessling et al., 2018).

The capacity to gather and focus light in local hotspots is not simply determined by irradiating and photographing a single valve along its optical axis. Digital holography allows for restructuring of the optical field associated with the valve after recording the hologram of a single valve. The diatom valve holograms were produced using a traditional Mach-Zehnder interferometer. A polarized beam splitter was utilized to separate a 633 nm polarized laser source into two beams, the reference and the object beams. In order to concentrate light on a single valve that was mounted on a glass slide and to capture light that was transmitted by the valve itself, microscope objectives were positioned all along the course of the object beam. After sending the object beam as well as the reference beam via a second recombining beam splitter, an interference pattern was obtained by employing a charge-coupled device plane array to gather the data (Caprio et al., 2014; Ferrara et al.,

2014). Following the acquisition of the interference pattern, the complex transmitted field may be recovered using the operator algebra, which substitutes the Fresnel-Kirchhoff diffraction integral and the lens transfer factor with algebraic operators, bypassing the time-consuming integral calculus (Nazarathy & Shamir, 1980).

It is possible to recover light propagation in a medium other than air when using operator algebra in digital holograms; all that has to be done is input the proper refractive index into the reconstruction procedure. This is one of the many advantages of utilizing operator algebra. At this stage, it has been shown that there is a connection between the photonic properties of diatom frustules and the spatial distribution of chloroplasts inside the cytoplasm of a living cell. It is particularly intriguing to explore how chloroplasts migrate throughout the cell in response to lighting conditions. For centric diatoms and under low-light situations, chloroplasts prefer to move near to the frustule walls, effectively connecting to modes governed by girdles and valves. During high irradiation, chloroplasts move to the nucleus (Furukawa et al., 1998) or the core of the cell, where visible light is captured mostly through the diffractive process. This migration is amplified for red light and therefore has been reported in the spectral range (Shihira-Ishikawa et al., 2007). It has been discovered that light transmission via valves is improved when compared to other wavelengths and when particular chlorophyll absorption maxima are located (Noyes et al., 2008; Romann et al., 2015).

5.4 PHOTOLUMINESCENCE OF DIATOM BIOSILICA

Marine and freshwater diatoms communicate in response to environmental changes periodically by secreting chemical signals that are amplified and spread by other individual cells in the population. Interestingly, some species of diatoms is capable of using light in the form of diffraction and luminescence for intercellular communication and social behavior (Font-Muñoz et al., 2021). Photoluminescence is the phenomenon of light emission from any matter after absorption of photons (Tebyetekerwa et al., 2020). Porous silicon is luminescent effectively in the spectrum of visible light when irradiated with ultraviolent light (Townley, 2011). The photoluminescent properties of diatom are species dependent, attributed to the optical properties of frustule structure and the conditions of the surrounding environment (De Stefano et al., 2005). The photoluminescence observed from diatoms is mainly within the blue visible spectrum region when irradiated with ultraviolet light (Ghobara et al., 2019). For example, *Thalassiosira rotula* Meuinier luminescent multiband photoluminescent spectrum is in the range of 450–690 nm at room temperature (De Stefano et al 2005) while blue photoluminescence is observed from *Cyclotella* sp. when excited at 300 nm wavelength (Mazumder et al., 2010).

Despite the fact that the bulk amorphous silica does not allow any photoluminescence emission, the various surface defects specifically present in nanostructured silica allow the presence of photoluminescent after excitation at different wavelengths (Tommasi, 2016; Qin et al., 2008). Furthermore, it is interesting to note that the photoluminescence features of diatoms are also capable of converting harmful UV radiation into blue and red radiation where the photosynthesis has its maximal activity (Tommasi, 2016). It is hence proposed that diatom frustules may not only function to protect the diatoms from predation, but also represent a mechanism to enhance photosynthetic efficiency and protect living diatom cells against UV radiation, which is detrimental to DNA integrity (Ellegaard et al., 2016). In general, there are two main sources of photoluminescence in diatoms: (1) frustule photoluminescence and (2) autofluorescence from chloroplasts and lipid layers (Tommasi, 2016).

5.4.1 Frustule Photoluminescence

Diatom frustules are also regarded as natural microlenses or photonic crystals. These biosilicas exhibit brilliant colors via diffraction of light and possess the ability to change the state of polarization of the incident radiation, the porous silicon photoluminescent when exposed to photoluminescent quenching gaseous substances (De Stefano et al., 2005; Tommasi et al., 2021). For example, blue photoluminescence peak centered at 440 nm is emitted from frustules of *Cyclotella* species, a freshwater centric diatom, when excited at 300 nm, 370 nm, and 380 nm (Mazumder et al., 2010). The photoluminescence is attributed to the geometrical features of nanostructures and their refractive index in respective of the environment (Font-Muñoz et al., 2021).

5.4.2 Autofluorescence

Diatoms have adapted to a variable light environment and successfully evolved to possess high photosynthetic productivity. This is especially the case for marine diatoms which often exhibit effective light-harvesting ability in oligotrophic oceans. The chloroplast autofluorescence derived from a portion of the absorbed energy from the more penetrating visible spectrum of blue-green light (450–550 nm) is reemitted to the surroundings in the red to far red spectrum (Font-Muñoz et al., 2021). These UV-excited photoluminescent features are commonly present in a large number of aquatic organisms especially marine organisms (Orellana et al., 2004). Autofluorescence of cellular pigments such as chlorophyll, fucoxanthin, or lipid cell layers in living diatom cells can be differentiated from the photoluminescent emitted from diatom biosilica by their distinguished peak spectra (Orellana et al., 2004). A larger and stronger peak red spectrum centered at around 680 nm is derived from the photosynthetic pigments. Meanwhile, the biosilica tends to emit luminescence with a peak spectrum within the blue-green region (400–510 nm) (LeDuff et al., 2016).

5.5 MODIFICATION AND FUNCTIONALIZATION OF FRUSTULES

Silicate phosphors are promising luminescent materials for biotechnological applications owing to their strong chemical stability, excellent moisture resistance, and low cost (Townley, 2011). The abilities to alter nanostructured diatom biosilica and control the luminescent properties offer a route to produce a biosensing platform. Currently, integrating functional molecules into the frustule by formulating specific growth conditions or by modifying nanocomposites after collection of frustules are the most commonly used methods to modify or enhance the functionality of biosilica. The two main approaches to these biosilica modification methods are integration of molecules into the architecture of nanostructured biosilica *in vivo* and attachment of functional molecules to the surface of biosilica frustules (Ragni et al., 2016).

5.5.1 Frustule Metabolic Doping

One example of functionalization of the silica surface is incorporating the intrinsic photoluminescent properties of the biosilica either by attaching nucleophilic moieties, organic molecules, or inorganic chemical compounds to semiconductors or photoluminescence surfaces with nanotopology to achieve increased photoluminescent (Townley, 2011). It is known that diatom biosilica exhibits visible photoluminescence when exposed to UV light, with a peak in the blue spectrum. The presence of silanol groups (=Si-OH) and organic residuals embedded in the frustule matrix are primarily responsible for this (Qin et al., 2008). *Pinnularia* sp. frustules with metabolically inserted germanium exhibit more intense photoluminescence emission than non-doped frustules (Jeffryes et al., 2008). Indeed, blue photoluminescence is induced by Ge-O flaws in amorphous silica, which superimposes atop that of diatom biosilica. It is advised that the initial Si:Ge molar ratio in culture medium needs to be at least 14:1 in order to introduce silicon and germanium into their frustules. It has been revealed that the germanium was alloyed into the silica matrix of the frustule as Ge-oxide nanoclusters or germanosilicates, rather from being adsorbed or randomly precipitated onto cell surfaces (Si-O-Ge). The frustule silica thickened owing to germanium metabolic insertion, however the shape of the areolae altered due to the thickening of the frustule caused by the presence of germanium oxides. As a result, metabolic insertion of germanium may be used to control pore morphology (Jeffryes et al., 2008). Ge-doped frustules, in addition to exhibiting photoluminescence, also display electroluminescence with narrow peaks that correlate with the predicted resonance modes of the diatom frustule represented as a photonic crystal slab (Rogato & De Tommasi, 2020). Other metal and semimetal ions investigated in metabolic integration include aluminum in *Stephanopixis turris*, where the Al/Si ratio in the cell walls could be raised to roughly 1:15 (Machill et al., 2013).

Titanium dioxide (TiO_2) plating improves the semiconducting properties of diatom frustules so much that they can be used as solar cells and in other devices. Sonochemistry is an effective and efficient way to put mesoporous titania into the macropores of certain species of diatoms while keeping the natural 3D shape and geometry of the pores. Sonochemistry greatly facilitated the creation of mesostructured TiO_2 because the unique properties of ultrasound hastened the condensation and polymerization activities (Mao et al., 2014). Metabolic implantation has enhanced the method even further. Titanium nanotubes are predominantly produced by the metabolic integration of TiO_2 nanoparticles in diatom frustules. Using a two-stage cultivation method, TiO_2 was metabolically delivered into the diatom frustules. It was revealed that diatom frustules metabolically implanted with nanostructured TiO_2 during two phases of growth in F/2 medium may replace a nanostructured surface doped with titanium in dye-sensitized solar cells (DSSCs) for heat and electricity production. The power efficiency of the DSSC constructed from titanium-doped diatom frustules is almost twice (9.45%) that of the ordinary DSSC (4.20%) that does not include any diatom frustules metabolically treated with TiO_2 (Gautam et al., 2016). When excited at 394 nm, europium in *Navicula* species makes photoluminescent frustule biosilica that gives off red light (614 nm). Unique red phosphors are good for white LED lighting because they have small particles, a large surface area, and a wider excitation band in the near-UV range (Zhang et al., 2013).

A dye is capable of stimulated emission in solution if it has a high-fluorescence quantum yield and negligible reabsorption. The incorporation of rhodamine dyes into the cell walls of the diatoms *Coscinodiscus granii* and *Coscinodiscus wailesii*, which allows for the production of luminous hybrid nanostructures, as well as the relationship with chemical structure, has been observed. Using free acids rather than esters is the key to getting significant self-reproducing structures. By altering the amino groups in the conjugated chromophore, fluorescent compounds with a wide spectrum ranging from green to red may be created (Kucki & Fuhrmann-Lieker, 2012). Dye photoluminescence is affected, both generally and specifically, by the regular structure of the frustule as well as its photonic properties.

Slab photonic crystals (PhCs) are photonic structures used in a wide range of modern optical technologies. Nanofabrication methods and cleanroom facilities are necessary for the creation of these components, which are both costly and harmful to the environment. Diatoms are often identified as the source of one of the natural structures known to exhibit PhC properties. The frustule has a unique structure. Valve and girdle components are built of a repetitive pattern of holes in a silica matrix. Certain species may display PhC characteristics because of their highly organized porosity network. The valves and girdles that are seen in diatom frustules are, in point of fact, PhC slabs that are made up of a recurring pattern of holes in a silica matrix. PhC slabs, when properly dimensioned, have the ability to couple and steer light of particular wavelengths (resonant modes) and/or prevent the transmission of light whose wavelength falls within photonic band-gaps. Both of these effects are dependent on the slab's thickness (Goessling et al., 2020; Li et al., 2006; Yamanaka et al., 2008).

5.5.2 Protein Immobilization on Diatom Frustules

Using immobilized proteins in industry and medicine has a number of benefits, such as making it easier to handle in continuous operations or preventing product contamination. Immobilizing proteins on a solid support material improves their activity by keeping their natural structure. Enzyme catalysis, detection, and drug administration are just a few of the various uses for this material. Proteins can be attached to the surfaces of polymers in a number of ways, such as by adsorption, entrapment, or covalent bonding. There are four ways to keep enzymes from moving around, including non-covalent adsorption and deposition, physical trapping, and chemical bonding (Eş et al., 2015; Homaei et al., 2013). In 2009, it was demonstrated that covalently attached antibody rabbit IgG to diatom biosilica of *Cyclotella* species is capable of amplifying the photoluminescence intensity (Gale et al., 2009). Arsenic-treated tropical freshwater diatoms, *Cosmioneis reimeri*, also depicted the enhancement of photoluminescent peak and distinguished X-ray diffraction peaks due to the presence of arsenic triocide and arsenic crystals (Chetia et al., 2021).

When utilized as a support material, diatom biosilica, in particular, has various advantages: Its porosity, which is characterized by high specific surface area values, it provides a large capacity for binding proteins, and a very effective interaction with a variety of adsorbates (De Stefano et al., 2009). The size of frustule pores is large enough to allow the incorporation of protein molecules with equivalent hydrodynamic diameters of the order of nanometers (Kröger et al., 2017). The most effective *in vitro* approaches for protein immobilization onto diatom biosilica are those based on chemical functional groups capable of permanently attaching proteins to the hydrogenated silica frustule by covalent binding.

As cross-linkers, chemically reactive functional species must be introduced to the frustule surface since the silanol (Si-OH) and silanolate (Si-O) groups are unable to form stable covalent connections with protein molecules. The most common method for replacing surface silanol groups (Si-OH) is to cover the crystalline and smooth surface with organofunctional alkoxysilane molecules. These molecules include 3-amino-propyl-triethoxysilane (APTES) and 3-amino-propyl-dimethyl-ethoxysilane (APDMES) (Terracciano et al., 2013). As a consequence, amine-terminated surfaces produced from diatom biogenic cell walls have the aptitude to covalently connect various molecular tools. In addition to the deposition of metal or the infiltration of polymer, diatom frustules may also be modified via the process of silanization or by a distinct design application for this particular nanomaterial (Mishra et al., 2017). Furthermore, for particular detection, functionalized diatom frustules may be cross-liked with another substrate. In particular, using a glass subtract with a cross-linked amine-functionalized diatom helps to enhance antibody molecules binding effectiveness and hence the intensity of photoluminescence in the detection of fungal pathogens (Mishra et al., 2020).

New microelectronics possibilities may arise when biosilica is converted to silicon without sacrificing its three-dimensional structure (Bao et al., 2007; Nassif & Livage, 2011). In the synthesis of nanomaterials, diatomic biosilica has shown promising results. Microchip-sized hybrid biosensors might be made possible by the ability of diatoms to hold proteins, enzymes, or antibodies (Brzozowska et al., 2020).

Although diatoms are often managed in aqueous solutions, frustules may be suspended in a colloidal-like solution and deposited on a dry, inert substrate. The use of gold electrodeposition to adhere antibody-functionalized diatoms to metal electrodes has been described (Leonardo et al., 2017). It is feasible to regulate the diatom immobilization orientation and yield by varying the electrodeposition parameters (applied voltage, duration, and gold concentration). Nanostructured electrodes with better electrochemical biosensing capabilities were created using frustules of varied sizes and shapes.

The promising results from these ultrasensitive sensing platforms possess a huge potential for biotechnology applications of diatom basilica such as diagnosis of microorganism infection, monitoring pathogens in food industries, and they even can be applied for ecosystem surveillance (Table 5.2).

5.5.3 Diatom Frustule as Substrate for SERS

Surface-enhanced Raman spectroscopy (SERS) is a popular detection technique in chemistry, materials science, and scientific diagnostics, with the potential to provide considerable sensing capabilities in the future. Using the vibrational absorption spectrum, it is possible to detect chemical

TABLE 5.2
Examples of Photoluminescent in Diatom Species

Species	Peaks or range of photoluminescent	Excitation Conditions	References
Thalassiosira rotula Meuinier	450–690 nm range	Irritated with 325 nm He-Cd laser	De Stefano et al. (2005)
Nitzschia frustulum	440–500 nm	Insertion of germanium into the diatom frustule	(T. Qin et al., 2008)
Cyclotella sp.	440 nm (Blue)	Excitement at 300 nm, 370 nm, and 380 nm	Mazumder et al. (2010)
Cyclotella meneghiniana	387 and 452 nm	Excitation wavelength at 274 nm	Goswami et al. (2011)
Amphora coffeaeformis	470, 480, 485 nm (major peaks), 498, 510, and 522 nm (minor peaks)	Crushed frustules under excitation wavelength 375 nm	Arteaga-Larios et al. (2014)
Gomphocymbella sp.	450, 463, 470, 485, and 490 nm	Crushed frustules under excitation wavelength 375 nm	Arteaga-Larios et al. (2014)
Thalassiosira pseudonana	450–650 nm range	Irritated with 325 nm He-Cd laser	LeDuff et al. (2016)
Navicula lundii	400 nm	Cross-linked AFD glass substrate with modified frustules	Manjita Mishra et al. (2020)
Pseudo-nitzschia delicatissima	450–490 nm range	Illuminating with excitation BL (470 nm) for 45 min in the pelagic environment	Font-Muñoz et al. (2021)
Amphora sp.	400–480 nm range	Cadmium sulfide nanoparticle (Cds-NP)/diatom biosilica excited with light source of 375 nm	González-Fortuna et al. (2021)

and biological molecules with exceptionally high sensitivity and specificity. However, creating and producing a trustworthy SERS substrate with high sensitivity and outstanding reproducibility remains a substantial challenge. They have been used directly as SERS substrates by coating Ag/Au nanoparticles on diatom frustules (Onesto et al., 2018; Pannico et al., 2016). Silver nanoparticles (Ag-NPs) self-assembled on diatom frustules' surfaces may be used as an integration platform to increase localized surface plasmon resonances. Deposition of silver nanoparticle-modified diatoms directly on the biosilica surface could be the ideal approach in sensing. Raman scattering of Rhodmine 6G from nanoparticles on diatoms showed an optical extinction two times larger than the NPs-on-glass structure, and a four-fold increase in sensitivities. There are two benefits of using in situ grown diatom-Ag-NPs over self-assembled ones: (1) an increase in the amount of Ag-NPs on the surface, which would lead to an increase in the number of hot spots, and (2) more Ag-NPs in the pores of diatom frustules, enabling a greater yield of SERS amplification (Ren et al., 2013).

Despite its improved detection sensitivity, a simple but cost-effective method for constructing SERS substrates with high SERS performance, such as sensitivity and repeatability, may be achieved by nanoimprinting regular micro-/nano-patterns of diatom frustules onto polydimethylsiloxane (PDMS) film (Losic et al., 2007; Qin et al., 2010).

In most practical SERS sensing applications, the target molecules are dispersed in solutions and free to penetrate into the liquid volume. Loading CTAB-coated gold nanorods (GNR) onto poly(L-lactic acid) nanofibrous paper yields a unique paper-based plasmonic SERS substrate with focusing effect. The immobilization of GNRs onto PLLA nanofibrous paper produces a hydrophobic paper

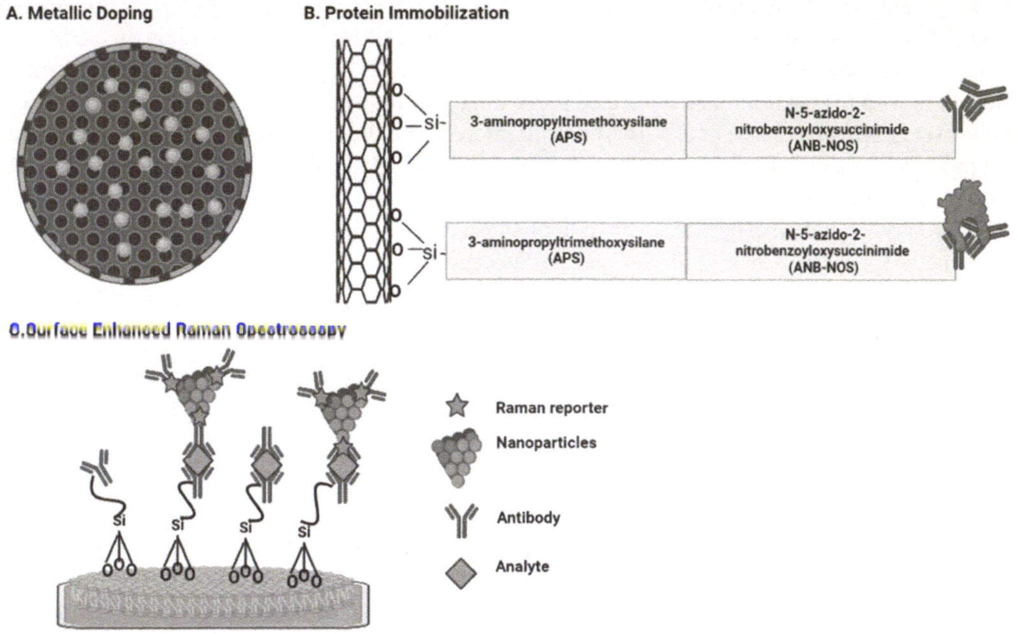

FIGURE 5.5 Frustule surface modification: (A) metallic doping, (B) protein, antibody or enzyme immobilization, (C) surface enhanced Raman spectroscopy (Rogato & De Tommasi, 2020).

substrate with a locally hydrophilic SERS active zone that may confine analyte molecules and inhibit random molecule diffusion. Due to this concentrating effect, the GNRs-PLLA SERS substrate has been shown to be exceptionally sensitive in the detection of R6G and malachite green at very low concentrations (0.1 nM), with excellent repeatability (8% RSD) and long-term stability. It is also less costly and does not need lengthy production processes (Shao et al., 2015). SERS active metallic colloids with hydrophilic surfaces were also employed to confine the analyte molecules after they were spotted onto a hydrophobic substrate. However, in order to modify the surface of the SERS substrates, these methods need certain manufacturing processes such as optical lithography and ion etching, as well as functional polymers (Kong et al., 2017) (Figure 5.5).

5.6 PHOTONIC CRYSTAL

Structured coloration and animal reflectors are examples of photonic crystals (PhCs) seen in nature. They may also be created artificially and employed in a variety of ways. PhCs feature periodic dielectric structures intended to generate the energy band structure for photons, which either permits or inhibits electromagnetic waves of particular frequencies from passing through for light-harvesting applications (Joannopoulos et al., 2008; Joannopoulos et al., 1997).

The girdle of *Coscinodiscus granii* has been described as a well-defined and well-preserved PhC. Under laboratory growth conditions, the nanometric morphological properties are very reproducible, which offers up a lot of opportunities for technical applications that need industrial upscaling (Goessling et al., 2020).

Research into frustules has focused on their well-organized porosity network, which indicates that certain species exhibit PhC features (Dossou, 2017; Fuhrmann et al., 2004). Despite several studies of frustule optical properties in different diatoms, additional experimental evidence of frustule PhC functions is necessary. It has been argued that since the frustule is unique to diatoms, the organism's potential biological utility is still being debated and remains elusive (Ghobara et al., 2019).

One potential role is to protect the cell against micro-grazers (Hamm et al., 2003). By functioning as a diffusion barrier, the frustule also promotes photosynthesis in watery conditions by facilitating cellular conversion of bicarbonate to CO_2 (Milligan & Morel, 2002). The pores allow for chemical interaction with the environment while also preventing potentially harmful agents from entering the cell via size exclusion (Losic et al., 2006). Furthermore, since the structures' period has dimensions that promote contact with photosynthetically beneficial sunlight, it is widely assumed that frustule perforation's ultimate role is to modulate light–cell interactions (Mishler, 2015).

Chlorophyll, which is plentiful, is capable of substantially absorbing the blue region of the electromagnetic spectrum. Chen et al. devised the incorporation of naturally existing diatom-derived chlorophyll as a light absorber component for the fabrication of solid-state solar cells in their research. The photosensitive chlorophyll components recovered from diatoms are placed on the surface of textured silicon solar cells as a layer of spin-coating anti-reflection film. The extracted chlorophyll's significant photon down-conversion action and overall reduction of light reflection increased solar cell efficiency by around 10%. It is also worth noting that the chlorophyll extracts have high thermal stability up to 90°C, making them a straightforward, low-cost, and environmentally friendly production option (Chen et al., 2015).

A layer of a 1D photonic crystal is a thin layer of dielectric film with a thickness of $\lambda/(4n)$, where λ is the wavelength of the incident light that corresponds to the center wavelength of the bandgap and n is the refractive index of the film. When light strikes the film, part of it bounces back and some of it passes through. The light then bounces back at the interface between the film and the substrate. Wavelengths focusing on interfere destructively as they reflect off the films and constructively as they transmit through the movies. This is due to the fact that the optical path difference of one round trip through the film is half the wavelength. The two reflected waves are half a wavelength out of phase and hence destructively interfere, while all other wavelengths pass through. Increasing the number of quarter-wavelength dielectric layers increases the quantity of reflected waves and increases destructive interference. This produces a 1D pattern of dielectric material, which is referred to as a 1D photonic crystal. This indicates that certain wavelengths are not transmitted by the photonic crystal (Mishler, 2015).

Many of the distinguishing features of diatoms are considered to influence their optical properties. Furthermore, diatoms contain multiple layers of periodic patterns, each with a different thickness and pattern. The nanostructured silica in 2D photonic crystals with structured pore arrays may be considered as a photonic crystal slab waveguide with a low refractive-index contrast (Kraai et al., 2020). The valve may have a hexagonal configuration of pores with a big lattice constant, whereas the girdle may have a square layout of holes with a small lattice constant (Mishler, 2015).

It was discovered in 2004 that the pore on the surface of the cell wall of *Coscinodiscus granii* acts as a photonic crystal, preferentially boosting light transport towards the chloroplasts in the cell (Fuhrmann et al., 2004). In 2010, it was suggested that in *Coscinodiscus wailesii*, the nanostructured silica contributes to light diffraction. Light is limited to the margins of the pores based on its wavelength, according to the results of modeling and simulation. Kieu et al. studied the relevance of the placement of the pores in the same species of diatoms. The photonic structure in this system was referred to as "quasi periodic" because of the pattern's inaccurate regularity (Kieu et al., 2014). 2D photonic crystal lattices consist of a periodic plane and are extruded or non-periodic in their third dimension (McCall et al., 1991), common examples include dielectric rods arranged periodically in air, and spaced air holes in a dielectric block (Solli & Hickmann, 2011) (Figure 5.6).

The dielectric constant of a 3D photonic crystal is designed to change periodically in three dimensions, resulting in a photonic bandgap that inhibits electromagnetic transmission and substantially impacts dispersion around a certain wavelength region. 3D photonic crystals and their variants with specifically designed defect structures to further modify their electromagnetic properties continue to attract intense interest in the quest to realize efficient dispersion engineering,

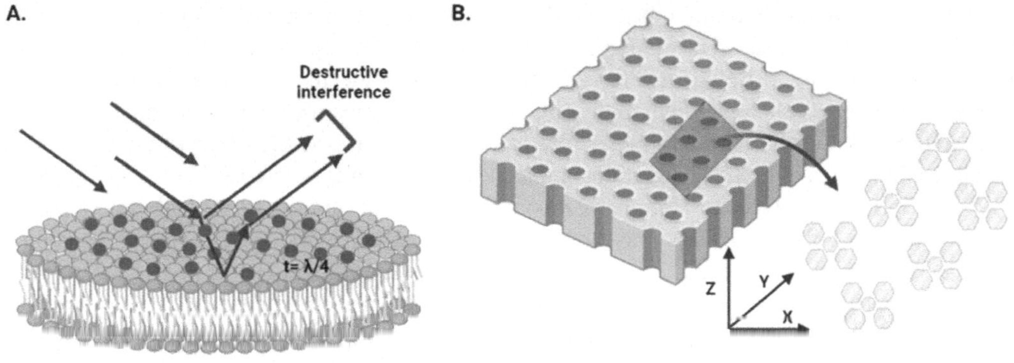

FIGURE 5.6 Diatoms are often referred to as "living photonic crystals." (A) A simplified illustration that explains the mechanics of photonic crystals. Incident light reflects off a stacking of thin sheets. In the model, the silica slab with air holes is indefinitely periodic along the x-axis and is approximated to be infinitely periodic along the y-axis by using a large number of periods. The photonic bandgap is the region of the electromagnetic spectrum where there is no transmission across the photonic crystal (Mishler, 2015). (B) *Pinnularia* species diatom frustules, with periodic unit cells in 2D. It is made up of regular arrays of sub-micron primary pores that are bordered with a thin layer of biosilica that contains numerous nanopores.

micro-integrated circuits, mirrorless lasers, and other advanced photonic applications. One of the greatest challenges in producing 3D photonic crystals for optical applications at sub-micrometer wavelengths is the successful fabrication of large-dimension crystals (>1000-unit cells) with sufficient refractive index difference. Many attempts have been undertaken to construct huge periodic nanostructures, including layer-by-layer photolithography, colloidal self-assembly, direct laser writing, and holographic lithography. All methods and approaches used so far have faced various technical and/or economic challenges, especially in cases where photonic crystals are intended to perform in the ultraviolet–visible range (Arsenault et al., 2004; Blanco et al., 2000; Von Freymann et al., 2010)

Some researchers have targeted *Coscinodiscus* species, most likely because of their comparatively large cell size in the range of 100–200 μm width. *Coscinodiscus* has a finely organized frustule in both two and three dimensions (Aitken et al., 2016; Xing et al., 2017). The porous biosilica nanostructured patterns are particularly fascinating for optical and photonics research. Diatom frustules are appealing model systems for biomimetic manufacturing of materials and electronics due to their various 3D regular silica architectures and nano- to micrometer widths. The whole non-destructive morphological information of the diatom frustule was collected utilizing high spatial resolution nano X-ray computed tomography. This diatom frustule structure was used as an input for the 3D printing of an up-scaled, artificial, self-similar item created utilizing additive manufacturing technologies (Zglobicka et al., 2019).

5.7 OPTICAL PROPERTIES-ASSOCIATED APPLICATION

Due to their optical qualities, diatom frustules have been employed in a range of interesting applications. Chemical and mechanical treatments may be used to regulate the synthesis of nanostructured silica. It is feasible to create label-free analytical equipment or increase the catalytic capabilities of enzymes, DNA, and antibodies using this nanopore structure. Enzyme conjugation with oligonucleotides has also been shown. Biomineralization enzymes found in sponges and diatoms have enabled prototyping and 3D bioprinting of customized structures (Lopez et al., 2005; Schröder et al., 2016).

Since nucleic acid- or protein-based biomacromolecular receptors play key roles in various biological processes, such as mediating and regulating strict recognition and chemical reactions within cells, they are ideal platforms for the identification of specific molecules. When compared to other detection technologies including optical, electrical, electrochemical, thermal, and magnetic, fluorescence detection offers better physical qualities, such as high sensitivity, outstanding spatial resolution, good tissue penetration, and a cheap detection system cost (Nakata et al., 2011). DNA is an intriguing molecular building component for a broad variety of nanoscale structures and applications. DNA nanostructures may be modified with other molecules to generate potentially unlimited functional DNA-based structures. Due to the variety of tasks that proteins may perform; they are an intriguing alternative for functionalizing DNA nanostructures. DNA nanostructures have a lot of promise for usage in life and material science applications (Ngo et al., 2019). Diatom frustules are attractive candidates for lab-on-particle applications because of their high sensitivity, low cost, and broad availability. It enables the study of the molecular event of antibody–antigen identification in diatoms using fluorophores with increased stability (De Stefano et al., 2008; De Stefano et al., 2009).

5.7.1 Application as an Analytical Tool

Diatoms are very sensitive to environmental conditions, the most noticeable of which are the wide range of diatom species found in almost any aquatic habitat, as well as the ecological preferences of many species, which can be seen in their abundance and variety. Even in microhabitats that only get water intermittently (discharge variations) or on a regular basis (spray zones), diatoms are present. Hydroperiod and flow variability and persistence may be inferred from diatom assemblage composition, which is particularly valuable in exploratory hydrogeological investigations (Cantonati et al., 2012).

The potential of diatom red lists has been emphasized, which characterize the ecological integrity and diatom diversity of inland water habitats and assess their vulnerability. In addition, it aids in the tracking of stressors and changes in the natural environment. The cumulative fraction of diatom species in risk categories is an efficient predictor of the ecological integrity of the spring habitat because of changes in groundwater nitrates or spring shape (Cantonati et al., 2022).

Natural or anthropogenic copper, zinc, and antimony pollution may be detected in the widespread species *Achnanthidium minutissimum* (Kütz.) Czarn. by studying the valve abnormalities (Cantonati et al., 2014). A study of diatom *Taxocenosis* at Santa Rosalia (Port) revealed lower abundance, distribution, and species richness, as well as an unusually high frequency of deformed diatom valves, which indicates that the diatoms are responding to metal pollution, despite the typical association structure for non-contaminated environments (Martínez et al., 2021).

In forensic practice, determining whether a dead person in the water drowned or was disposed of after death is challenging. Despite its time-consuming and labor-intensive nature, diatom sampling is now a significant adjunct technique for locating drowning deaths and locales. This uses genetic markers to differentiate between species. Flora profiles derived from diatoms could be used in forensic investigations. With almost 2000 recognized species, it is a remarkably varied microorganism. Chemical digestion approaches are less sensitive and more difficult to apply than gene sequence-based diatom testing. Other phytoplankton in water may be recognized in human tissue samples from drowning cases by amplifying certain DNA sequences that are not found in the human genome. These locations include the ribosomal small subunit 16s/18s rRNA gene and the cytochrome c oxidase subunit I (cox I) gene (Abe et al., 2003; Suto et al., 2009; Uchiyama et al., 2012).

5.7.2 Application as a Biosensor

Diatom frustules provide a versatile platform for sensitive and specific immunocomplex biosensing. The platform's key benefits are the nanostructured biosilica's natural blue photoluminescence emission and the ease with which this material may be antibody-functionalized. Diatom frustules are gaining appeal as biosensors because they are photoluminescent and formed of nanostructured amorphous silica, the surface of which may be easily functionalized and modified for specific uses. Diatomite natural deposits or freshly created diatom frustules might be converted into 3D porous silica nanostructures known as diatomite nanoparticles using low-cost processes. The optical characteristics of these nanostructured materials make them suitable for application in semiconductors and optoelectronics. The use of silica frustules in the production of these materials would increase the overall durability and utility of nanotechnology. There are still some issues to work out before the application can be improved further.

Another significant hurdle that must be overcome by any medical point of care technology is the complexity of physiological fluid composition. Because of this, only a few studies have shown that biomarkers and ions in physiological fluids may be detected directly. Because the concentrations of many target analytes in physiological fluids are so low, they are vulnerable to interference from other substances and cannot be adequately purified. In order to remain relevant, point-of-care devices must continue to increase their specificity and reduce their detection limits. Smart materials and the incorporation of lab-on-a-chip sample pretreatment and analyte enrichment features may enhance PhC designs in the future.

Nanomaterials in biosensors provide exciting opportunities for the design of novel biosensor technology. Nanomaterials improve the mechanical, electrochemical, optical, and magnetic properties of biosensors, opening the way for single-molecule and high-throughput biosensor arrays. Biological molecules have distinct structures and activities, and it remains a significant challenge to find out how to combine the structure and function of nanomaterials with biomolecules to create single-molecule multifunctional nanocomposites, nanofilms, and nanoelectrodes. Current techniques confront significant challenges in processing, characterization, interface difficulties, the availability of high-quality nanomaterials, nanomaterial tailoring, and the principles governing the behavior of these nanoscale composites on the surface of electrodes.

Probe preimmobilization on frustules may offer great specificity for target analyte binding for lattice spacing, enhanced fluorescence, and configuration transition-based sensing. The approach of probe immobilization is expected to make SERS-based sensing more difficult, since the additional signal complexity caused by the biomolecular binding element makes discerning the target's individual Raman peaks even more difficult. SERS nanotags may be placed into probes or targets to ensure specificity for a wide range of complicated biological components. However, using this nanotag to monitor dynamic processes in biological systems would be problematic.

The advancements in fluorescence spectroscopy and SERS sensing principles may allow for the creation of biosensors with high selectivity, sensitivity, and affinity. However, expensive instruments or facilities are needed to detect the optical signal, restricting efforts to combine both technologies into a small device. Indirectly, this highlights the importance of spectrometer-free sensing approaches and lower detection limits with excellent specificity (Table 5.3).

5.8 CONCLUSION AND PERSPECTIVES

Diatoms may be found all over the planet, in both saltwater and freshwater habitats. These characteristics ensure that this potentially lucrative material is accessible, affordable, and useful in biomolecular nanotechnology and biomineralization. The girdle sustains the most distinguishing aspects of PhCs. These characteristics are highly repeatable in individual specimens and among diatom species. As a result, a broad variety of bandgap configurations are available for various applications. Diatom-based biosensors have extremely promising future applications in clinical diagnosis, food analysis, process control, and environmental monitoring. There remain many unsolved questions about the economic potential of diatoms. Optimization of analyte surface coverage and surface functionalization is crucial before scaling up the process for large-scale production, which

TABLE 5.3
Reported Application of Diatom Frustule as Biosensor

No.	Application	Detected concentration	Species	References
1	3-Aminopropyltriethoxysilane (APES) modification for BSA protein detection	3×10^{-5} M	*Amphora* species	(Viji et al., 2014)
2	SERS, to detect melamine and xylene	1000 ppm to 0.1 ppm for melamine, 1 mM for xylene	*Pinnularia* species	(Kong et al., 2016)
3	Conjugation with self-assembled gold nanoparticles for detecting malachite green and fluoride	1 nM malachite green and 100 nM fluoride	*Cosmioneis hawaiiensis*	(Chamuah et al., 2017)
4	SERS immunoassay with anti-IL-8 antibodies, detection in blood plasma	6.2 pg/mL	*Pseudostaurosira trainorii*	(Kamińska et al., 2017)
5	Antibody immobilization to detect *Salmonella typhi*	10 pg	*Amphora* species	(Selvaraj et al., 2018)
6	Aminopropyltriethoxylsilane (APTES) modification to target mouse IgG	10^{-16} M–10^{-15} M	Pinnularia species	(Squire et al., 2018)
7	Diatom-ZrO2 nano composites to detect methyl parathion	54.3 pM	*Phaeodactylum tricornutum*	(Gannavarapu et al., 2019)
8	Antibody immobilization to detect *Bacillus anthracis* Sterne	Not reported	*Thalassiosira pseudonana*	(Ford et al., 2020)
9	SERS, pesticide carbendazim detection	10^{-11} mol/L	*Coscinodiscus* sp.	(Cai et al., 2021)

is dependent on numerous micro- and macro-factors such as growing, harvesting, genetic manipulation, genomic, proteomic, and metabolic information. This is a significant challenge for diatom-based research. Future research should also emphasize understanding the mechanism of interaction between nanomaterials and biomolecules on the surface of electrodes or nanofilms, as well as leveraging their unique properties to create a new generation of biosensors.

REFERENCES

Abe, S., Suto, M., Nakamura, H., Gunji, H., Hiraiwa, K., Suzuki, T., Itoh, T., Kochi, H., & Hoshiai, G.-I. (2003). A novel PCR method for identifying plankton in cases of death by drowning. *Medicine, Science and the Law* 43(1): 23–30. DOI:10.1258/rsmmsl.43.1.23

Aitken, Z. H., Luo, S., Reynolds, S. N., Thaulow, C., & Greer, J. R. (2016). Microstructure provides insights into evolutionary design and resilience of *Coscinodiscus* sp. frustule. *Proceedings of the National Academy of Sciences* 113(8): 2017–2022. DOI:10.1073/pnas.1519790113

Aravindu, K., Mass, O., Vairaprakash, P., Springer, J. W., Yang, E., Niedzwiedzki, D. M., Kirmaier, C., Bocian, D. F., Holten, D., & Lindsey, J. S. (2013). Amphiphilic chlorins and bacteriochlorins in micellar environments. Molecular design, *de novo* synthesis, and photophysical properties. *Chemical Science* 4(9): 3459–3477. DOI:10.1039/c3sc51335a

Arsenault, A., Fournier-Bidoz, S., Hatton, B., Miguez, H., Tétreault, N., Vekris, E., Wong, S., Kitaev, V., & Ozin, G. A. (2004). Towards the synthetic all-optical computer: science fiction or reality? *Journal of Materials Chemistry* 14(5): 781–794. DOI:10.1039/b314305h

Arteaga-Larios, N. V., Nahmad, Y., Navarro-Contreras, H. R., Encinas, A., & Viridiana García-Meza, J. (2014). Photoluminescence shift in frustules of two pennate diatoms and nanostructural changes to their pores. *Luminescence* 29(8): 969–976. DOI:10.1002/bio.2646

Ballottari, M., Girardon, J., Dall'Osto, L., & Bassi, R. (2012). Evolution and functional properties of photosystem II light harvesting complexes in eukaryotes. *Biochimica et Biophysica Acta (BBA)-Bioenergetics* 1817(1): 143–157. DOI:10.1016/j.bbabio.2011.06.005

Bao, Z., Weatherspoon, M. R., Shian, S., Cai, Y., Graham, P. D., Allan, S. M., Ahmad, G., Dickerson, M. B., Church, B. C., & Kang, Z. (2007). Chemical reduction of three-dimensional silica micro-assemblies into microporous silicon replicas. *Nature* 446(7132): 172–175. DOI:10.1038/nature05570

Bedoshvili, Y. D., & Likhoshway, Y. V. (2019). Cellular mechanisms of diatom valve morphogenesis. *Diatoms: Fundamentals and Applications* 99–114. DOI: 10.1002/9781119370741.ch5

Blanco, A., Chomski, E., Grabtchak, S., Ibisate, M., John, S., Leonard, S. W., Lopez, C., Meseguer, F., Miguez, H., & Mondia, J. P. (2000). Large-scale synthesis of a silicon photonic crystal with a complete three-dimensional bandgap near 1.5 micrometres. *Nature* 405(6785): 437–440. DOI:10.1038/35013024

Brzozowska, W., Sprynskyy, M., Wojtczak, I., Dąbek, P., Witkowski, A., & Buszewski, B. (2020). "Outsourcing" diatoms in fabrication of metal-doped 3D biosilica. *Materials* 13(11): 2576. DOI:10.3390/ma13112576

Cai, J., Wang, Z., Wang, M., & Zhang, D. (2021). Au nanoparticle-grafted hierarchical pillars array replicated from diatom as reliable SERS substrates. *Applied Surface Science* 541: 148374. DOI:10.1016/j.apsusc.2020.148374

Cantonati, M., Angeli, N., Bertuzzi, E., Spitale, D., & Lange-Bertalot, H. (2012). Diatoms in springs of the Alps: spring types, environmental determinants, and substratum. *Freshwater Science* 31(2): 499–524. DOI:10.1899/11-065.1

Cantonati, M., Angeli, N., Virtanen, L., Wojtal, A. Z., Gabrieli, J., Falasco, E., Lavoie, I., Morin, S., Marchetto, A., & Fortin, C. (2014). *Achnanthidium minutissimum* (Bacillariophyta) valve deformities as indicators of metal enrichment in diverse widely-distributed freshwater habitats. *Science of the Total Environment* 475: 201–215. DOI:10.1016/j.scitotenv.2013.10.018

Cantonati, M., Hofmann, G., Spitale, D., Werum, M., & Lange-Bertalot, H. (2022). Diatom Red Lists: Important tools to assess and preserve biodiversity and habitats in the face of direct impacts and environmental change. *Biodiversity and Conservation* 1–25. DOI:10.1007/s10531-021-02339-9

Chamuah, N., Chetia, L., Zahan, N., Dutta, S., Ahmed, G. A., & Nath, P. (2017). A naturally occurring diatom frustule as a SERS substrate for the detection and quantification of chemicals. *Journal of Physics D: Applied Physics* 50(17): 175103. DOI:10.1088/1361-6463/aa63b0

Chen, C.-T., Hsu, F.-C., Huang, J.-Y., Chang, C.-Y., Chang, T.-Y., Lin, H.-M., Lin, T.-Y., & Chen, Y.-F. (2015). Effects of a thermally stable chlorophyll extract from diatom algae on surface textured Si solar cells. *RSC Advances* 5(44): 35302–35306. DOI:10.1039/C5RA00265F

Chen, X., Wang, C., Baker, E., & Sun, C. (2015). Numerical and experimental investigation of light trapping effect of nanostructured diatom frustules. *Scientific Reports* 5(1): 1-9. DOI: 10.1038/srep11977

Chetia, L., Kalita, D., Gogoi, A., & Ahmed, G. A. (2021). Enhanced photoluminescence properties of arsenic-treated diatom frustules of *Cosmioneis reimeri*. *MRS Advances* 7(13–14): 284–288. DOI: 10.1557/s43580-021-00131-z

Croce, R., & van Amerongen, H. (2011). Light-harvesting and structural organization of photosystem II: from individual complexes to thylakoid membrane. *Journal of Photochemistry and Photobiology B: Biology* 104(1–2): 142–153. DOI: 10.1016/j.jphotobiol.2011.02.015

Croce, R., & Van Amerongen, H. (2014). Natural strategies for photosynthetic light harvesting. *Nature Chemical Biology* 10(7): 492–501. DOI: 10.1038/nchembio.1555

De Stefano, L., Lamberti, A., Rotiroti, L., & De Stefano, M. (2008). Interfacing the nanostructured biosilica microshells of the marine diatom *Coscinodiscus wailesii* with biological matter. *Acta Biomaterialia* 4(1): 126–130. DOI:10.1016/j.actbio.2007.09.003

De Stefano, L., Maddalena, P., Moretti, L., Rea, I., Rendina, I., De Tommasi, E., Mocella, V., & De Stefano, M. (2009). Nano-biosilica from marine diatoms: A brand new material for photonic applications. *Superlattices and Microstructures* 46(1–2): 84–89. DOI:10.1016/j.spmi.2008.10.031

De Stefano, L., Rea, I., Rendina, I., De Stefano, M., & Moretti, L. (2007). Lensless light focusing with the centric marine diatom *Coscinodiscus walesii*. *Optics Express* 15(26): 18082–18088. DOI:10.1364/OE.15.018082

De Stefano, L., Rendina, I., De Stefano, M., Bismuto, A., & Maddalena, P. (2005). Marine diatoms as optical chemical sensors. *Applied Physics Letters* 87(23): 233902. DOI:10.1063/1.2140087

De Stefano, L., Rotiroti, L., De Stefano, M., Lamberti, A., Lettieri, S., Setaro, A., & Maddalena, P. (2009). Marine diatoms as optical biosensors. *Biosensors and Bioelectronics* 24(6): 1580–1584. DOI:10.1016/j.bios.2008.08.016

De Tommasi, E. (2016). Light manipulation by single cells: the case of diatoms. *Journal of Spectroscopy* 2016: 1–13. DOI:10.1155/2016/2490128

De Tommasi, E., Esposito, E., Romano, S., Crescitelli, A., Di Meo, V., Mocella, V., Zito, G., & Rendina, I. (2021). Frontiers of light manipulation in natural, metallic, and dielectric nanostructures. *La Rivista del Nuovo Cimento* 44(1): 1–68. DOI:10.1007/s40766-021-00015-w

Demmig-Adams, B., Garab, G., Adams III, W., & Govindjee, U. (2014). *Non-Photochemical Quenching and Energy Dissipation in Plants, Algae and Cyanobacteria* (Vol. 40). Springer.

Depauw, F. A., Rogato, A., Ribera d'Alcalá, M., & Falciatore, A. (2012). Exploring the molecular basis of responses to light in marine diatoms. *Journal of Experimental Botany* 63(4): 1575–1591. DOI:10.1093/jxb/ers005

Di Caprio, G., Coppola, G., Stefano, L. D., Stefano, M. D., Antonucci, A., Congestri, R., & Tommasi, E. D. (2014). Shedding light on diatom photonics by means of digital holography. *Journal of Biophotonics* 7(5); 341–350. DOI:10.1002/jbio.201200198

Dossou, K. B. (2017). Large field enhancement obtained by combining Fabry–Perot resonance and Rayleigh anomaly in photonic crystal slabs. *Journal of Optics* 19(4); 045101. DOI:10.1088/2040-8986/aa608b

Ellegaard, M., Lenau, T., Lundholm, N., Maibohm, C., Friis, S. M. M., Rottwitt, K., & Su, Y. (2016). The fascinating diatom frustule—can it play a role for attenuation of UV radiation? *Journal of Applied Phycology 28*(6): 3295–3306. DOI: 10.1007/s10811-016-0893-5

Eş, I., Vieira, J. D. G., & Amaral, A. C. (2015). Principles, techniques, and applications of biocatalyst immobilization for industrial application. *Applied Microbiology and Biotechnology* 99(5): 2065–2082. DOI: 10.1007/s00253-015-6390-y

Falkowski, P., Chen, Y., Green, B., & Parson, W. (2003). *Light-Harvesting Antennas in Photosynthesis.* Kluwer Dordrecht.

Fawley, M. W. (1989). A new form of chlorophyll c involved in light-harvesting. *Plant Physiology* 91(2): 727–732. DOI:10.1104/pp.91.2.727

Ferrara, M. A., Dardano, P., De Stefano, L., Rea, I., Coppola, G., Rendina, I., Congestri, R., Antonucci, A., De Stefano, M., & De Tommasi, E. (2014). Optical properties of diatom nanostructured biosilica in *Arachnoidiscus* sp: micro-optics from mother nature. *PLoS One* 9(7): e103750. DOI:10.1371/journal.pone.0103750

Fiedor, L., Zbyradowski, M., & Pilch, M. (2019). Tetrapyrrole pigments of photosynthetic antennae and reaction centers of higher plants: Structures, biophysics, functions, biochemistry, mechanisms of regulation, applications. *Advances in Botanical Research* 90: 1–33). Elsevier. DOI:10.1016/bs.abr.2019.04.001

Flory, F., Escoubas, L., & Berginc, G. (2011). Optical properties of nanostructured materials: a review. *Journal of Nanophotonics* 5(1): 052502. DOI: 10.1117/1.3609266

Font-Muñoz, J. S., Sourisseau, M., Cohen-Sánchez, A., Tuval, I., & Basterretxea, G. (2021). Pelagic diatoms communicate through synchronized beacon natural fluorescence signaling. *Science Advances* 7(51): eabj5230–eabj5230. DOI:10.1126/sciadv.abj5230

Ford, N. R., Xiong, Y., Hecht, K. A., Squier, T. C., Rorrer, G. L., & Roesijadi, G. (2020). Optimizing the design of diatom biosilica-targeted fusion proteins in biosensor construction for *Bacillus anthracis* detection. *Biology* 9(1): 14. DOI:10.3390/biology9010014

Friedrichs, L., Hörnig, M., Schulze, L., Bertram, A., Jansen, S., & Hamm, C. (2013). Size and biomechanic properties of diatom frustules influence food uptake by copepods. *Marine Ecology Progress Series* 481: 41–51. DOI:10.3354/meps10227

Fuhrmann, T., Landwehr, S., Rharbi-Kucki, E., & Sumper, M. (2004). Diatoms as living photonic crystals. *Applied Physics B* 78(3): 257–260. DOI:10.1007/s00340-004-1419-4

Furukawa, T., Watanabe, M., & Shihira-Ishikawa, I. (1998). Green-and blue-light-mediated chloroplast migration in the centric diatom *Pleurosira laevis*. *Protoplasma* 203(3): 214–220. DOI:10.1007/BF01279479

Gale, D. K., Gutu, T., Jiao, J., Chang, C.-H., & Rorrer, G. L. (2009). Photoluminescence detection of biomolecules by antibody-functionalized Diatom Biosilica. *Advanced Functional Materials* 19(6): 926–933. DOI:10.1002/adfm.200801137

Ganapathy, S., Oostergetel, G. T., Wawrzyniak, P. K., Reus, M., Chew, A. G. M., Buda, F., Boekema, E. J., Bryant, D. A., Holzwarth, A. R., & De Groot, H. J. (2009). Alternating syn-anti bacteriochlorophylls form concentric helical nanotubes in chlorosomes. *Proceedings of the National Academy of Sciences* 106(21); 8525–8530. DOI: 10.1073/pnas.0903534106

Gannavarapu, K. P., Ganesh, V., Thakkar, M., Mitra, S., & Dandamudi, R. B. (2019). Nanostructured diatom-ZrO_2 composite as a selective and highly sensitive enzyme free electrochemical sensor for detection of methyl parathion. *Sensors and Actuators B: Chemical* 288: 611–617. DOI:10.1016/j.snb.2019.03.036

Gautam, S., Kashyap, M., Gupta, S., Kumar, V., Schoefs, B., Gordon, R., Jeffryes, C., Joshi, K. B., & Vinayak, V. (2016). Metabolic engineering of TiO_2 nanoparticles in *Nitzschia palea* to form diatom nanotubes: an ingredient for solar cells to produce electricity and biofuel. *RSC Advances* 6(99): 97276–97284. DOI:10.1039/C6RA18487A

Ghobara, M. M., Ghobara, M. M., Mazumder, N., Vinayak, V., Reissig, L., Gebeshuber, I. C., Tiffany, M. A., Gordon, R., & Gordon, R. (2019). On light and diatoms: A photonics and photobiology review. *Diatoms: Fundamentals and Applications* 129–189. DOI: 10.1002/9781119370741.ch7

Goessling, J. W., Cartaxana, P., & Kühl, M. (2016). Photo-protection in the centric diatom *Coscinodiscus granii* is not controlled by chloroplast high-light avoidance movement. *Frontiers in Marine Science* 2: 115. DOI:10.3389/fmars.2015.00115

Goessling, J. W., Su, Y., Cartaxana, P., Maibohm, C., Rickelt, L. F., Trampe, E. C., Walby, S. L., Wangpraseurt, D., Wu, X., & Ellegaard, M. (2018). Structure-based optics of centric diatom frustules: modulation of the in vivo light field for efficient diatom photosynthesis. *New Phytologist* 219(1); 122–134. DOI:10.1111/nph.15149

Goessling, J. W., Su, Y., Maibohm, C., Ellegaard, M., & Kühl, M. (2019). Differences in the optical properties of valve and girdle band in a centric diatom. *Journal of the Royal Society Interface Focus* 9(1); 20180031. DOI:10.1098/rsfs.2018.0031

Goessling, J. W., Wardley, W. P., & Lopez-Garcia, M. (2020). Highly reproducible, bio-based slab photonic crystals grown by diatoms. *Advanced Science* 7(10); 1903726. DOI:10.1002/advs.201903726

González-Fortuna, G., Arteaga-Larios, N., Nahmad, Y., Navarro-Contreras, H. R., & García-Meza, J. V. (2021). Frustules of *Amphora* sp. as a photonic crystal with photoluminescent CdS nanoparticles. *Luminescence* 36(3): 788–794. DOI:10.1002/bio.4003

Goss, R., & Lepetit, B. (2015). Biodiversity of NPQ. *Journal of Plant Physiology* 172: 13–32. DOI:10.1016/j.jplph.2014.03.004

Goswami, B., Choudhury, A., & Buragohain, A. K. (2011). Luminescence properties of a nanoporous freshwater diatom. *Luminescence* 27(1): 16–19. DOI:10.1002/bio.1315

Green, B., Parson, W. W., & Parson, W. W. (2003). *Light-Harvesting Antennas in Photosynthesis* (Vol. 13). Springer Science & Business Media.

Grønning, J., & Kiørboe, T. (2020). Diatom defence: Grazer induction and cost of shell-thickening. *Functional Ecology* 34(9): 1790–1801. DOI:10.1111/1365-2435.13635

Gundermann, K., Büchel, C., & Hohmann-Marriott, M. (2014). *The Structural Basis of Biological Energy Generation.* Springer, New York, 21–37.

Hamm, C. E., Merkel, R., Springer, O., Jurkojc, P., Maier, C., Prechtel, K., & Smetacek, V. (2003). Architecture and material properties of diatom shells provide effective mechanical protection. *Nature* 421(6925): 841–843. DOI:10.1038/nature01416

Heintze, C., Formanek, P., Pohl, D., Hauptstein, J., Rellinghaus, B., & Kröger, N. (2020). An intimate view into the silica deposition vesicles of diatoms. *BMC Materials* 2(1): 1–15. DOI:10.1186/s42833-020-00017-8

Hoek, C., Mann, D., Jahns, H. M., & Jahns, M. (1995). *Algae: An Introduction to Phycology.* Cambridge University Press.

Homaei, A. A., Sariri, R., Vianello, F., & Stevanato, R. (2013). Enzyme immobilization: an update. *Journal of Chemical Biology* 6(4): 185–205. DOI:10.1007/s12154-013-0102-9

Jeffryes, C., Gutu, T., Jiao, J., & Rorrer, G. L. (2008). Two-stage photobioreactor process for the metabolic insertion of nanostructured germanium into the silica microstructure of the diatom *Pinnularia* sp. *Materials Science and Engineering: C* 28(1): 107–118. DOI:10.1016/j.msec.2007.01.002

Jeffryes, C., Solanki, R., Rangineni, Y., Wang, W., Chang, C. h., & Rorrer, G. L. (2008). Electroluminescence and photoluminescence from nanostructured diatom frustules containing metabolically inserted germanium. *Advanced Materials* 20(13): 2633–2637. DOI: 10.1002/adma.200800292

Joannopoulos, J. D., Johnson, S. G., Winn, J. N., & Meade, R. D. (2008). *Molding the Flow of Light*. Princeton Univ. Press, Princeton, NJ.

Joannopoulos, J. D., Villeneuve, P. R., & Fan, S. (1997). Photonic crystals: putting a new twist on light. *Nature* 386(6621): 143–149. DOI:10.1038/386143a0

Johnsen, S., & Sosik, H. (2004). Shedding light on light in the ocean. *Oceanus Magazine* 43(2); 1–5.

Kamińska, A., Sprynskyy, M., Winkler, K., & Szymborski, T. (2017). Ultrasensitive SERS immunoassay based on diatom biosilica for detection of interleukins in blood plasma. *Analytical and Bioanalytical Chemistry* 409(27): 6337–6347. DOI: 10.1007/s00216-017-0566-5

Kieu, H., Pan, T., Wang, Z., Le, M., Nguyen, H., & Vo, M. (2014). Accurate 3D shape measurement of multiple separate objects with stereo vision. *Measurement Science and Technology* 25(3): 035401. DOI:10.1088/0957-0233/25/3/035401

Kong, X., Squire, K., Li, E., LeDuff, P., Rorrer, G. L., Tang, S., Chen, B., McKay, C. P., Navarro-Gonzalez, R., & Wang, A. X. (2016). Chemical and biological sensing using diatom photonic crystal biosilica with in-situ growth plasmonic nanoparticles. *IEEE Transactions on Nanobioscience* 15(8): 828–834. DOI:10.1109/TNB.2016.2636869

Kong, X., Xi, Y., Le Duff, P., Chong, X., Li, E., Ren, F., Rorrer, G. L., & Wang, A. X. (2017). Detecting explosive molecules from nanoliter solution: A new paradigm of SERS sensing on hydrophilic photonic crystal biosilica. *Biosensors and Bioelectronics* 88: 63–70. DOI:10.1016/j.bios.2016.07.062

Kraai, J. A., Wang, A. X., & Rorrer, G. L. (2020). Photonic crystal enhanced SERS detection of analytes separated by ultrathin layer chromatography using a diatom frustule monolayer. *Advanced Materials Interfaces* 7(11): 2000191. DOI:10.1002/admi.202000191

Kröger, N., Dubey, N., & Kumari, E. (2017). Immobilization of proteins on diatom biosilica. *Diatom Nanotechnology: Progress and Emerging Applications* 126. DOI:10.1039/9781788010160-00126

Kröger, N., & Poulsen, N. (2008). Diatoms—from cell wall biogenesis to nanotechnology. *Annual Review of Genetics* 42: 83–107. DOI:10.1146/annurev.genet.41.110306.130109

Kucki, M., & Fuhrmann-Lieker, T. (2012). Staining diatoms with rhodamine dyes: control of emission colour in photonic biocomposites. *Journal of the Royal Society Interface* 9(69): 727–733. DOI:10.1098/rsif.2011.0424

Kuczynska, P., Jemiola-Rzeminska, M., & Strzalka, K. (2015). Photosynthetic pigments in diatoms. *Marine Drugs* 13(9): 5847–5881. DOI:10.3390/md13095847

LeDuff, P., Roesijadi, G., & Rorrer, G. L. (2016). Micro-photoluminescence of single living diatom cells. *Luminescence* 31(7): 1379–1383. DOI: 10.1002/bio.3118

Leonardo, S., Garibo, D., Fernández-Tejedor, M., O'Sullivan, C., & Campàs, M. (2017). Addressed immobilization of biofunctionalized diatoms on electrodes by gold electrodeposition. *Biofabrication* 9(1): 015027. DOI:10.1088/1758-5090/aa6400

Leonardo, S., Prieto-Simón, B., & Campàs, M. (2016). Past, present and future of diatoms in biosensing. *TrAC Trends in Analytical Chemistry* 79: 276–285. DOI:10.1016/j.trac.2015.11.022

Li, J., Sun, G., & Chan, C. T. (2006). Optical properties of photonic crystals composed of metal-coated spheres. *Physical Review B* 73(7): 075117. DOI:10.1103/PhysRevB.73.075117

Lopez, P. J., Descles, J., Allen, A. E., & Bowler, C. (2005). Prospects in diatom research. *Current Opinion in Biotechnology* 16(2): 180–186. DOI:10.1016/j.copbio.2005.02.002

Losic, D., Mitchell, J. G., Lal, R., & Voelcker, N. H. (2007). Rapid fabrication of micro-and nanoscale patterns by replica molding from diatom biosilica. *Advanced Functional Materials* 17(14): 2439–2446. DOI: 10.1002/adfm.200600872

Losic, D., Rosengarten, G., Mitchell, J. G., & Voelcker, N. H. (2006). Pore architecture of diatom frustules: potential nanostructured membranes for molecular and particle separations. *Journal of Nanoscience and Nanotechnology* 6(4): 982–989. DOI:10.1166/jnn.2006.174

Lowenstam, H. A. (1989). Biomineralization processes. In: *On Biomineralization*. Oxford University Press. DOI:10.1093/oso/9780195049770.001.0001

Machill, S., Köhler, L., Ueberlein, S., Hedrich, R., Kunaschk, M., Paasch, S., Schulze, R., & Brunner, E. (2013). Analytical studies on the incorporation of aluminium in the cell walls of the marine diatom *Stephanopyxis turris*. *BioMetals* 26(1): 141–150. DOI: 10.1007/s10534-012-9601-3

Mao, L., Liu, J., Zhu, S., Zhang, D., Chen, Z., & Chen, C. (2014). Sonochemical fabrication of mesoporous TiO_2 inside diatom frustules for photocatalyst. *Ultrasonics Sonochemistry* 21(2): 527–534. DOI:10.1016/j.ultsonch.2013.09.001

Marella, T. K., López-Pacheco, I. Y., Parra-Saldívar, R., Dixit, S., & Tiwari, A. (2020). Wealth from waste: Diatoms as tools for phycoremediation of wastewater and for obtaining value from the biomass. *Science of the Total Environment* 724: 137960. DOI:10.1016/j.scitotenv.2020.137960

Martínez, Y. J., Siqueiros-Beltrones, D. A., & Marmolejo-Rodríguez, A. J. (2021). Response of benthic diatom assemblages to contamination by metals in a marine environment. *Journal of Marine Science and Engineering* 9(4): 443. DOI:10.3390/jmse9040443

Mazumder, N., Gogoi, A., Kalita, R. D., Ahmed, G. A., Buragohain, A. K., & Choudhury, A. (2010). Luminescence studies of fresh water diatom frustules. *Indian Journal of Physics* 84(6): 665–669. DOI:10.1007/s12648-010-0068-1

McCall, S., Platzman, P., Dalichaouch, R., Smith, D., & Schultz, S. (1991). Microwave propagation in two-dimensional dielectric lattices. *Physical Review Letters* 67(15): 2017. DOI:10.1103/PhysRevLett.67.2017

Mcheik, A., Cassaignon, S., Livage, J., Gibaud, A., Berthier, S., & Lopez, P. J. (2018). Optical properties of nanostructured silica structures from marine organisms. *Frontiers in Marine Science* 5: 123. DOI:10.3389/fmars.2018.00123

Milligan, A. J., & Morel, F. M. (2002). A proton buffering role for silica in diatoms. *Science* 297(5588): 1848–1850. DOI:10.1126/science.1074958

Mishler, J. H. (2015). *Modeling Diatom Frustules as Photonic Crystals* (Doctoral dissertation, Honors Thesis). William Fulbright College of Arts and Sciences. The University of Arkansas, Arkansas, USA.

Mishra, M., Arukha, A. P., Bashir, T., Yadav, D., & Prasad, G. (2017). All new faces of diatoms: potential source of nanomaterials and beyond. *Frontiers in Microbiology* 8: 1239. DOI:10.3389/fmicb.2017.01239

Mishra, M., Singh, S. K., Bhardwaj, A., Kumar, L., Singh, M. K., & Sundaram, S. (2020). Development of a diatom-based photoluminescent immunosensor for the early detection of Karnal Bunt disease of wheat crop. *ACS Omega* 5(14): 8251–8257. DOI:10.1021/acsomega.0c00551

Mitchell, J. G., Seuront, L., Doubell, M. J., Losic, D., Voelcker, N. H., Seymour, J., & Lal, R. (2013). The role of diatom nanostructures in biasing diffusion to improve uptake in a patchy nutrient environment. *PLoS One* 8(5): e59548. DOI:10.1371/journal.pone.0059548

Nakata, E., Liew, F., Nakano, S., & Morii, T. (2011). Recent progress in the construction methodology of fluorescent biosensors based on biomolecules. *Biosensors–Emerging Materials and Applications* 123. DOI:10.5772/17724

Nassif, N., & Livage, J. (2011). From diatoms to silica-based biohybrids. *Chemical Society Reviews* 40(2): 849–859. DOI: 10.1039/c0cs00122h

Nazarathy, M., & Shamir, J. (1980). Fourier optics described by operator algebra. *JOSA*, 70(2): 150–159. DOI: 10.1364/JOSA.70.000150

Ngo, T. A., Dinh, H., Nguyen, T. M., Liew, F. F., Nakata, E., & Morii, T. (2019). Protein adaptors assemble functional proteins on DNA scaffolds. *Chemical Communications* 55(83); 12428–12446. DOI:10.1039/C9CC04661E

Noyes, J., Sumper, M., & Vukusic, P. (2008). Light manipulation in a marine diatom. *Journal of Materials Research* 23(12): 3229–3235. DOI:10.1557/JMR.2008.0381

Onesto, V., Villani, M., Coluccio, M., Majewska, R., Alabastri, A., Battista, E., Schirato, A., Calestani, D., Coppedé, N., & Cesarelli, M. (2018). Silica diatom shells tailored with Au nanoparticles enable sensitive analysis of molecules for biological, safety and environment applications. *Nanoscale Research Letters* 13(1): 1–9. DOI:10.1186/s11671-018-2507-4

Orellana, M. V., Petersen, T. W., & Van Den Engh, G. (2004). UV-excited blue autofluorescence of *Pseudo-nitzschia multiseries* (Bacillariophyceae). *Journal of Phycology* 40(4), 705–710. DOI:10.1111/j.1529-8817.2004.03226.x

Pannico, M., Rea, I., Chandrasekaran, S., Musto, P., Voelcker, N. H., & De Stefano, L. (2016). Electroless gold-modified diatoms as surface-enhanced Raman scattering supports. *Nanoscale Research Letters* 11(1). 1–6. DOI: 10.1186/s11671-016-1539-x

Qin, D., Xia, Y., & Whitesides, G. M. (2010). Soft lithography for micro-and nanoscale patterning. *Nature Protocols* 5(3): 491. DOI:10.1038/nprot.2009.234

Qin, T., Gutu, T., Jiao, J., Chang, C.-H., & Rorrer, G. L. (2008). Photoluminescence of silica nanostructures from bioreactor culture of marine diatom *Nitzschia frustulum*. *Journal of Nanoscience and Nanotechnology* 8(5): 2392–2398. DOI:10.1166/jnn.2008.241

Ragni, M., & d'Alcalà, M. R. (2007). Circadian variability in the photobiology of *Phaeodactylum tricornutum*: pigment content. *Journal of Plankton Research* 29(2): 141–156. DOI:10.1093/plankt/fbm002

Ragni, R., Cicco, S., Vona, D., Leone, G., & Farinola, G. M. (2016). Biosilica from diatoms microalgae: smart materials from bio-medicine to photonics. *Journal of Materials Research* 32(2): 279–291. DOI:10.1557/jmr.2016.459

Reddy, K. R., Jiang, J., Krayer, M., Harris, M. A., Springer, J. W., Yang, E., Jiao, J., Niedzwiedzki, D. M., Pandithavidana, D., & Parkes-Loach, P. S. (2013). Palette of lipophilic bioconjugatable bacteriochlorins for construction of biohybrid light-harvesting architectures. *Chemical Science* 4(5); 2036–2053. DOI:10.1039/C3SC22317E

Ren, F., Campbell, J., Wang, X., Rorrer, G. L., & Wang, A. X. (2013). Enhancing surface plasmon resonances of metallic nanoparticles by diatom biosilica. *Optics Express* 21(13): 15308–15313. DOI: 10.1364/OE.21.015308

Rogato, A., & De Tommasi, E. (2020). Physical, chemical, and genetic techniques for diatom frustule modification: Applications in nanotechnology. *Applied Sciences* 10(23): 8738. DOI:10.3390/app10238738

Romann, J., Valmalette, J.-C., Chauton, M. S., Tranell, G., Einarsrud, M.-A., & Vadstein, O. (2015). Wavelength and orientation dependent capture of light by diatom frustule nanostructures. *Scientific Reports* 5(1): 1–6. DOI: 10.1038/srep17403

Romann, J., Valmalette, J.-C., Røyset, A., & Einarsrud, M.-A. (2015). Optical properties of single diatom frustules revealed by confocal microspectroscopy. *Optics Letters* 40(5): 740–743. DOI:10.1364/OL.40.000740

Sardo, A., Orefice, I., Balzano, S., Barra, L., & Romano, G. (2021). Mini-review: potential of diatom-derived silica for biomedical applications. *Applied Sciences* 11(10): 4533. DOI:10.3390/app11104533

Schliep, M., Cavigliasso, G., Quinnell, R. G., Stranger, R., & Larkum, A. W. (2013). Formyl group modification of chlorophyll a: a major evolutionary mechanism in oxygenic photosynthesis. *Plant, Cell & Environment* 36(3): 521–527. DOI: 10.1111/pce.12000

Schröder, H. C., Grebenjuk, V. A., Wang, X., & Müller, W. E. (2016). Hierarchical architecture of sponge spicules: biocatalytic and structure-directing activity of silicate in proteins as model for bioinspired applications. *Bioinspiration & Biomimetics* 11(4): 041002. DOI:10.1088/1748-3190/11/4/041002

Selvaraj, V., Muthukumar, A., Nagamony, P., & Chinnuswamy, V. (2018). Detection of typhoid fever by diatom-based optical biosensor. *Environmental Science and Pollution Research* 25(21): 20385–20390. DOI:10.1007/s11356-017-9362-1

Shao, J., Tong, L., Tang, S., Guo, Z., Zhang, H., Li, P., Wang, H., Du, C., & Yu, X.-F. (2015). PLLA nanofibrous paper-based plasmonic substrate with tailored hydrophilicity for focusing SERS detection. *ACS Applied Materials & Interfaces* 7(9): 5391–5399. DOI:10.1021/am508881k

Sharma, N., Simon, D. P., Diaz-Garza, A. M., Fantino, E., Messaabi, A., Meddeb-Mouelhi, F., Germain, H., & Desgagné-Penix, I. (2021). Diatoms biotechnology: various industrial applications for a greener tomorrow. *Frontiers in Marine Science* 8: 106. DOI: 10.3389/fmars.2021.636613

Shihira-Ishikawa, I., Nakamura, T., Higashi, S. i., & Watanabe, M. (2007). Distinct responses of chloroplasts to blue and green laser microbeam irradiations in the centric diatom Pleurosira laevis. *Photochemistry and Photobiology* 83(5): 1101–1109. DOI:10.1111/j.1751-1097.2007.00167.x

Solli, D., & Hickmann, J. (2011). Study of the properties of 2D photonic crystal structures as a function of the air-filling fraction and refractive index contrast. *Optical Materials* 33(3): 523–526. DOI:10.1016/j.optmat.2010.10.041

Songaila, E., Augulis, R. n., Gelzinis, A., Butkus, V., Gall, A., Büchel, C., Robert, B., Zigmantas, D., Abramavicius, D., & Valkunas, L. (2013). Ultrafast energy transfer from chlorophyll c 2 to chlorophyll a in fucoxanthin–chlorophyll protein complex. *The Journal of Physical Chemistry Letters* 4(21): 3590–3595. DOI: 10.1021/jz401919k

Squire, K., Kong, X., LeDuff, P., Rorrer, G. L., & Wang, A. X. (2018). Photonic crystal enhanced fluorescence immunoassay on diatom biosilica. *Journal of Biophotonics* 11(10); e201800009. DOI: 10.1002/jbio.201800009

Stomp, M., Huisman, J., Stal, L. J., & Matthijs, H. C. (2007). Colorful niches of phototrophic microorganisms shaped by vibrations of the water molecule. *The ISME Journal* 1(4): 271–282. DOI:10.1038/ismej.2007.59

Suto, M., Kato, N., Abe, S., Nakamura, M., Tsuchiya, R., & Hiraiwa, K. (2009). PCR detection of bacterial genes provides evidence of death by drowning. *Legal Medicine* 11: S354–S356. DOI:10.1016/j.legalmed.2009.01.062

Takahashi, F., Yamagata, D., Ishikawa, M., Fukamatsu, Y., Ogura, Y., Kasahara, M., Kiyosue, T., Kikuyama, M., Wada, M., & Kataoka, H. (2007). AUREOCHROME, a photoreceptor required for photomorphogenesis in stramenopiles. *Proceedings of the National Academy of Sciences* 104(49): 19625–19630. DOI:10.1073/pnas.0707692104

Tebyetekerwa, M., Zhang, J., Xu, Z., Truong, T. N., Yin, Z., Lu, Y., Ramakrishna, S., Macdonald, D., & Nguyen, H. T. (2020). Mechanisms and applications of steady-state photoluminescence spectroscopy in two-dimensional transition-metal dichalcogenides. *ACS Nano* 14(11): 14579–14604. DOI:10.1021/acsnano.0c08668

Terracciano, M., Rea, I., Politi, J., & De Stefano, L. (2013). Optical characterization of aminosilane-modified silicon dioxide surface for biosensing. *Journal of the European Optical Society-Rapid Publications* 8. DOI:10.2971/jeos.2013.13075

Townley, H. E. (2011). Diatom frustules: Physical, optical, and biotechnological applications. *The Diatom World* (pp. 273–289). Springer Netherlands. DOI: 10.1007/978-94-007-1327-7_12

Uchiyama, T., Kakizaki, E., Kozawa, S., Nishida, S., Imamura, N., & Yukawa, N. (2012). A new molecular approach to help conclude drowning as a cause of death: simultaneous detection of eight bacterioplankton species using real-time PCR assays with TaqMan probes. *Forensic Science International* 222(1–3): 11–26. DOI:10.1016/j.forsciint.2012.04.029

Viji, S., Anbazhagi, M., Ponpandian, N., Mangalaraj, D., Jeyanthi, S., Santhanam, P., Devi, A. S., & Viswanathan, C. (2014). Diatom-based label-free optical biosensor for biomolecules. *Applied Biochemistry and Biotechnology* 174(3): 1166–1173. DOI: 10.1007/s12010-014-1040-x

Von Freymann, G., Ledermann, A., Thiel, M., Staude, I., Essig, S., Busch, K., & Wegener, M. (2010). Three-dimensional nanostructures for photonics. *Advanced Functional Materials* 20(7): 1038–1052. DOI:10.1002/adfm.200901838

Wang, W., Zhao, S., Pi, X., Kuang, T., Sui, S. F., & Shen, J. R. (2020). Structural features of the diatom photosystem II–light-harvesting antenna complex. *The FEBS Journal* 287(11): 2191–2200. DOI:10.1111/febs.15183

Wang, Y., Cai, J., Jiang, Y., Jiang, X., & Zhang, D. (2013). Preparation of biosilica structures from frustules of diatoms and their applications: current state and perspectives. *Applied Microbiology and Biotechnology* 97(2): 453–460. DOI:10.1007/s00253-012-4568-0

Xing, Y., Yu, L., Wang, X., Jia, J., Liu, Y., He, J., & Jia, Z. (2017). Characterization and analysis of *Coscinodiscus* genus frustule based on FIB-SEM. *Progress in Natural Science: Materials International* 27(3): 391–395. DOI:10.1016/j.pnsc.2017.04.019

Yamanaka, S., Yano, R., Usami, H., Hayashida, N., Ohguchi, M., Takeda, H., & Yoshino, K. (2008). Optical properties of diatom silica frustule with special reference to blue light. *Journal of Applied Physics* 103(7): 074701. DOI: 10.1063/1.2903342

Yu, L.-J., Suga, M., Wang-Otomo, Z.-Y., & Shen, J.-R. (2018). Structure of photosynthetic LH1–RC supercomplex at 1.9 Å resolution. *Nature* 556(7700): 209–213. DOI:10.1038/s41586-018-0002-9

Zapata, M., Garrido, J. L., & Jeffrey, S. W. (2006). Chlorophyll c pigments: current status. *Chlorophylls and Bacteriochlorophylls* (pp. 39–53). Springer. DOI:10.1007/1-4020-4516-6_3

Zglobicka, I., Chmielewska, A., Topal, E., Kutukova, K., Gluch, J., Krüger, P., Kilroy, C., Swieszkowski, W., Kurzydlowski, K. J., & Zschech, E. (2019). 3D diatom–designed and selective laser melting (SLM) manufactured metallic structures. *Scientific Reports* 9(1): 1–9. DOI:10.1038/s41598-019-56434-7

Zhang, G., Jiang, W., Wang, L., Liao, X., Liu, P., Deng, X., & Li, J. (2013). Preparation of silicate-based red phosphors with a patterned nanostructure via metabolic insertion of europium in marine diatoms. *Materials Letters* 110: 253–255. DOI:10.1016/j.matlet.2013.08.045

6 Case Studies in Diatom Biorefinery

G. Flora[1], G. Mary Stephy[1], and Ashokkumar Veeramuthu[2,3]
[1]Department of Botany, St. Mary's College (Autonomous), Thoothukudi, Tamil Nadu, India
[2]Center of Excellence in Catalysis for Bioenergy and Renewable Chemicals (CBRC), Faculty of Science, Chulalongkorn University, Bangkok, Thailand
[3]Center for Transdisciplinary Research, Department of Pharmacology, Saveetha Dental College, Saveetha Institute of Medical and Technical Sciences, Saveetha University, Chennai, India

6.1 INTRODUCTION

Diatoms (Bacillariophyta) are the most prevalent phytoplankton species globally and play critical roles as primary producers. They are members of the Heteronkontophyta clade, which includes around 285 genera and over 12,500 identified species (Cavelier-Smith, 2018; Mann & Droop, 1996; Winter & Duthie, 2000), however, many of them have yet to be described. They capture solar energy to build quick biomass and commence the grazing food chain because they are the primary producers (Round et al., 1990). These algae groups are responsible for half of the organic compounds in the ocean and 20% of the oxygen on the planet (Falkowski et al., 1998; Scala & Bowler, 2001). Even in the most extreme climatic conditions, they can be found in both marine and freshwater environments (Choi et al., 2008). Diatoms were formed through an endosymbiotic interaction between red algae and heterotrophic flagellates about 240 million years ago (Karaman et al., 2019). Diatoms can produce more lipid and biomass than other algae groups (Marella & Tiwari, 2020). In the deep ocean, diatoms help to export carbon and play a great role in the biogeochemical cycles of macronutrients including nitrogen (N), phosphorus (P), and silicon (Si) (Seth et al., 2021; Buesseler, 1998; Kolber et al., 2001; Sarthou et al., 2005; Treguer et al., 1995). They are important keystone species that influence zooplankton grazing and recruitment (Isigoien et al., 2002; Ianora et al., 2004) at the bottom of the pelagic food chain. As a result, every change in diatom levels has a profound impact on the pelagic food web (Leterme, 2015).

Because diatoms are the only algae possessing silica in their cells, they are the most important and distinctive category of algae (Pawolski et al., 2018). The silica cell wall of diatoms allows for a wide range of shapes and sizes. Frustules are a type of silica cell wall that protects diatoms from photoinhibition, acts as a mechanical barrier against grazers, and regulates sinking rate (Zhang et al., 2017). Size, growth pattern, and environmental parameters including temperature, light, nutrition, salinity, and growth phase all influence their silica concentration (Saxena et al., 2020).

Diatoms produce eicosapentaenoic acid (EPA), docosahexaenoic acid (DHA), and other nutrients that can be used as a food source for zooplankton and fish. They remove excess chemical compounds and inorganic nutrients from water, such as carbon, phosphorus, and silica, and account for 40% of the oxygen in water bodies (Yi et al., 2017). Due to triacylglycerol (TAG) profiles that are abundant in medium-chain fatty acids, diatoms have emerged as a better feedstock for biofuel generation than chlorophytes (Zulu et al., 2018). Diatoms have the potential to absorb heavy metals from

DOI: 10.1201/9781003436553-6

wastewater, but they frequently endure morphological changes and stress in the process, causing them to overproduce secondary metabolites (Gautam et al., 2017; Lakshmi et al, 2020). Due to a large amount of fucoxanthin in diatoms, they have a distinctive golden-brown color (Seth et al., 2021). This fucoxanthin is of high demand in pharmaceutical and nutraceutical industries (Marella & Tiwari, 2020). In marine plants, diatoms are responsible for 25–40% of total primary production (Bertrand, 2010). They generate a variety of high-value products; hence this chapter aims to provide a clear perspective on diatom biorefinery and its challenges.

6.2 FEEDSTOCK FOR DIATOM BIOREFINERY

Microalgae are now thought to be superior to land plants because of their faster growth rate, microscopic nature, higher biomass and oil productivity, simple life cycle, ability to survive in a variety of ecological environments, and absence of competition with land for food (Hogan et al., 2021). Oil and microalgal biomass are appealing renewable feedstocks for the developing biotechnology and chemical industries, particularly in the production of biodiesel. Microalgal cultures have been tweaked to produce more oil per acre and in a shorter time frame than plants (Martin et al., 2018). Most Bacillariophyceae microalgae are extremely prolific and adaptable eukaryotic organisms with high lipid and fatty acid contents (4–28% fatty acids and 10–39% lipids of dry weight biomass) (Hogan et al., 2021).

Diatoms are microscopic eukaryotic protists that occur in oceans, rivers, and lakes. Diatoms convert sunlight, carbon dioxide, and water into biological energy through photosynthesis and chemosynthesis (Cheirsilp et al., 2020). Diatoms store carbon in triglycerides (neutral lipids), which is the optimum substrate for biodiesel production (Yi et al., 2017). The fatty acid profile is essential to understanding a diatom's viability as a biodiesel feedstock.

Compared to other microalga types, diatoms can accumulate a substantial amount of carbohydrates, proteins, lipids, and colors (Jensen et al., 2019).

They can accumulate neutral lipids like triacylglycerol (TAG) in cytosolic or plastidic lipid bodies, as well as polymeric carbohydrates like starch granules, which can be exploited as biofuel feedstocks (Singh et al., 2021). Biofuels, omega-3 fatty acids, carotenoids, surfactants, bio-lubricants, and other diatom-based products are critical in the biomanufacturing and energy industries (Karim et al., 2020). The nutrient availability, including carbon (C), silicate (Si), nitrogen (N), and others, as well as environmental factors like temperature and light, have a direct impact on diatom development and the content of storage molecules (Khan et al., 2021). Biogenic silica is a compound present in diatom cell wall, which they obtain in the form of silicic acid from the surrounding water (Saxena et al., 2021).

6.3 LIPIDS

The algae's lipid content or lipid productivity has the most significant impact on the final cost of biofuel generation (Joseph et al., 2017). One of the most critical factors in establishing whether a feedstock is acceptable for the production of biodiesel is the fatty acid methyl ester (FAME) profile (Saxena et al., 2021). FAME profiles (CFPP) are also used to identify key parameters impacting biodiesel quality and features such as the cetane number (CN), kinematic viscosity (KV), iodine value (IV), cold filter plugging point, and cloud point (CP) (Marella et al., 2019). Depending on the physiological status and environmental factors, the composition of fatty acids of microalgae can change significantly, both qualitatively and quantitatively. This makes identifying local microalgae species as a potential renewable feedstock more important because they have a competitive advantage in the local geographic, climatic, and biological conditions (Joseph et al., 2017).

Lipid bodies are specialized structures in diatoms that store lipids. The lipid body size varies by species and is regulated by the size of the cell (Saxena et al., 2021). Diatoms accumulate lipids based

TABLE 6.1
Percentage of Fatty Acids Obtained from Different Species of Diatoms

	% Total fatty acid		
Diatom species	**EPA**	**DHA**	**References**
Porosira glacialis	31	5.5	Svenning et al., 2019
Nitzschia laevis	3.43	–	Lu et al., 2019
Phaeodactylum tricomutum	33	–	Barka et al., 2016
Fistulifera solaris	14.2	–	Tanaka et al., 2017
Aulacoseira baicalensis	11.2	10.4	Shishlvannikov et al., 2018
Synedra acus	14.6	3.5	Shishlvannikov et al., 2018
Gomphonema gracile	26.1	–	Demailly et al., 2019
Diatom-dominated biofilm	26.5	0.70	Marella et al., 2019

on C/N and C/Si ratios, where a lower ratio results in the availability of carbon resulting in storage lipids. In diatoms, silica deficiency causes fat buildup and nitrogen deficiency, in contrast to other species that do not have a need for silica to undergo cell division (Vella et al., 2019).

Diatoms store carbon in triglycerides form which is the optimum substrate for the production of biodiesel (Yi et al., 2017). EPA and DHA are PUFAs found in diatoms. In *Chaetoceros* sp., the DHA and EPA contents are 12.47% and 13.07%, respectively, but in *Thalassiosira* sp., they are 5.92% and 3.47%. *Thalassiosira* sp. and *Chaetoceros* sp. had 21.48% and 28.00% MUFA contents, respectively (Bhattacharjya et al., 2020). During diatom cultivation, species-specific alterations and physico-chemical conditions influence the PUFA content (Lin et al., 2018). Phospholipids are significant elements of the membrane lipid fraction, which contain PUFA as a primary component. Diatoms tend to move a portion of their phospholipids into stored lipids, which are primarily neutral lipids when they are under nutrient stress or nutrient deficiency. As a result, variations in nutrition availability affect diatoms' FAME profiles (Table 6.1).

6.4 CARBOHYDRATES

First- and second-generation bioethanol production can benefit from a carbohydrate-based microalgal feedstock. Harun et al. (2010) examined the wasted biomass retrieved following supercritical carbon dioxide extraction of lipids from the microalga *Chlorococcum* sp. for suitability for conversion into bioethanol. In addition to primary biomass utilization, the data imply that algal biomass residues are a suitable fuel for bioethanol synthesis. The use of discarded diatom biomass for ethanol production is very interesting. Diatoms vary from other microalgae in that their storage carbohydrate, termed chrysolaminarin, is found within the chrysolaminarin vacuole. This is a soluble carbohydrate that has $\beta(1\rightarrow3)$- and $\beta(1\rightarrow6)$-linked glucose units in an 11:1 ratio (Hogan et al., 2021). As a result, the potential of diatom polysaccharides to ferment ethanol must be assessed within the context of the biorefinery idea, which is viewed as a long-term solution for biomass energy (Hogan et al., 2021).

Carbon partitioning causes carbohydrate generation in diatoms, with chrysolaminarin as the soluble and lipid as the storage molecule (Jensen et al., 2019). Carbohydrate and fat concentrations rise during stressful situations, such as nutrient depletion or harsh physical conditions. Bioreactors have been used to increase biomass and metabolite production by altering growth conditions and allowing for greater storage compound generation (Yin et al., 2020). Diatoms are attractive species for industrial applications because of their effective nutrient acquisition mechanisms and faster growth rate, as well as their high storage chemical content (Suparmaniam et al., 2019). *Thalassiosira*

TABLE 6.2
Characteristics of Diatom Feedstock

Proximate analysis		Ultimate analysis	
Properties	**Quantities (%)**	**Properties**	**Quantities (%)**
Ash content	32.8	C	31.9
Moisture content	11.45	H	4.26
Crude fat	3.70	N	5.96
Crude protein	38.7	Si	6.76
Acid detergent fiber	17.2	Miscellaneous	9.02
Neutral detergent fiber	21.0	O	42.1
Lignin	6.53		

Source: adopted from Wang & Seibert (2017).

pseudonana and *Phaeodactylum tricornutum* have shown that amino acids and cellular proteins are low when nitrogen levels are low, resulting in photosynthetic downregulation and an increase in fatty acid production. Because nitrogen, phosphorous, and silica play such important roles in cellular activities, any nutritional deficiency can severely impact growth and metabolite levels (Cointet et al., 2019) (Table 6.2).

6.5 PRETREATMENT TECHNIQUES FOR DIATOM BIOMASS

Diatoms were first cultivated on a large scale in the early nineteenth century, owing to their economic and ecological benefits and the future need for food, fuel, and feed (Bekirogullari et al., 2020). However, due to low biomass yields in current culture systems, the potential benefits of these organisms have not been completely investigated. In addition, the nutrient availability, such as carbon (C), silicate (Si), nitrogen (N), and others, as well as environmental conditions like temperature and light, have a direct impact on diatom growth and storage molecule content (Khan et al., 2021).

Concerns over global warming caused by fossil fuels and natural resource depletion have prompted a current push toward biodiesel research (Gavrilescu & Chisti, 2015). Biodiesel is a renewable, non-toxic, biodegradable, and environmentally beneficial alternative to petroleum. Despite these benefits, the expense of producing microalgal biomass is now substantially greater than the cost of delivering biomass from other energy crops such as canola, soybean, rapeseed, palm, and jatropha (Chisti, 2007). As a result, it is critical to establish an energy- and cost-effective production strategy that may help achieve competitive biodiesel production. Diatoms are more productive in lipid and biomass than other algal groups (Marella & Tiwari, 2020). A report from the Aquatic Species Program revealed 50 microalgal strains for biofuel generation, of these 60% are diatoms (Sheehan et al., 1998). Diatoms are a prominent group of microalgal species with enormous potential in producing lipids and carotenoids for the biofuel and nutraceutical industries (Schiano di Visconte, 2019).

The microalgal lipid extraction methods are suboptimal (Otero et al., 2017), and laboratory-scale extractions typically rely on harmful organic solvents such as hexane, methanol, and chloroform. However, supercritical fluid extraction using CO_2 is currently a standard method for extracting essential biodiesel lipids and omega fatty acids on an industrial scale (Herrero & Ibanez, 2015). This green technique replaces conventional organic solvent-based lipid extraction from microalgae because supercritical fluid extraction is more efficient than using hazardous solvents such as hexane, methanol, and chloroform and has a low environmental impact, making it ideal for food applications.

Due to their great mechanical strength and chemical resistance, the cell membranes and cell walls of many microalgae, particularly the frustules of *Amphora* sp., may interfere with the effectiveness of lipid extractions (Kim et al., 2013). As a result, biomass pretreatment has been explored for optimal biomolecule extraction from microalgae and other organisms (Yellapu et al., 2016).

6.6 BIOMASS PRETREATMENT

Chemical methods can disturb microalgal cells, such as treatment with alkalis, acids, and surfactants. Chemical treatments induce hydrolysis of the cell membrane (Liang et al., 2015). The biogenic silica of diatom cell wall is obtained from surrounding water (Guillard, 1973). When acids, alkalis, or detergents are used, there is a risk of permanent contamination of the target product as well as destruction (such as protein denaturing) (Middelberg, 1995). Industrial use for the recovery of high-value goods is doubtful due to these issues (Liang et al., 2015).

Lyophilized diatom biomass was homogeneously combined with 0.5 M HCl solution for acid pretreatment. While in base pretreatment, anionic detergent pretreatment, and non-ionic detergent pretreatment, the biomass was combined with 0.5% sodium dodecyl sulfate, 0.5 M NaOH, and 0.5% Tween-20, respectively. Before supercritical fluid extraction of lipids, it was agitated in a shaker at 120 rpm for 4 hours at 20°C. The SFE method is a fluid extraction method that uses fluids at temperatures above their optimum temperature. The density of the supercritical fluid is similar to that of liquids, while its viscosity is equivalent to that of a gas (Abhari & Khaneghah, 2020). Then the biomass samples were centrifuged, and the biomass pellets were rinsed thoroughly with distilled water and centrifuged again. The pellets were dried and used for lipid extraction using supercritical fluids (Hogan et al., 2021). Fatty acid LC-MS investigations indicate that each pretreatment has a distinct effect on the fatty acid profile. Comparatively, the pretreated biomass showed better yield than untreated biomass (Hogan et al., 2021).

Diatom photosynthetic activity was found to be significantly impacted by high irradiation and nutritional constraint, resulting in increased lipid buildup (Cointet et al., 2019). Along with light intensity, the wavelength can influence PUFA content; for example, in the marine diatom *Fistulifera solaris*, higher UV-C wavelength induced increased PUFA content (Arakaki et al., 2017). Blue light, even at a high light intensity, had no deleterious impact on growth but did increase lipid content. Although many diatoms have been reported to have high PUFA content, the composition of PUFA and fatty acids is species-specific and strain-specific. For example, the EPA level in *Amphora* sp. was 3.9% fatty acid (TFA), whereas it was 14.2% in *Amphora coffeaeformis* (Indrayani et al., 2020) (Table 6.3).

6.7 PROCESSING OF BIOPRODUCTS FROM DIATOMS

The world is placing greater emphasis on finding sustainable and renewable energy sources that can help lower the planet's carbon impact (Yew et al., 2019). Biofuels are being explored as potential energy supplies to replace fossil fuels. Diatoms are dominant phytoplankton communities and have a growing biotechnological use (George et al., 2020). Upstream processing focuses on microalgal cultivation and biomass output maximization, whereas downstream processing focuses on bioproduct extraction and purification from microalgal biomass (Manirafasha et al., 2016).

6.7.1 Upstream Processing

In microalgae research, upstream processes are considered the baseline. These procedures are crucial both technically and commercially because they directly impact the quality and quantity of the microalgal biomass produced. Upstream processes have numerous primary steps that should be examined in microalgae research (e.g. cultivation media preparation, bioreactor design, CO_2 supplementation, and modification and control of environmental parameters, etc.) (Manirafasha et al.,

TABLE 6.3
Untreated and Pretreated Biomass of *Amphora* sp. Relative Percentage of Extractable Fatty Acids

Fatty acid name	Control	Untreated biomass	0.5M HCl_ ASN-III biomass	0.5M NaOH_ ASN-III biomass	0.5% SDS_ ASN-III biomass	0.5% Tween-20_ASN-III biomass
EPA (C20:05)	53.9	53.1	14.8	35.5	41.7	16.1
DHA (C22:06)	0.6	0.3	0.2	0.7	1.1	0.2
Lauric (C12.00)	0.1	0.1	0.2	0.1	0.3	21.2
Myristic (C14:00)	8.6	9.4	1.7	4.3	6.3	8.2
Palmitoleic (C16:01)	33.2	32.0	17.0	20.2	21.7	0.8
Oleic/Elaidic (C18:01)	0.1	0.1	6.7	8.9	9.8	7.5
Nonadecylic (C19:00)	ND	ND	3.7	1.1	0.3	4.3
Pelargonic (C9:00)	0.7	1.6	2.9	1.6	1.0	1.2
Capric (C10:00)	0.1	ND	0.6	0.2	ND	0.1

Adopted from Hogan et al. (2021). (Note: ND, most of these fatty acids were found below the detection limit of the equipment employed or were detected in such little amounts that the relative % could not be reported in the table within one decimal point.)

2016). Aside from choosing the right microalgae species, the growth circumstances, such as pH, temperature, nutrient availability, and light intensity, may help microalgae cells accumulate specific useful components (Yew et al., 2019).

The cultivation method is determined by the type of microalgae strain (Chew et al., 2018). Diatoms and brown seaweeds have been extensively used to produce FX in vitro and in vivo utilizing various strains (Bauer et al., 2019). However, few marine diatom species have been studied yet for their commercial potentials in FX production, such as *Cyclotella cryptic*, *Odontella aurita*, and *Phaeodactylum tricornutum*. The increased efficiency of carotenoid production in *Odontella aurita*, *Phaeodactylum tricornutum*, *Nitzschia laevis*, and *Isochrysis galbana* (up to 15–40 mg FX g^{-1} DW) when cultured in standardized conditions in photobioreactors, fermenters, and open ponds, results in higher FX production in diatoms (1–2.5% of DW) (Gerin et al., 2020; McClure et al., 2018; Mao et al., 2020; Zarekarizi et al., 2019; Ishika et al., 2019; Xia et al., 2018). To improve FX production, a variety of techniques should be applied, including strain selection, genetic engineering, directed gene editing, physical and chemical mutagenesis, system biology, metabolic engineering, cell disruption, culturing, and harvesting (Seth et al., 2021).

Microalgae can be used to make high-value goods such as triterpenoids (betulin and lupeol), commercial pigments (FX and astaxanthin), and polyhydroxy-alkanoates (Butlet et al., 2020). A microalgal-based biorefinery worked on two concepts: low-volume–high-value and large-volume–poor-value applications (Gholkar et al., 2021). The major goal is to expand the quantity of goods while keeping the process as cost-effective as possible. However, selecting or identifying a diatom strain that can produce a stable biomass and biomolecules under different conditions in outdoor cultivation is one of the most difficult tasks (Cho et al., 2017). *P. tricornutum* and *Tisochrysis lutea* were successfully grown in tubular photobioreactors for 170 days as an incubation period of semi-continuous culture conditions and observed maximum fucoxanthin contents of 0.6% 0.7% mg g^{-1} DW, revealing that the intensity of light plays a crucial role in fucoxanthin production (Pereira et al., 2021a).

6.7.2 Downstream Processing

Harvesting, purification, product extraction, and product finalizing are some of the sequential processes involved in microalgae downstream processing. Sixty percent of the overall biodiesel production cost is compensated by algae-based biodiesels from downstream processing (Kim et al., 2012). The nature of the microalgal strain cell walls will influence downstream processing. Previously, organic solvents such as acetone, n-hexane, ethanol, and others were often used, but in a recent work, FX and lipids were extracted from *Chlorella vulgaris* wet biomass and *Phaeodactylum tricornutum* wet biomass using a pressured subcritical extraction solvent technique. Grinding is used to break down cells before extraction. To standardize the extraction conditions, three different solvents (ethyl acetate, ethanol, and n-hexane) and three subcritical temperatures (150, 100, and 50°C) were used. Pressure is used to extract bioactive molecules more efficiently, shortening the time required for temperature-sensitive FX to be treated at higher temperatures. This procedure is significant because it avoids drying algal biomass, saving both energy and time (Derwenskus et al., 2019). However, due to the environmental risk and chemical toxicity associated with organic solvent usage during bioactive molecule extraction, new techniques have recently been developed that take into account cost, time, and energy, and that are environmentally friendly. Higher temperature conditions combined with an organic solvent have been shown to alter the chemical moiety of FX in earlier studies. Furthermore, the lack of light during the extraction method, presence of ascorbic acid, and alkaline pH are all favorable for FX stability. Furthermore, the nursing of diatoms/microalgae is a new strategy that proposes extracting bioactive molecules without damaging the biological biomass, allowing for long-term milking of important chemicals (Vinayak et al., 2015).

6.8 BIOPRODUCTS FROM DIATOM BIOREFINING

Chlorophylls, fucoxanthin (FX), and eicosapentaenoic acid (EPA) are extremely valued marketable products with prospective applications in human health, aquaculture, pharmaceuticals, foods, and cosmetics, and are abundant in diatoms. Several diatoms, such as *O. aurita*, *Thalassiosira pseudonana*, and *P. tricornutum*, can simultaneously accumulate large amounts of Chrl, EPA, and Fx, and these species could be regarded as candidate producers for these high-value compounds (Yang et al., 2020). The most important byproducts or value metabolites from microalgae are carotenoids and lipids, which are high-value metabolites (HVMs) and low-value metabolites (LVMs), respectively (Gateau et al., 2017; Mimouni et al., 2015). Compared to chlorophytes, diatoms have emerged as preferable feedstocks for biofuel production due to abundant triacylglycerol (TAG) profiles in medium-chain fatty acids. Diatoms are also high in the fatty acids eicosapentaenoic acid (EPA) and docosahexaenoic acid (DHA), and are regarded as a viable alternative supply of these nutrients (Zulu et al., 2018). Compared to bacteria and yeast species, *Phaeodactylum tricornutum* has the potential for heterologous terpenoid production (Vavitsas et al., 2018). It is a promising alternative substrate. To meet global market demands, it is, therefore, critical to develop efficient technology to manufacture these high-value diatom products (Yang et al., 2020).

6.8.1 Fucoxanthin (FX)

Fucoxanthin ($C_{42}H_{58}O_6$) is extracted from diatoms and marine brown algae, and is a commercially important carotenoid (Su et al., 2019). FX is a keto carotenoid with an allenic link, a conjugated carbonyl, and a 5,6-monoepoxide in its chemical formula. This pigment has a unique structural property in that it has both an allene and an epoxide moiety in its molecule, altering its electrical properties dramatically (Buchel, 2003). FX is a potent antioxidant that has been shown to have anti-obesity, anti-inflammatory, anti-diabetic, anti-cancer, antihypertensive, and anti-osteoporotic activities (Martínez-Ruiz et al., 2022) and to suppress the inflammatory response (Kim et al., 2010). The carotenoids loroxanthin and fucoxanthin, generated mostly by marine diatoms *Phaeodactylum* and *Isochrysis*, are powerful antioxidants (Gupta et al., 2021; Yadavalli et al., 2020; Bajpai & Bajpai, 1993) (Table 6.4).

TABLE 6.4
Amount of EPA and FX Obtained from Various Diatom Species

		EPA		FX		
Strain	**Biomass (g L^{-1})**	**Productivity (mg L^{-1}.d^{-1})**	**Content (% DW)**	**Productivity (mg L^{-1}.d^{-1})**	**Content (% DW)**	**References**
C. cryptic	1.25	33.90	37.83	NA	NA	Clement et al., 2017
Phaeodactylum tricornutum	3.44	9.58	1.67	4.73	0.83	Yodsuwan et al., 2017
Bellerochea sp.	0.57	0.35	1.17	NA	NA	Li & Oliver, 2008
Odontella aurita	2.78	2.08	1.68	1.30	0.59	Vilchez et al., 2011
Nitzschia brevirostris	3.16	NA	5.28	NA	0.08	Remmers et al., 2017
N. laevis	6.48	19.73	2.74	NA	NA	Wang et al., 2018

6.8.2 Lipids

Lipids come in various forms, including triacylglyceride (TAG), fatty acids, chrysolaminarin, and others, and are used to store energy. These chemical compounds are commercially feasible, as they are used to make fatty acids, green fuel, glycerol, and a variety of other valuable chemicals. Myristic acid, palmitoleic acid, EPA, and DHA are the major fatty acids detected, which are widely employed in the health and biofuel industries (Yi et al., 2017; Yadavalli et al., 2020).

6.8.3 EPA

EPA is basically found in phytoplankton and marine animals, and is a PUFA (polyunsaturated fatty acid) (Bajpai & Bajpai, 1993). *T. pseudonana* accumulates up to 26% EPA and slightly more DHA than *P. tricornutum* (30% EPA and DHA in trace amounts) (Zulu et al., 2018).

6.8.4 Biofuel

Diafuel, a biofuel made from diatoms, is one of the promising bioenergy sources that could potentially replace future oil needs (Vinayak et al., 2019). Diatoms primarily create 13–21 carbon fatty acids (FAs), primarily saturated and MUFAs, which yield more energy when oxidized than polyunsaturated FAs (PUFAs) with the same number of carbons. In marine and freshwater diatoms, the two most common FAs are 16:1 and 16:0, which together account for up to 70% of the cell's lipid profile (Orcutt & Patterson, 1975). The qualities of the FAs used to make biodiesel directly impact its properties. Biodiesel properties like cold flow, cetane number, emissions level, viscosity, oxidative stability, and lubricity are all affected by the length and saturation level of the FAs (Knothe, 2005; Durrett et al., 2008). Therefore, using diatoms in a microbial fuel cell (MFC) can result in a low-cost Diafuel harvest. This is because diatoms not only reabsorb CO_2, but they also replenish oil tanks, as diatoms produce the majority of extra crude oil (Vinayak et al., 2015). However, because diatoms are inflexible due to silica shell technologies, they should be enhanced in MFC, also known as photo MFC, to harvest Diafuel rather than extracting it (Khan et al., 2021).

6.8.5 Value-Added Metabolites Extracted from Diatoms

Microalgae have received more interest in recent years as a viable and alternative renewable feedstock for the long-term generation of carbon-based fuels. This has been exacerbated by their high-fat content and rapid growth rate (Bayu et al., 2020). Other algal classes can achieve higher lipid and biomass productivity than diatoms (Marella & Tiwari, 2020). Diatoms have a high nutritional

FIGURE 6.1 Schematic representation of value-added products obtained from diatoms.

value, which can be exploited to make new compounds, including animal feed, antioxidants, vegetarian protein supplements, and vitamins (Sharma et al., 2021). Silica and pigments are diatom components that should be used to their full potential because they are both economically valuable and in high demand (Bayu et al., 2020). Diatom frustules, which are formed of silica and have hierarchical nanopore sizes, show promise in nanobiotechnological silica-based enterprises (Sharma et al., 2021). The genomes of diatoms were disclosed due to better molecular technologies, and always creates interest to humans to study them as model organisms using various culture conditions and customized photobioreactors to better comprehend the practical applications of diatoms and their valuable products. As a result, diatoms have found a niche for their multiscale biorefinery approaches (Mal et al., 2022). The synthesis of bio-products by diatoms can increase the economic viability of the biorefinery idea while also adding value to the service (Marella & Tiwari, 2020). Figure 6.1 shows a list of value-added products obtained from diatoms.

6.8.6 Pigments

Chlorophyll and carotenoids are the most common pigments found in diatoms. Chlorophyll a and c are the most important photosynthetic pigments (Mal et al., 2022). Diatoms are used to extract the commercially important substances β-carotene, fucoxanthin, astaxanthin, lycopene, lipid droplets, and proteins, with bioactivities that benefit human nutrition (Maeda et al. 2017; Pereira et al., 2021b; Leyland et al., 2020). Su et al. (2019) estimate that fucoxanthin ($C_{42}H_{58}O_6$) costs around 30,000 US dollars per gram and it has various potential health benefits (Mohamadnia et al., 2020; Pereira et al., 2021b).

6.8.7 Silica

The silica cell wall provides both mechanical and biological protection against microorganisms such as viruses and bacteria. Silaffins, pleuralins, silacidins, long-chain polyamines, and other proteins make up the majority of it. Because it creates highly pure porous nanospheres of various architectural shapes, biosilica has great potential as a biotechnological material. Bioactive glasses, drug-delivery systems, ceramics, and a variety of other microdevices are all made with diatom biosilica (Terracciano et al., 2015; Maher et al., 2018; Terracciano et al., 2018).

6.8.8 Biohydrocarbons

Diatoms are a good source of oil for making biodiesel. The presence of fatty acids such as oleic acid, palmitic acid, and triacylglycerides (TAGs) such as dilinoleoyloleoylgylcerol in *Navicula cryptocephala* is important for biofuels. *Phaeodactylum tricornutum*, on the other hand, has revealed a greater TAG content in cells when nitrogen and phosphorus were depleted. This could be because of a protective mechanism in which cell division slows and fatty acids are stored as TAGs (Rodolfi et al., 2017).

6.8.9 Antiviral Potency

Haslea ostrearia contains marennine, a blue pigment that has antiviral activity (Gastineau et al., 2012; Berge et al., 1999). In addition, naviculan, a sulfated polysaccharide in *Navicula directa*, has shown antiviral activity against herpesviruses (HSV-1, HSV-2) and influenza virus (Ahmadi et al., 2015).

6.8.10 Aquafeed

Diatoms are regarded as the optimum food for fisheries and hatcheries. Under typical laboratory circumstances, *Chaetoceros* sp. is thought to be the best choice for bivalve rearing. Secondary metabolites such as PUFAs, EPA, and DHA are abundant in these raw materials, resulting in a high yield and market value (Mal et al., 2022).

6.9 NANOBIOTECHNOLOGY

Diatoms' intricate structure and fascinating patterns are frequently employed in nanotechnology to create nanomaterials. Biosilica nanoparticles, nanofibers, and nanopolymers are widely used because of their unique mechanical, optical, thermostable, and chemical properties. Also, *Thalassiosira*, *Coscinodiscus*, and *Cyclotella* species produce biosilica microcapsule structures that are changed at the nanoscale to build natural living nanodevices (Ozkan & Rorrer, 2017; Maher et al., 2018).

A phototropism-based artificial selection technique (micro-electrochemical system) leads diatoms toward the chemostat template. This method could be used to mass-produce a 3D cloth with matching patterns at the nanoscale (Adams et al., 2017; Drum & Gordon, 2003). Many folds can be increased in solar efficiency when thin-film solar cells are coated with diatoms, which trap light inside tiny pores (Johnson, 2009) (Table 6.5).

6.10 CHALLENGES AND PERSPECTIVES OF DIATOM BIOREFINERY

In oceans, freshwaters, and rivers, diatoms are among the most opaque photosynthetic microorganisms. Because they fix more than 25% of the carbon dioxide in the atmosphere, they

TABLE 6.5
Value-added Chemicals Obtained from Different Diatom Species

Diatom species	Potential chemicals	Applications	Reference
Thalassiosira sp.	Lipids (24–39%)	Aquaculture feed	Botte et al., 2018 Mishra et al., 2017
Chaetoceros gracilis	High lipid content (30–70%)	Lipid-based fuel production, aquaculture feed	Wang & Seibert, 2017
Phaeodactylum tricornutum	EPA (24–39%)	Lipid producers, Functional foods	Mishra et al., 2017 Rodolfi et al., 2017
Coscinodiscus sp.	Nanoporous silica	Biosensors	Mishra et al., 2017
Chaetoceros calcitrans	Fucoxanthin (5.2 mg g^{-1})	Fucoxanthin production	Foo et al., 2015
Nitzschia sp.	EPA (1.9–4.7%), arachidonic acid (0.6–4.7%)	Food supplement	Mishra et al., 2017
Cyclotella sp.	Lipids (42%)	Lipid-based biofuels	Botte et al., 2018
Skeletonema costatus	Bioactive compound	Antiphatogenic secondary metabolite	Hildebrand et al., 2012

are critical in reducing global warming. They represent a vast reservoir of untapped potential with a number of uses, including wastewater treatment, CO_2 reduction, and as primary producers in the aquatic food chain (Tiwari et al., 2022). Microalgae biofuel is one such source with various advantages, such as being less harmful, biodegradable, sulfur-free, and a very appealing alternative for fossil fuel (Živković & Veljković, 2018). Microalgae have emerged as a viable option for wastewater treatment since they can handle a wide range of wastewater effluents, including human waste, industrial waste, agricultural waste, dairy waste, slaughterhouse waste, and so on (Adesra et al., 2021; Ekambaram et al., 2020; Shindhal et al., 2021). Diatoms have the potential to absorb heavy metals from wastewater, but they frequently endure morphological changes and stress in the process, causing them to overproduce secondary metabolites (Gautam et al., 2017; Lakshmi et al., 2020).

Diatoms' high operational costs do not encourage researchers in economically depressed areas (Khan et al., 2021). The biggest roadblock to developing diatom-based biofuels is the low return on investment (ROI), although the biochemical components in are is predicted to boost the process' revenue, making biofuel production more competitive. The potential economic value of diatom-derived biochemical products is being researched, particularly in the food, nutraceuticals, pharmaceuticals, nano-biomaterials, and biofuels industries. Some of the key limitations of diatoms for fucoxanthin (FX) production include identifying and isolating the optimum diatom strains for FX production, standardization of methods to obtain pure cultures, and appropriate nutritional needs in a photobioreactor-based production system. In addition, more research is needed to focus on lowering input costs during downstream processing in order to obtain high-quality and high-quantity FX at a fair price and purity. Aside from efficient growing methods, crucial information on the pathways involved in the production of bioactive algal metabolites is required, as is the discovery of genes that could be used for genetic engineering (Seth et al., 2021). The most challenging element of diatoms is that foreign DNA insertion into the genome occurs at random, resulting in transgenic lines with different degrees of expression despite being transformed with the same design (Chu et al., 2016). Loss-of-function studies are also limited by the random insertions into diatom genomes, as total gene deletion is exceedingly uncommon. The continual increase in water salinity owing to evaporation is a barrier to outdoor microalgae cultivation for FX production (Seth et al., 2021).

The cost of microalgal fuel is expected to rise in the near future. As a result, commercial algal fuel production from species with high lipid contents is promoted. Diatom solar panels must be tested on the ground because diatoms are more viable possibilities for biofuel production than any other algae. Diatoms could be a spark to boost India's economy, especially in light of the current COVID-19 outbreak. Because diatom solar panels only require sunshine, CO_2, water, and a few macro- and micronutrients, they would be affordable. Furthermore, adopting pretreatment processes like ultrasonication to recover diatoms grown in wastewater could lower the cost of operating these diatom solar panels (Khan et al., 2021). However, employing algae such as diatoms in a microbial fuel cell (MFC) paired PFC/PEC would produce Diafuel and bioelectricity to power hybrid electric vehicles in the near future (Vinayak et al., 2019).

6.11 CONCLUSIONS

Diatoms are a form of microalgae that have a high lipid content. They are a promising feedstock for making biofuels and other high-value products. As a result, the diatom to biorefinery technique has gotten a lot of interest from researchers because it yields more biofuels than other algal sources. The current review has focused on the current opportunities for diatoms to be used in various value-added bioproducts to improve the circular bioeconomy. Implementing this at a commercial scale could reduce the usage of fossil fuels and also help to enhance the national economy.

REFERENCES

Abhari, K. and Khaneghah, A.M. (2020). Alternative extraction techniques to obtain, isolate and purify proteins and bioactive from aquaculture and by-products. *Advances in Food and Nutrition Research* 92, 35–52.

Adams, L.A., Essien, E.R., Adesalu, A.T. and Julius, M.L. (2017). Bioactive glass 45S5 from diatom biosilica. *Journal of Science: Advanced Materials and Devices* 2(4), 476–482.

Adesra, A., Srivastava, V.K. and Varjani, S. (2021). Valorization of dairy wastes: integrative approaches for value added products. *Indian Journal of Microbiology 61*(3), 270–278.

Ahmadi, A., Zorofchian Moghadamtousi, S., Abubakar, S. and Zandi, K. (2015). Antiviral potential of algae polysaccharides isolated from marine sources: a review. *BioMed Research International* 2015, 825203.

Arakaki, A., Matsumoto, T., Tateishi, T., Matsumoto, M., Nojima, D., Tomoko, Y. and Tanaka, T. (2017). UV-C irradiation accelerates neutral lipid synthesis in the marine oleaginous diatom Fistulifera solaris. *Bioresource Technology* 245, 1520–1526.

Bajpai, P. and Bajpai, P.K. (1993). Eicosapentaenoic acid (EPA) production from microorganisms: a review. *Journal of Biotechnology 30*(2), 161–183.

Barka, F., Angstenberger, M., Ahrendt, T., Lorenzen, W., Bode, H.B. and Büchel, C. (2016). Identification of a triacylglycerol lipase in the diatom Phaeodactylum tricornutum. *Biochimica et Biophysica Acta (BBA) – Molecular and Cell Biology of Lipids* 1861(3), 239–248.

Bauer, C.M., Schmitz, C., Corrêa, R.G., Herrera, C.M., Ramlov, F., Oliveira, E.R., Pizzato, A., Varela, L.A.C., Cabral, D.Q., Yunes, R.A. and Lopes, R.G. (2019). In vitro fucoxanthin production by the *Phaeodactylum tricornutum* diatom. *Studies in Natural Products Chemistry* 63, 211–242.

Bayu, A., Rachman, A., Noerdjito, D.R., Putra, M.Y. and Widayatno, W.B. (2020). March. High-value chemicals from marine diatoms: a biorefinery approach. *IOP Conference Series: Earth and Environmental Science* 460(1), 012012.

Bekirogullari, M., Figueroa-Torres, G.M., Pittman, J.K. and Theodoropoulos, C. (2020). Models of microalgal cultivation for added-value products-A review. *Biotechnology Advances* 44, 107609.

Bergé, J.P., Bourgougnon, N., Alban, S., Pojer, F., Billaudel, S., Chermann, J.C., Robert, J.M. and Franz, G. (1999). Antiviral and anticoagulant activities of a water-soluble fraction of the marine diatom *Haslea ostrearia. Planta Medica* 65(07), 604–609.

Bertrand, M. (2010). Carotenoid biosynthesis in diatoms. *Photosynthesis Research* 106(1), 89–102.

Bhattacharjya, R., Marella, T.K., Tiwari, A., Saxena, A., Singh, P.K. and Mishra, B. (2020). Bioprospecting of marine diatoms *Thalassiosira, Skeletonema* and *Chaetoceros* for lipids and other value-added products. *Bioresource Technology* 318, 124073.

Botte, P., d'Ippolito, G., Gallo, C., Sardo, A. and Fontana, A. (2018). Combined exploitation of CO2 and nutrient replenishment for increasing biomass and lipid productivity of the marine diatoms Thalassiosira weissflogii and Cyclotella cryptica. *Journal of Applied Phycology* 30(1), 243–251.

Büchel, C. (2003). Fucoxanthin-chlorophyll proteins in diatoms: 18 and 19 kDa subunits assemble into different oligomeric states. *Biochemistry* 42(44), 13027–13034.

Buesseler, K.O. (1998). The decoupling of production and particulate export in the surface ocean. *Global Biogeochemical Cycles*, 12(2), 297–310.

Butler, T., Kapoore, R.V. and Vaidyanathan, S. (2020). *Phaeodactylum tricornutum*: a diatom cell factory. *Trends in Biotechnology* 38(6), 606–622.

Cavalier-Smith, T. (2018). Kingdom Chromista and its eight phyla: a new synthesis emphasising periplastid protein targeting, cytoskeletal and periplastid evolution, and ancient divergences. *Protoplasma* 255(1), 297–357.

Cheirsilp, B., Srinuanpan, S. and Mandik, Y.I. (2020). Efficient harvesting of microalgal biomass and direct conversion of microalgal lipids into biodiesel. In: *Microalgae Cultivation for Biofuels Production*, pp. 83–96. Academic Press.

Chew, K.W., Chia, S.R., Show, P.L., Yap, Y.J., Ling, T.C. and Chang, J.S. (2018). Effects of water culture medium, cultivation systems and growth modes for microalgae cultivation: A review. *Journal of the Taiwan Institute of Chemical Engineers* 91, 332–344.

Chisti, Y. (2007). Biodiesel from microalgae. *Biotechnology Advances* 25(3), 294–306.

Cho, D.H., Choi, J.W., Kang, Z., Kim, B.H., Oh, H.M., Kim, H.S. and Ramanan, R., (2017). Microalgal diversity fosters stable biomass productivity in open ponds treating wastewater. *Scientific Reports*, *7*(1), 1–11.

Choi, H.G., Joo, H.M., Jung, W., Hong, S.S., Kang, J.S. and Kang, S.H. (2008). Morphology and phylogenetic relationships of some psychrophilic polar diatoms (Bacillariophyta). *Nova Hedwigia*, 133, 7–30.

Chu, L., Ewe, D., Bártulos, C.R., Kroth, P.G. and Gruber, A. (2016). Rapid induction of GFP expression by the nitrate reductase promoter in the diatom *Phaeodactylum tricornutum*. *PeerJ* 4, e2344.

Clement, R., Jensen, E., Prioretti, L., Maberly, S.C. and Gontero, B. (2017). Diversity of CO_2-concentrating mechanisms and responses to CO2 concentration in marine and freshwater diatoms. *Journal of Experimental Botany* 68(14), 3925–3935.

Cointet, E., Wielgosz-Collin, G., Bougaran, G., Rabesaotra, V., Gonçalves, O. and Méléder, V. (2019). Effects of light and nitrogen availability on photosynthetic efficiency and fatty acid content of three original benthic diatom strains. *PLoS One* 14(11), e0224701.

Demailly, F., Elfeky, I., Malbezin, L., Le Guédard, M., Eon, M., Bessoule, J.J., Feurtet-Mazel, A., Delmas, F., Mazzella, N., Gonzalez, P. and Morin, S. (2019). Impact of diuron and S-metolachlor on the freshwater diatom *Gomphonema gracile*: Complementarity between fatty acid profiles and different kinds of ecotoxicological impact-endpoints. *Science of the Total Environment* 688, 960–969.

Derwenskus, F., Metz, F., Gille, A., Schmid-Staiger, U., Briviba, K., Schließmann, U. and Hirth, T. (2019). Pressurized extraction of unsaturated fatty acids and carotenoids from wet Chlorella vulgaris and *Phaeodactylum tricornutum* biomass using subcritical liquids. *Gcb Bioenergy* 11(1), 335–344.

Drum, R.W. and Gordon, R. (2003). Star Trek replicators and diatom nanotechnology. *Trends in Biotechnology* 21(8), 325–328.

Durrett, T.P., Benning, C. and Ohlrogge, J. (2008). Plant triacylglycerols as feedstocks for the production of biofuels. *The Plant Journal* 54(4), 593–607.

Ekambaram, N., Varjani, S., Goswami, S., Singh, U. and Kapoor, T. (2020). Chitosan-based silver nanocomposite for hexavalent-chromium removal from tannery industry effluent using a packed-bed reactor. *Journal of Environmental Engineering* 146(6), 04020032.

Falkowski, P.G., Barber, R.T. and Smetacek, V. (1998). Biogeochemical controls and feedbacks on ocean primary production. *Science* 281(5374), 200–206.

Foo, S.C., Yusoff, F.M., Ismail, M., Basri, M., Chan, K.W., Khong, N.M. and Yau, S.K. (2015). Production of fucoxanthin-rich fraction (FxRF) from a diatom, *Chaetoceros calcitrans* (Paulsen) Takano 1968. *Algal Research* 12, 26–32.

Gastineau, R., Pouvreau, J.B., Hellio, C., Morançais, M., Fleurence, J., Gaudin, P., Bourgougnon, N. and Mouget, J.L. (2012). Biological activities of purified marennine, the blue pigment responsible for the greening of oysters. *Journal of Agricultural and Food Chemistry* 60(14), 3599–3605.

Gateau, H., Solymosi, K., Marchand, J. and Schoefs, B. (2017). Carotenoids of microalgae used in food industry and medicine. *Mini Reviews in Medicinal Chemistry* 17(13), 1140–1172.

Gautam, S., Pandey, L.K., Vinayak, V. and Arya, A. (2017). Morphological and physiological alterations in the diatom *Gomphonema pseudoaugur* due to heavy metal stress. *Ecological Indicators* 72, 67–76.

Gavrilescu, M. and Chisti, Y. (2005). Biotechnology—a sustainable alternative for chemical industry. *Biotechnology Advances* 23(7–8), 471–499.

George, J., Kahlke, T., Abbriano, R.M., Kuzhiumparambil, U., Ralph, P.J. and Fabris, M. (2020). Metabolic engineering strategies in diatoms reveal unique phenotypes and genetic configurations with implications for algal genetics and synthetic biology. *Frontiers in Bioengineering and Biotechnology* 8, 513.

Gérin, S., Delhez, T., Corato, A., Remacle, C. and Franck, F. (2020). A novel culture medium for freshwater diatoms promotes efficient photoautotrophic batch production of biomass, fucoxanthin, and eicosapentaenoic acid. *Journal of Applied Phycology* 32(3), 1581–1596.

Gholkar, P., Shastri, Y. and Tanksale, A. (2021). Renewable hydrogen and methane production from microalgae: A techno-economic and life cycle assessment study. *Journal of Cleaner Production* 279, 123726.

Guillard, R.R. (1973). Division rates. *Handbook of Phycological Methods: Culture Methods and Growth Measurements*. Cambridge University Press, London, pp. 2289–2311.

Gupta, A.K., Seth, K., Maheshwari, K., Baroliya, P.K., Meena, M., Kumar, A. and Vinayak, V. (2021). Biosynthesis and extraction of high-value carotenoid from algae. *Frontiers in Bioscience-Landmark* 26(6), 171–190.

Harun, R., Danquah, M.K. and Forde, G.M. (2010). Microalgal biomass as a fermentation feedstock for bioethanol production. *Journal of Chemical Technology & Biotechnology* 85(2), 199–203.

Herrero, M. and Ibanez, E. (2015). Green processes and sustainability: An overview on the extraction of high added-value products from seaweeds and microalgae. *The Journal of Supercritical Fluids* 96, 211–216.

Hildebrand, M., Davis, A.K., Smith, S.R., Traller, J.C. and Abbriano, R. (2012). The place of diatoms in the biofuels industry. *Biofuels* 3(2), 221–240.

Hogan, P., Otero, P., Murray, P. and Saha, S.K. (2021). Effect of biomass pre-treatment on supercritical CO_2 extraction of lipids from marine diatom Amphora sp. and its biomass evaluation as bioethanol feedstock. *Heliyon* 7(1), e05995.

Ianora, A., Turner, J.T., Esposito, F., Carotenuto, Y., Romano, G., Fontana, A., Guisande, C. and Miralto, A. (2004). Copepod egg production and hatching success is reduced by maternal diets of a non-neurotoxic strain of the dinoflagellate *Alexandrium tamarense*. *Marine Ecology Progress Series* 280, 199–210.

Indrayani, I., Moheimani, N.R., de Boer, K., Bahri, P.A. and Borowitzka, M.A. (2020). Temperature and salinity effects on growth and fatty acid composition of a halophilic diatom, *Amphora* sp. MUR258 (Bacillariophyceae). *Journal of Applied Phycology* 32(2), 977–987.

Irigoien, X., Harris, R.P., Verheye, H.M., Joly, P., Runge, J., Starr, M., Pond, D., Campbell, R., Shreeve, R., Ward, P. and Smith, A.N. (2002). Copepod hatching success in marine ecosystems with high diatom concentrations. *Nature* 419(6905), 387–389.

Ishika, T., Laird, D.W., Bahri, P.A. and Moheimani, N.R. (2019). Co-cultivation and stepwise cultivation of *Chaetoceros muelleri* and *Amphora* sp. for fucoxanthin production under gradual salinity increase. *Journal of Applied Phycology* 31(3), 1535–1544.

Jensen, E.L., Yangüez, K., Carrière, F. and Gontero, B. (2019). Storage compound accumulation in diatoms as response to elevated CO_2 concentration. *Biology* 9(1), 5.

Johnson, R.C. (2009). Diatoms could triple solar cell efficiency. *EE Times*.

Joseph, M.M., Renjith, K.R., John, G., Nair, S.M. and Chandramohanakumar, N. (2017). Biodiesel prospective of five diatom strains using growth parameters and fatty acid profiles. *Biofuels* 8(1), 81–89.

Karaman, E.S., Wang, Z., Di Benedetto, G., Zunino III, J.L., Meng, X. and Mitra, S. (2019). Fabrication of supercapacitors and flexible electrodes using biosilica from cultured diatoms. *Materials Today Energy* 11, 166–173.

Karim, A., Islam, M.A., Khalid, Z.B., Faizal, C.K.M., Khan, M.M.R. and Yousuf, A. (2020). Microalgal cell disruption and lipid extraction techniques for potential biofuel production. In: *Microalgae Cultivation for Biofuels Production*, pp. 129–147. Academic Press.

Khan, M.J., Bawra, N., Verma, A., Kumar, V., Pugazhendhi, A., Joshi, K.B. and Vinayak, V. (2021). Cultivation of diatom *Pinnularia saprophila* for lipid production: a comparison of methods for harvesting the lipid from the cells. *Bioresource Technology* 319, 124129.

Kim, J., Yoo, G., Lee, H., Lim, J., Kim, K., Kim, C.W., Park, M.S. and Yang, J.W. (2013). Methods of downstream processing for the production of biodiesel from microalgae. *Biotechnology Advances* 31(6), 862–876.

Kim, K.N., Heo, S.J., Yoon, W.J., Kang, S.M., Ahn, G., Yi, T.H. and Jeon, Y.J. (2010). Fucoxanthin inhibits the inflammatory response by suppressing the activation of NF-κB and MAPKs in lipopolysaccharide-induced RAW 264.7 macrophages. *European Journal of Pharmacology* 649(1–3), 369–375.

Kim, S.M., Kang, S.W., Kwon, O.N., Chung, D. and Pan, C.H. (2012). Fucoxanthin as a major carotenoid in Isochrysis aff. galbana: Characterization of extraction for commercial application. *Journal of the Korean Society for Applied Biological Chemistry* 55(4), 477–483.

Knothe, G. (2005). Dependence of biodiesel fuel properties on the structure of fatty acid alkyl esters. *Fuel Processing Technology* 86(10), 1059–1070.

Kolber, Z.S., Gerald, F., Plumley, Lang, A.S., Beatty, J.T., Blankenship, R.E., VanDover, C.L., Vetriani, C., Koblizek, M., Rathgeber, C. and Falkowski, P.G. (2001). Contribution of aerobic photoheterotrophic bacteria to the carbon cycle in the ocean. *Science* 292(5526), 2492–2495.

Lakshmi, S., Suvedha, K., Sruthi, R., Lavanya, J., Varjani, S. and Nakkeeran, E. (2020). Hexavalent chromium sequestration from electronic waste by biomass of Aspergillus carbonarius. *Bioengineered* 11(1), 708–717.

Leterme, S.C. (2015). The oil production capacity of diatoms. *Annals of Aquaculture and Research* 2(1), 1007.

Leyland, B., Boussiba, S. and Khozin-Goldberg, I. (2020). A review of diatom lipid droplets. *Biology* 9(2), 38.

Liang, Y., Kashdan, T., Sterner, C., Dombrowski, L., Petrick, I., Kroger, M., Hofer, R. (2015). Chapter 2 Algal Biorefineries. In: *Industrial Bioregineries and White Biotechnology*, pp. 35–90. Elsevier, Amsterdam, Netherlands.

Lin, M. and Oliver, D.J. (2008). The role of acetyl-coenzyme a synthetase in Arabidopsis. *Plant Physiology* 147(4), 1822–1829.

Lin, Q., Zhuo, W.H., Wang, X.W., Chen, C.P., Gao, Y.H. and Liang, J.R. (2018). Effects of fundamental nutrient stresses on the lipid accumulation profiles in two diatom species *Thalassiosira weissflogii* and *Chaetoceros muelleri*. *Bioprocess and Biosystems Engineering* 41(8), 1213–1224.

Lu, X., Liu, B., He, Y., Guo, B., Sun, H. and Chen, F. (2019). Novel insights into mixotrophic cultivation of *Nitzschia laevis* for co-production of fucoxanthin and eicosapentaenoic acid. *Bioresource Technology* 294, 122145.

Maeda, Y., Nojima, D., Yoshino, T. and Tanaka, T. (2017). Structure and properties of oil bodies in diatoms. *Philosophical Transactions of the Royal Society B: Biological Sciences* 372(1728), 20160408.

Maher, S., Kumeria, T., Aw, M.S. and Losic, D. (2018). Diatom silica for biomedical applications: Recent progress and advances. *Advanced Healthcare Materials* 7(19), 1800552.

Mal, N., Srivastava, K., Sharma, Y., Singh, M., Rao, K.M., Enamala, M.K., Chandrasekhar, K. and Chavali, M. (2022). Facets of diatom biology and their potential applications. *Biomass Conversion and Biorefinery* 12(5), 1959–1975.

Manirafasha, E., Ndikubwimana, T., Zeng, X., Lu, Y. and Jing, K. (2016). Phycobiliprotein: Potential microalgae derived pharmaceutical and biological reagent. *Biochemical Engineering Journal* 109, 282–296.

Mann, D.G. and Droop, S.J.M. (1996). Biodiversity, biogeography and conservation of diatoms. In: *Biogeography of Freshwater Algae*, pp. 19–32. Springer, Dordrecht.

Mao, X., Chen, S.H.Y., Lu, X., Yu, J. and Liu, B. (2020). High silicate concentration facilitates fucoxanthin and eicosapentaenoic acid (EPA) production under heterotrophic condition in the marine diatom *Nitzschia laevis*. *Algal Research*, 52, 102086.

Marella, T.K. and Tiwari, A. (2020). Marine diatom *Thalassiosira weissflogii* based biorefinery for co-production of eicosapentaenoic acid and fucoxanthin. *Bioresource Technology*, 307, 123245.

Marella, T.K., Datta, A., Patil, M.D., Dixit, S. and Tiwari, A. (2019). Biodiesel production through algal cultivation in urban wastewater using algal floway. *Bioresource Technology* 280, 222–228.

Martin, L.A., Popovich, C.A., Martínez, A.M., Bilbao, P.G.S., Damiani, M.C. and Leonardi, P.I. (2018). Hybrid two-stage culture of *Halamphora coffeaeformis* for biodiesel production: Growth phases, nutritional stages and biorefinery approach. *Renewable Energy* 118, 984–992.

Martínez-Ruiz, M., Martínez-González, C.A., Kim, D.H., Santiesteban-Romero, B., Reyes-Pardo, H., Villaseñor-Zepeda, K.R., Meléndez-Sánchez, E.R., Ramírez-Gamboa, D., Díaz-Zamorano, A.L., Sosa-Hernández, J.E. and Coronado-Apodaca, K.G. (2022). Microalgae bioactive compounds to topical applications products—A review. *Molecules*, 27(11), 3512.

McClure, D.D., Luiz, A., Gerber, B., Barton, G.W. and Kavanagh, J.M. (2018). An investigation into the effect of culture conditions on fucoxanthin production using the marine microalgae *Phaeodactylum tricornutum*. *Algal Research* 29, 41–48.

Middelberg, A.P. (1995). Process-scale disruption of microorganisms. *Biotechnology Advances* 13(3), 491–551.

Mimouni, V., Ulmann, L., Haimeur, A., Guéno, F., Meskini, N. and Tremblin, G. (2015). Marine microalgae used as food supplements and their implication in preventing cardiovascular diseases. *OCL* 22(4), D409.

Mishra, M., Arukha, A.P., Bashir, T., Yadav, D. and Prasad, G.B.K.S. (2017). All new faces of diatoms: potential source of nanomaterials and beyond. *Frontiers in Microbiology* 8, 1239.

Mohamadnia, S., Tavakoli, O., Faramarzi, M.A. and Shamsollahi, Z. (2020). Production of fucoxanthin by the microalga *Tisochrysis lutea*: A review of recent developments. *Aquaculture* 516, 734637.

Orcutt, D.M. and Patterson, G.W. (1975). Sterol, fatty acid and elemental composition of diatoms grown in chemically defined media. *Comparative Biochemistry and Physiology Part B: Comparative Biochemistry* 50(4), 579–583.

Otero, P., Saha, S.K., Gushin, J.M., Moane, S., Barron, J. and Murray, P. (2017). Identification of optimum fatty acid extraction methods for two different microalgae *Phaeodactylum tricornutum* and *Haematococcus pluvialis* for food and biodiesel applications. *Analytical and Bioanalytical Chemistry* 409(19), 4659–4667.

Ozkan, A. and Rorrer, G.L. (2017). Effects of light intensity on the selectivity of lipid and chitin nanofiber production during photobioreactor cultivation of the marine diatom *Cyclotella* sp. *Algal Research* 25, 216–227.

Pawolski, D., Heintze, C., Mey, I., Steinem, C. and Kröger, N. (2018). Reconstituting the formation of hierarchically porous silica patterns using diatom biomolecules. *Journal of Structural Biology* 204(1), 64–74.

Pereira, A.G., Otero, P., Echave, J., Carreira-Casais, A., Chamorro, F., Collazo, N., Jaboui, A., Lourenço-Lopes, C., Simal-Gandara, J. and Prieto, M.A. (2021a). Xanthophylls from the sea: algae as source of bioactive carotenoids. *Marine Drugs* 19(4), 188.

Pereira, H., Sá, M., Maia, I., Rodrigues, A., Teles, I., Wijffels, R.H., Navalho, J. and Barbosa, M. (2021b). Fucoxanthin production from *Tisochrysis lutea* and *Phaeodactylum tricornutum* at industrial scale. *Algal Research* 56, 102322.

Remmers, I.M., Martens, D.E., Wijffels, R.H. and Lamers, P.P. (2017). Dynamics of triacylglycerol and EPA production in *Phaeodactylum tricornutum* under nitrogen starvation at different light intensities. *PloS One* 12(4), e0175630.

Rodolfi, L., Biondi, N., Guccione, A., Bassi, N., D'Ottavio, M., Arganaraz, G. and Tredici, M.R. (2017). Oil and eicosapentaenoic acid production by the diatom *Phaeodactylum tricornutum* cultivated outdoors in Green Wall Panel (GWP®) reactors. *Biotechnology and Bioengineering* 114(10), 2204–2210.

Round, F.E., Crawford, R.M. and Mann, D.G. (1990). *Diatoms: Biology and Morphology of the Genera*. Cambridge University Press.

Sarthou, G., Timmermans, K.R., Blain, S. and Tréguer, P. (2005). Growth physiology and fate of diatoms in the ocean: a review. *Journal of Sea Research* 53(1–2), 25–42.

Saxena, A., Marella, T.K., Singh, P.K. and Tiwari, A. (2021). Indoor mass cultivation of marine diatoms for biodiesel production using induction plasma synthesized nanosilica. *Bioresource Technology* 332, 125098.

Saxena, A., Prakash, K., Phogat, S., Singh, P.K. and Tiwari, A. (2020). Inductively coupled plasma nanosilica based growth method for enhanced biomass production in marine diatom algae. *Bioresource Technology* 314, 123747.

Scala, S. and Bowler, C. (2001). Molecular insights into the novel aspects of diatom biology. *Cellular and Molecular Life Sciences CMLS* 58(11), 1666–1673.

Schiano di Visconte, G., Spicer, A., Chuck, C.J. and Allen, M.J. (2019). The microalgae biorefinery: a perspective on the current status and future opportunities using genetic modification. *Applied Sciences* 9(22), 4793.

Seth, K., Kumar, A., Rastogi, R.P., Meena, M. and Vinayak, V. (2021). Bioprospecting of fucoxanthin from diatoms—Challenges and perspectives. *Algal Research*, 60, 102475.

Sharma, N., Simon, D.P., Diaz-Garza, A.M., Fantino, E., Messaabi, A., Meddeb-Mouelhi, F., Germain, H. and Desgagné-Penix, I. (2021). Diatoms biotechnology: Various industrial applications for a greener tomorrow. *Frontiers in Marine Science* 8, 636613.

Sheehan, J., Dunahay, T., Benemann, J. and Roessler, P. (1998). A look back at the US Department of Energy's aquatic species program: biodiesel from algae. *National Renewable Energy Laboratory* 328, 1–294.

Shindhal, T., Rakholiya, P., Varjani, S., Pandey, A., Ngo, H.H., Guo, W., Ng, H.Y. and Taherzadeh, M.J. (2021). A critical review on advances in the practices and perspectives for the treatment of dye industry wastewater. *Bioengineered* 12(1), 70–87.

Shishlyannikov, S.M., Nikonova, A.A., Bukin, Y.S. and Gorshkov, A.G. (2018). Fatty acid trophic markers in Lake Baikal phytoplankton: A comparison of endemic and cosmopolitan diatom-dominated phytoplankton assemblages. *Ecological Indicators* 85, 878–886.

Singh, P.K., Bhattacharjya, R., Saxena, A., Mishra, B. and Tiwari, A. (2021). Utilization of wastewater as nutrient media and biomass valorization in marine Chrysophytes – *Chaetoceros* and *Isochrysis*. *Energy Conversion and Management: X* 10, 100062.

Su, J., Guo, K., Huang, M., Liu, Y., Zhang, J., Sun, L., Li, D., Pang, K.L., Wang, G., Chen, L. and Liu, Z. (2019). Fucoxanthin, a marine xanthophyll isolated from *Conticribra weissflogii* ND-8: Preventive anti-inflammatory effect in a mouse model of sepsis. *Frontiers in Pharmacology* 10, 906.

Suparmaniam, U., Lam, M.K., Uemura, Y., Lim, J.W., Lee, K.T. and Shuit, S.H. (2019). Insights into the microalgae cultivation technology and harvesting process for biofuel production: A review. *Renewable and Sustainable Energy Reviews* 115, 109361.

Svenning, J.B., Dalheim, L., Eilertsen, H.C. and Vasskog, T. (2019). Temperature dependent growth rate, lipid content and fatty acid composition of the marine cold-water diatom *Porosira glacialis*. *Algal Research* 37, 11–16.

Tanaka, T., Yabuuchi, T., Maeda, Y., Nojima, D., Matsumoto, M. and Yoshino, T. (2017). Production of eicosapentaenoic acid by high cell density cultivation of the marine oleaginous diatom *Fistulifera solaris*. *Bioresource Technology* 245, 567–572.

Terracciano, M., De Stefano, L. and Rea, I. (2018). Diatoms green nanotechnology for biosilica-based drug delivery systems. *Pharmaceutics* 10(4), 242.

Terracciano, M., Shahbazi, M.A., Correia, A., Rea, I., Lamberti, A., De Stefano, L. and Santos, H.A. (2015). Surface bioengineering of diatomite based nanovectors for efficient intracellular uptake and drug delivery. *Nanoscale* 7(47), 20063–20074.

Tiwari, A., Marella, T.K. and Saxena, A. (2022). Diatom biorefinery: From carbon mitigation to high-value products. *Biomass, Biofuels, Biochemicals* 401–420.

Treguer, P., Nelson, D.M., Van Bennekom, A.J., DeMaster, D.J., Leynaert, A. and Quéguiner, B. (1995). The silica balance in the world ocean: a reestimate. *Science* 268(5209), 375–379.

Vavitsas, K., Fabris, M. and Vickers, C.E. (2018). Terpenoid metabolic engineering in photosynthetic microorganisms. *Genes* 9(11), 520.

Vella, F.M., Sardo, A., Gallo, C., Landi, S., Fontana, A. and d'Ippolito, G. (2019). Annual outdoor cultivation of the diatom *Thalassiosira weissflogii*: productivity, limits and perspectives. *Algal Research 42,* 101553.

Vílchez, C., Forján, E., Cuaresma, M., Bédmar, F., Garbayo, I. and Vega, J.M. (2011). Marine carotenoids: biological functions and commercial applications. *Marine Drugs* 9(3), 319–333.

Vinayak, V., Joshi, K.B. and Sarma, P.M. (2019). Diafuel™ (diatom biofuel) vs electric vehicles, a basic comparison: a high potential renewable energy source to make India energy independent. *Diatoms: Fundamentals and Applications*, 537–582.

Vinayak, V., Manoylov, K.M., Gateau, H., Blanckaert, V., Hérault, J., Pencréac'h, G., Marchand, J., Gordon, R. and Schoefs, B. (2015). Diatom milking: a review and new approaches. *Marine Drugs* 13(5), 2629–2665.

Wang, H., Zhang, Y., Chen, L., Cheng, W. and Liu, T. (2018). Combined production of fucoxanthin and EPA from two diatom strains *Phaeodactylum tricornutum* and *Cylindrotheca fusiformis* cultures. *Bioprocess and Biosystems Engineering* 41(7), 1061–1071.

Wang, J.K. and Seibert, M. (2017). Prospects for commercial production of diatoms. *Biotechnology for Biofuels* 10(1), 1–13.

Winter, J.G. and Duthie, H.C. (2000). Stream epilithic, epipelic and epiphytic diatoms: habitat fidelity and use in biomonitoring. *Aquatic Ecology* 34(4), 345–353.

Xia, S., Gao, B., Fu, J., Xiong, J. and Zhang, C. (2018). Production of fucoxanthin, chrysolaminarin, and eicosapentaenoic acid by *Odontella aurita* under different nitrogen supply regimes. *Journal of Bioscience and Bioengineering* 126(6), 723–729.

Yadavalli, R., Ratnapuram, H., Motamarry, S., Reddy, C.N., Ashokkumar, V. and Kuppam, C. (2020). Simultaneous production of flavonoids and lipids from *Chlorella vulgaris* and *Chlorella pyrenoidosa*. *Biomass Conversion and Biorefinery* 12(3): 683–691.

Yang, R., Wei, D. and Xie, J. (2020). Diatoms as cell factories for high-value products: chrysolaminarin, eicosapentaenoic acid, and fucoxanthin. *Critical Reviews in Biotechnology* 40(7), 993–1009.

Yellapu, S.K., Bezawada, J., Kaur, R., Kuttiraja, M. and Tyagi, R.D. (2016). Detergent assisted lipid extraction from wet yeast biomass for biodiesel: A response surface methodology approach. *Bioresource Technology* 218, 667–673.

Yew, G.Y., Lee, S.Y., Show, P.L., Tao, Y., Law, C.L., Nguyen, T.T.C. and Chang, J.S. (2019). Recent advances in algae biodiesel production: from upstream cultivation to downstream processing. *Bioresource Technology Reports* 7, 100227.

Yi, Z., Xu, M., Di, X., Brynjolfsson, S. and Fu, W. (2017). Exploring valuable lipids in diatoms. *Frontiers in Marine Science* 4, 17.

Yin, Z., Zhu, L., Li, S., Hu, T., Chu, R., Mo, F., Hu, D., Liu, C. and Li, B. (2020). A comprehensive review on cultivation and harvesting of microalgae for biodiesel production: Environmental pollution control and future directions. *Bioresource Technology* 301, 122804.

Yodsuwan, N., Sawayama, S. and Sirisansaneeyakul, S. (2017). Effect of nitrogen concentration on growth, lipid production and fatty acid profiles of the marine diatom *Phaeodactylum tricornutum*. *Agriculture and Natural Resources* 51(3), 190–197.

Zarekarizi, A., Hoffmann, L. and Burritt, D. (2019). Approaches for the sustainable production of fucoxanthin, a xanthophyll with potential health benefits. *Journal of Applied Phycology* 31(1), 281–299.

Zhang, S., Liu, H., Ke, Y. and Li, B. (2017). Effect of the silica content of diatoms on protozoan grazing. *Frontiers in Marine Science* 4, 202.

Živković, S. and Veljković, M. (2018). Environmental impacts the production and use of biodiesel. *Environmental Science and Pollution Research*, 25(1), 191–199.

Zulu, N.N., Zienkiewicz, K., Vollheyde, K. and Feussner, I. (2018). Current trends to comprehend lipid metabolism in diatoms. *Progress in Lipid Research* 70, 1–16.

7 Extraction Technologies from Diatoms

Hao Yi Gay[1], Wen Yi Chia[2], Doris Ying Ying Tang[2], Wee Jun Ong[1], and Kit Wayne Chew[3]
[1]School of Energy and Chemical Engineering, Xiamen University Malaysia, Malaysia
[2]Department of Chemical and Environmental Engineering, Faculty of Science and Engineering, University of Nottingham Malaysia, Selangor, Malaysia
[3]School of Chemistry, Chemical Engineering and Biotechnology, Nanyang Technological University, Singapore

7.1 INTRODUCTION

Ever since the discovery of fossil fuels back in the 1760s, humankind was thought to be blessed by mother nature to be given a powerful energy source, which would then go on to spur the first Industrial Revolution in that period, and subsequently the development of human civilization in the following centuries (Hsu and Robinson, 2017). However, the combustion of such fuels results in massive carbon dioxide (CO_2) emissions, which is the most abundant greenhouse gas (GHG), that results in global warming via radiative forcing, as well as ocean acidification through the dissolution of CO_2 into ocean waters to form weak carbonic acid. Fast forward to the world we know today, approximately 85% of national energy requirements are provided by fossil fuels, amounting to annual CO_2 emissions of 29 gigatons in recent years and a cumulative emission of over 1.5 trillion tons since the introduction of fossil fuel combustion to the year 2017 (Ritchie, 2019). Despite the severe consequences of anthropogenic CO_2 pollution, the greater concern surrounding fossil fuels for the rising global population is their depletion, with coal and oil reserves expected to be exhausted in 110 and 50 years, respectively, due to the increasingly rapid consumption relative to the time required to produce such fossil fuels which is virtually eternal (Day and Day, 2017).

Hence, humans would eventually be forced to make their own improvements in the form of renewable and sustainable energy sources, namely solar, wind, hydropower, and geothermal, which also significantly reducing GHG emissions. Another form of sustainable energy source on the rise is bioenergy or biofuels, such as biochar, biodiesel, bioethanol, and biohydrogen, which are mainly derived from plant-based products. According to the biomass feedstock, these biofuels and their production can be categorized into four generations, in which the first generation of biofuels is produced from the oils of edible seeds, crops, and animal fats. This method of sourcing biofuels is relatively unsustainable due to deforestation to occupy precious soil resources and arable land, which presents the issue of the food versus fuel crisis since it sacrifices edible food sources and the resources required to produce food crops (Sajjadi et al., 2018). Conversely, second-generation biofuels utilize mostly non-edible and lower-value biomass such as food crops residues, food waste, and cooking oil from the general population as well as lignocellulosic feedstock like sugarcane bagasse. For example, due to the abundance of oil palm wastes in Malaysia and Indonesia, oil palm

DOI: 10.1201/9781003436553-7

processing facilities have subjected oil palm residues to combustion for electricity generation for their own use. Oil palm leaves, palm kernel shells, and empty fruit bunches could also be pelletized or briquetted to be made into biocoal and enhance their energy density, to obtain an average calorific value of 18.41 MJ/kg (Kaniapan et al., 2021), which is comparable to lower-rank coals at around 23 MJ/kg (Speight, 1990). Next, third-generation biofuel sources mainly are comprised of algae or diatoms, which involves the cultivation and harvesting of algae cells, extraction of algae lipids, followed by transesterification of the lipids to form fatty acid methyl esters (FAMEs) that are the foundation of biofuels or biodiesels. Diatoms can also produce bioethanol upon extraction of their fermentable sugars. Lastly, fourth-generation biofuels involve advanced biotechnology such that the bio-engineered plants or algae undergo carbon capture and storage, in which the carbon-rich bodies could be used as feedstock for biofuel generation (Saha et al., 2018).

Among the four generations of biofuels, the third is the focus of this study since diatoms are the most promising source for biofuel production. They have the potential to strike a balance between increasing demands for food, bioenergy, and valuable chemicals production by being able to produce various carbon compounds such as carbohydrates or polysaccharides, pigments, lipids, proteins, vitamins, and antioxidants for bio-products, as well as lipids and fermentable sugars for biofuels and bioethanol synthesis as mentioned. Diatoms are one of the most suitable solutions to the topic of the food and bioenergy supply trade-off in the agriculture sector, as they can grow rapidly without the need for soil, excessive land area, excessive freshwater volume, fertilizers, or pesticides, hence they will not be contesting for food resources (Khan, et al., 2018). Owing to their rapid growth, they can be harvested daily and can produce up to 10 times more oils or lipids on the same land area as a typical terrestrial vegetable oil crop such as oil palm, olives, nuts, and others, as diatoms can contain around 30% lipid with respect to their dry biomass weight and even up to 70% lipid in high oil content strains and by optimizing the growing conditions. From another point of view, the growing area of diatoms could be significantly less than that of crops by 49 to 132 times, since diatom cells can double in size in 24 hours or even 13 hours in some strains like *Scenedesmus* sp. Moreover, the chemical and physical properties of the microalgal biofuels are more environmentally friendly compared to conventional fossil fuels since they contain no sulfur with less CO_x, SO_x, and hydrocarbon emissions, while having comparable performance to their conventional counterparts (Khoo et al., 2020).

As mentioned, the procedures for obtaining biofuels from diatoms are mainly comprised of diatom strain isolation to select the most suitable strain for lipid production, diatom cultivation to grow the biomass and produce the lipids, cell harvesting to concentrate the diatom culture and drying for increasing storage stability, pretreatment of diatoms for cell wall and membrane lysis, extraction of diatom lipids, and finally lipid transesterification to transform them into FAMEs. Currently, the scale-up of this biofuel production system remains a challenge in terms of economic viability due to the sheer amount of energy required in the culturing, harvesting, drying, and pretreatment steps, as well as the very high medium and solvent volume requirements for the culturing and extraction steps (Rashid et al., 2014). Hence, any major breakthrough in just one of these steps would bring us one step closer to realizing the commercialization of diatom biofuels.

Herein, this chapter focuses on the integration of diatom pretreatment and lipid extraction into a one-step, simultaneous process that can simplify the extraction process as well as increasing the extraction efficiency of diatom lipids. The proposed pretreatment and extraction steps would be electrolysis and liquid biphasic flotation (LBF), respectively, with novel integration of nanobubbling into the flotation system in the LBF which would normally use microbubbles around the size of 5–20 μm. This proposed technique will be referred to as electrolysis-assisted liquid biphasic nanobubble flotation (ELBNF) from this time forward. This review begins with the characteristics, lipids, and cell wall of the diatom, followed by cell wall pretreatment techniques and electrolysis, as well as a comparison of different microalgal lipid extraction techniques and LBF. The chapter further discusses the characteristics of nanobubbles and their proposed effect, with critical analysis of previous research into electrolysis-assisted LBF (or liquid biphasic electric flotation, LBEF), and finally the factors that govern the performance of this ELBNF system.

7.2 LIPIDS IN DIATOMS

A study conducted to assess the most promising diatom species for biofuel production determined that diatoms species like *Navicula cincta*, *N. gregaria*, and *H. coffeaeformis* have higher overall scores than chlorophytes like *C. vulgaris* in terms of biomass and oil production requirements, triacylglyceride (TAG) content, and biodiesel properties (Martín et al., 2020). Despite that, *C. vulgaris* remains the highlighted species in this study owing to its extensive research and applications and its ease of procurement. Besides, typical *Chlorella* strains still contain 28–32% oil content (dry weight biomass basis) which is significantly higher than vegetable oil crops as previously mentioned (Aguoru and Okibe, 2015). The lipid composition in *C. vulgaris* can be categorized into neutral lipids, which are mono-/di/tri-acylglycerol stored in the cytoplasm and polar lipids like phospholipids and glycolipids which are normally found in cell walls, cellular membranes, and organellar membranes, as illustrated in Figure 7.1.

According to a study carried out on the analysis of *C. vulgaris* lipid content and composition, it has been found that the main fatty acid chains that comprise the phospholipid content in this species are palmitic acid (C16:0) and oleic acid (C18:1) ranging from 18.82–24.58% and 60.98–73.22%, respectively, depending on nitrogen availability (Aguoru and Okibe, 2015). Some diatom strains are also mainly constituted of C16:0 and C18:1 fatty acids. Hence, *C. vulgaris*, albeit containing trace amounts of palmitoleic acid (C16:1), is a suitable diatom strain since monounsaturated fatty acids (MUFAs), especially C16:1 and C18:1 fatty acids, have been proven to be ideal lipids for biofuel production (Ge et al., 2016). The reason behind this statement was that a high composition of these fatty acids will strike a balance between desirable high cetane number (CN), cold flow properties, and oxidative stability. Generally, a higher CN will give better ignition properties and combustion efficiency, which is achieved with increasing chain length and chain saturations

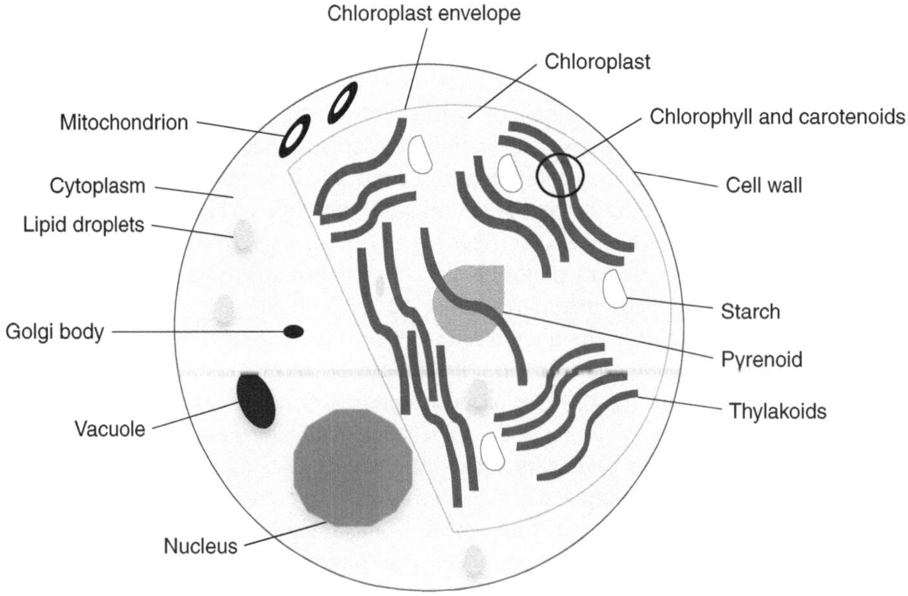

FIGURE 7.1 Schematic structure of *Chlorella vulgaris* with its different organelles (Safi et al., 2014).

TABLE 7.1
Lipid Content and Important Biofuel MUFAs Make-up for Several Diatom Strains

Diatom strain	Lipid content (% dry wt.)	C16:0 FA make-up (%)	C16:1 FA make-up (%)	C18:1 FA make-up (%)	References
Botryococcus braunii	25–75	20–30	–	44–53	(Aguoru and Okibe, 2015; Sadeghin et al., 2018)
Chlorella sorokiniana	21–26	19–30	–	12–21	(Piligaev et al., 2018)
Chlorella sp.	28–32	20–24	12–16	17–23	(Anto, et al., 2019; Aguoru and Okibe, 2015)
Chlorella vulgaris	25	19–25	–	61–73	(Aguoru and Okibe, 2015)
Chlorella vulgaris BDUG 91771	22–27.5	29	12	22	(Mathimani, et al., 2018)
Chlorella vulgaris UV715-EMS25	41	25	–	50	(Sarayloo et al., 2018)
Chlorella vulgaris SAG 211-11b	53	15	–	60	(Sakarika and Kornaros, 2016)
Crypthecodinium cohnii	20–48	23–32	–	7–18	(Aguoru and Okibe, 2015; Liu et al., 2017; Couto et al., 2010)
Isochrysis galbana	22–42	12–19	12–26	–	(Fidalgo et al., 1998)
Micractinium sp.	28–36	21–32	1–3	14–21	(Piligaev et al., 2018)
Nannochloropsis sp.	25–31	24–32	25–28	8–10	(Wahidin, et al., 2013; Hu and Gao, 2003)
Neochloris oleoabundans	35–54	22–24	2–4	27–32	(Aguoru and Okibe, 2015; Sun et al., 2014)
Phaeodactylum tricornutum	20–30	20–25	33–46	2	(Aguoru and Okibe, 2015; Yang et al., 2017)
Tetraselmis suecica	15–23	9	–	11	(Aguoru and Okibe, 2015; Almutairi and Toulibah, 2017)

in the FAMEs; cold-flow properties refer to the temperature that biofuels start to crystallize and precipitate to not be able to pass through an oil filter, hence having higher degrees of unsaturation will improve the cold-flow properties; oxidative stability, on the other hand, accentuates the disadvantages of increasing double bonds in FAMEs as they are more prone to autoxidation, which greatly reduces the biofuel quality (Stansell, et al., 2011). Table 7.1 shows the lipid content, C16:0, C16:1, and C18:1 fatty acid make-up in total fatty acids for some microalgal strains. Furthermore, the linolenic acid found in *C. vulgaris* is an effective antibacterial agent for *Staphylococcus aureus* and *Salmonella typhi*, which are common causes of skin infections and typhoid, respectively, in humans (Tiong et al., 2020).

After the lipids or fatty acids have been extracted from the green diatoms, the last step required is the conversion of such fatty acids into the biofuels that are desired. The main process that takes place is the transesterification reaction between the fatty acids and an alcohol (commonly methanol or ethanol) to produce fatty acid methyl ester biodiesel or biofuel, with the help of a catalyst (van Gerpen et al., 2004). Hence, numerous researches have compared and studied the ideal catalysts for the transesterification process, mainly to maximize the biodiesel conversion and yield.

Conventionally, these catalysts consist of homogeneous or heterogeneous acid and base catalysts, in which some of the examples are sulfuric, hydrochloric, phosphoric, or organic sulfonic acids and boron trifluoride, BF_3, for acids; and sodium or potassium hydroxide for bases (Cheirsilp, et al., 2020). Recently, there has been wider acceptance of an enzyme called lipase as an environmentally friendly and economically feasible alternative as it delivers higher purity biodiesel, with the absence of acidic/alkaline wastewater and saponification side reaction, as well as efficient separation and recovery of the enzyme (Hossain et al., 2020).

7.3 DIATOM CELL WALL DISRUPTION TECHNIQUES

One of the challenges surrounding diatoms as an alternative feedstock for biofuel production is the pretreatment process which involves the degradation of the diatom cell wall to liberate the intracellular lipids for efficient subsequent extraction. It is also crucial to determine the optimum cell wall disruption techniques based on the characterization of cell walls of different diatom strains. For example, species containing the highly resistant trilaminar algaenan matrix, as previously mentioned, require more vigorous techniques. A study that carried out a comparison of different pretreatment methods for lipid recovery from *Scenedesmus* sp. was conducted and found that microwave and bead-milling techniques lead to better lipid recovery improvements by 5.5 and 4.5 times, respectively, while autoclaving and sonication methods resulted in improvements by factors of 2.5 and 3, respectively, when compared to untreated biomass (Lee et al., 2010). The use of surfactants such as 3-DAPS, MTAB, and SDS also improved the FAME recovery of *Scenedesmus* sp. up to 72%, which was 10- to 16-fold higher than the recoveries of the untreated biomass, although the study also took into account the concentration of surfactants and stage of diatom growth phase (Lai, Francesco, et al., 2016).

C. vulgaris, being one of the most researched strains for biofuel production, also has been used with numerous pretreatment methods. The use of cationic surfactant hexadecyltrimethylammonium bromide (CTAB) efficiently disrupted the cell wall structures to achieve nearly 90% lipid recovery, so long as the concentration of CTAB exceeded the critical micelle concentration to form surfactant-complex micelles with the cell wall for effective cell disruption (Lai, Zhou, et al., 2016). Moreover, the application of surfactants offers another advantage of additional recovery since flotation techniques will recover both the surfactant and the diatom cells altogether, thereby decreasing the environmental discharge of the surfactant while increasing its concentration by recycling it back into subsequent treatment cycles. The lower energy input requirement also further solidifies surfactant-pretreated diatoms as an effective strategy to improve diatom biofuel economics (Coward, et al., 2014). However, disposal of biomass systems with high surfactant concentrations can be harmful to aquatic life, and this technique generally requires longer durations per cycle. Nevertheless, there are six categories (mechanical, chemical, thermal, electromagnetic, biological, and electrical) of cell wall disruption methods as stated in Kapoore et al. (2018). Both conventional and recently advanced pretreatment techniques for *C. vulgaris* have varying lipid recoveries, advantages, and disadvantages, as summarized in Table 7.2, although their performances cannot be directly compared with each other due to the different lipid extraction techniques adopted and different *C. vulgaris* strains across different studies.

7.4 DIATOM LIPID EXTRACTION TECHNIQUES

After the diatom pretreatment or the cell wall disruption step, the next challenge is to extract the intracellular lipids from the biomass and accumulate the lipids in such a way that they are easily collected. To date, only solvents are used during this step owing to the "like-dissolves-like" mechanism, where, for example, non-polar lipids or polar solvents will dissolve polar lipids (Halim, et al., 2012). In solvent recovery systems, a mixture of polar and non-polar solvents with certain

TABLE 7.2
A Comparison of Some of the Studied *Chlorella Vulgaris* (Otherwise Stated) Cell Wall Disruption Pretreatment Techniques

Pretreatment techniques (category)	Operating conditions	Lipid recovery (% biomass)	Advantages	Disadvantages	References
Ultrasonication (mechanical)	24 kHz, 400 W, 60 s	35.5	Minimal maintenance cost, rapid, no hazardous substances	High energy input and thus operational cost	(Hattab and Ghaly, 2015; Rokicka et al., 2020)
Microwave (electromagnetic)	2.45 GHz, 400 W, 60 s	41.3	Relatively low energy input, decreased solvent usage, rapid heating	High maintenance cost due to heat generated	
Grinding (mechanical)	With quartz sand, dehydrated for 8 h	29	Quick at laboratory scale	Potential degradation of intracellular compounds and time consuming	(Hattab and Ghaly ,2015; Zheng et al., 2011)
Bead milling (mechanical)	1500 rpm, 20 min	10	Effective cell wall disruption and quick extraction	Requires additional step for microbeads separation, hence high energy input and costs in total	
Cellulose (biological)	55°C, 10 h	24	High selectivity (effective towards cell walls but mild towards cells)	Time-consuming treatment, high enzyme costs, inability to recycle enzymes	
Lysozyme (biological)	55°C, 10 h	22			
Pulsed electric field (electrical)	20 kV/cm, 130 Hz, 6 μs pulse width	12.8 (from *Chlorella pyrenoidosa*)	High selectivity, relatively low energy input	Advancements in this field still in infancy	(Han et al., 2019)
Steam explosion (thermal)	Target 157.5°C and approx. 130 psi at 300 s	19.9	Low maintenance costs and energy input, relatively suitable for commercial application	Efficiency widely varies across different strains, decreasing efficiency with increased biomass concentration	(Aguirre and Bassi, 2014; Hattab and Ghaly, 2015)
Silver nanoparticles (AgNP) (chemical)	150 μg AgNP/g biomass, 100 rpm, 40 min	18–21	Environmentally friendly	Additional separation process required for nanoparticles, advancements in infancy	(Abdul Razack, et al., 2016; Seo et al., 2016)
[Emim]DEP IL (chemical)	5 wt.% C. vulgaris loading in IL, 120°C, 2 h	25	Environmentally friendly, highly stable, ability to tune anions and cations	High cost, toxic reagents in the synthesis of ILs	(Choi et al., 2014)

ratios is used in the lipid recovery process. The solvent molecules will penetrate the pretreated porous cell walls, where the non-polar solvent will form lipid–solvent complexes with the neutral lipid droplets in the cytoplasm, then the complexes will diffuse out from the diatom cell walls due to the concentration gradient of such complexes between the intracellular matrix and the bulk solvent. Nonetheless, there will be certain fractions of neutral lipids that are attached to the polar lipids found in the cell membranes via hydrogen bonding, the function of the polar solvent is to liberate the neutral lipids by forming its own hydrogen bonds with the polar solvent to separate the neutral–polar lipid complex. Hence, this phenomenon results in the neutral and polar lipids leaving the diatoms into the bulk solvent. Finally, the recovery of lipids is achieved via evaporation or distillation of the solvents. Hence, the complications in solvent extractions are to select solvents, or combination of solvents, that have high selectivity toward the lipids, that are insoluble in water when wet biomass is used, with suitable properties to penetrate the microalgal cell wall, as well as being volatile and non-hazardous to improve the quality during the lipid recovery process (Karim et al., 2020). The solvents also have to be inexpensive as lipid recovery operations generally comprise 30–40% of the total cost in the entire diatom biofuel production process due to the large volume of solvents used in the extraction process (Kumar et al., 2018).

7.5 CONVENTIONAL SOLVENT EXTRACTION TECHNIQUES

Soxhlet extraction is one of the conventional techniques where the hot solvent comes into contact with the target lipid in the thimble to undergo extraction, and the lipid-carrying solvent will siphon down to the flask. Generally, Soxhlet extraction is a simple technique that ensures complete or near-to-complete lipid recovery from the diatoms, while providing itself with a built-in solvent recovery system. However, temperature control is a delicate factor in this technique to prevent lipid oxidation which can happen above 120°C (Chen et al., 2020), while high energy demand for heating and water consumption for cooling make it difficult for sustainable scale-up for industrial application. The Folch method has also been extensively applied in lipid extraction from biological sources and diatoms are no exception. Although the Folch method represents a more efficient and rapid technique for microalgal lipid recovery, as well as the benefit of eliminating high-temperature conditions, unlike Soxhlet extraction, the extensive use and exposure to chloroform will result in acute toxicity toward living organisms as well as the environment.

Another conventional method that has been extensively applied in various studies is the Bligh and Dyer method, which is a modification of the Folch method. This method homogenizes an initial chloroform/methanol/water ratio of 1:2:0.8 (taking into account the water content in the biomass), followed by the addition of chloroform and water to achieve a final ratio of 2:2:1.8. The several advantages of the Bligh and Dyer method over the Folch method include time consumption, where the former can be completed within an hour while the latter takes about 24 hours. In terms of solvent consumption, the Bligh and Dyer method only uses a total of five parts solvent (chloroform + methanol) to one part biomass, while the Folch method uses four times more solvent than the former with the same biomass volume (Folch, Lees, and Stanley, 1957; Bligh and Dyer, 1959). Moreover, the Bligh and Dyer method allows simultaneous lipid extraction and separation, where the lipid can be found at the chloroform upper phase (Khoo et al., 2020). Araujo et al. compared the lipid extraction performance of several methods and discovered that the Bligh and Dyer method yielded approximately three times more oil than the Folch method and 29 times more than Soxhlet method, at 52.5%, 16.1%, and 1.8%, respectively, from *C. vulgaris*, while all the methods were assisted by ultrasound (Araujo et al., 2013). Nevertheless, the conventional Bligh and Dyer method still presents toxicity issues due to the use of chloroform, hence many researchers have formulated a method by substituting the solvents while remaining with the same working principles. An example was by de Jesus et al., where they utilized 2-methyltetrahydrofuran:isoamyl alcohol (2:1 v/v) and cyclopentyl methyl ether:methanol (1:1.7) to obtain 95.7 and 89.4 mg lipid/g biomass, respectively.

However, the high costs of these solvents were a hindrance for industrial application as hexane prices remain cheaper, hence the hexane:isopropanol (3:2) system was the most cost-efficient Bligh and Dyer alternative (de Jesus et al., 2019).

7.6 DIATOM LIPID EXTRACTION TECHNIQUES

Due to the disadvantages of the techniques that involve environmentally and bodily harmful solvents as mentioned in the previous section, various green solvents have been proposed and experimented with to attempt to reduce the environmental impact upon solvent discharge and at the same time provide comparable or even better microalgal lipid extraction performance. One of these methods is extraction using supercritical fluids which are "solvents" or compounds that exist beyond the critical temperature and pressure, and at the same time, lower than the required pressure to be compressed into a solid. One way to visually indicate the existence of such compounds is the non-distinct liquid and gas phase interface as well as the increased turbidity of the compound. Supercritical fluids combine the benefits of both liquid and gaseous phases in such a way that they provide the solubility performance of liquids around 0.1–1 mL^{-1}, while at the same time overcoming the mass transfer limitations of liquids due to their viscosity and diffusion coefficients properties that lie between liquids and gases (Khoo et al., 2020). One of the common fluids used in supercritical fluid extraction is supercritical carbon dioxide (SC-CO_2), which possesses a critical temperature and pressure of 31.1°C and 7.28 MPa, respectively, hence high temperatures are not required for the extraction process which is highly beneficial for lipid extractions since they could be oxidized in higher temperatures as previously mentioned. Other rationales behind using CO_2 as the supercritical solvent include low cost and the absence of flammability and toxicity. Polar co-solvents are often used to accompany the SC-CO_2 as it is a non-polar compound, hence its lipid extraction efficiency is insufficient by itself (Khoo et al., 2020). Obeid et al. combined ethanol (10% v/v) into SC-CO_2 and managed to extract 97%, 25%, and 36% of total neutral lipids, glycolipids, and phospholipids, respectively, from *C. vulgaris* (Obeid et al., 2018). Also with *C. vulgaris*, Yang et al. demonstrated the high lipid yield from CO_2-expanded-methanol (CXM) of 84.7% total lipids within just 30 minutes of continuous extraction, and the extracted yield per ml of CXM (1.49%/mL) was better than for pressurized methanol, methanol, and CO_2-expanded-ethanol of 1.21, 1.06, and 0.87%/mL, respectively (Yang, et al., 2015). However, several drawbacks to SC-CO_2 include the low yield toward polar lipids as seen in Obeid et al.'s work as well as the high equipment costs for such pressurization, which will be difficult for industrial scale-up.

Another recently advanced green solvent for microalgal lipid extraction is ILs which are also eco-friendly and non-flammable, while being able to maintain at a liquid state in a wide range of temperatures from 0 to 140°C given that they are non-aqueous organic salt solutions (Vasistha et al., 2021). They are generally comprised of small, inorganic anions and large, organic cations; this combination of ions enables them to be fine-tuned depending on the required operation; other benefits of using ILs as a green solvent are negligible volatility and great recyclability (Khoo et al., 2020). Choi et al. studied the effect on imidazolium-based ILs on lipid extraction from *C. vulgaris*, and concluded that the lipid extraction yields of [Emim]Oac, [Emim]DEP, [Emim]BF_4, and [Emim]Cl of 22.4, 25.0, 22.0, and 23.5% wt. cell were higher than that of conventional hexane-methanol solvent of 18.5% wt. cell (Choi et al., 2014). A study using [Bmim][$MeSO_4$] on ultrasound-pretreated *C. vulgaris* also displayed a higher lipid yield of 47 mg/g cell, which was 2.2- and 1.6-fold greater than that of the Soxhlet and Bligh and Dyer methods, respectively (Kim et al., 2013). Despite the environmental friendliness of ILs, some of their synthesis processes were not as they involved toxic and volatile reagents, and another disadvantage is the very high cost of synthesizing such solvents. The lack of understanding in the mechanism of lipid extraction in ILs adds another element of uncertainty for potential commercial application (Khoo et al., 2020).

7.7 SOLVENT SUBLATION SEPARATION

This section provides some of the context that the to-be-discussed diatom biocompound extraction technique, LBF, is built upon, which is solvent sublation. It is essentially a separation technique involving the generation of bubbles in the liquid media where the adsorptive gas bubble surfaces separate the constituents of the solutions or suspensions from the bulk liquid (Sobianowska, et al., 2009).

The main components that make up the solvent sublation system are a column equipped with any kind of gas generation system, whose purpose is for generating numerous tiny gas bubbles which also determines the total interfacial contact area between the gas bubbles and the aqueous solution. There are three main methods to generate the bubbles for sublation, where the first and most common is the dispersed-gas generation system. A gas cylinder is connected to the column via a gas humidifier and gas sparger, in which its pore characteristics determine the size of the gas bubbles (Sobianowska, et al., 2009). The second method is via dissolved gas generation, where the dissolved air in the pressurized water (4–5 bar) in a tank is charged to the flotation column via a solenoid valve (Zouboulis, et al., 1990). The third method is via electrolysis, where fine bubbles of hydrogen and oxygen gas are liberated on the cathode and anode, respectively.

The operational parameters that directly govern the separation efficiency of the solvent sublation technique include temperature and gas flow rate. In higher temperature conditions, higher concentrations of surface-active components would adhere to the bubble surface since adsorption from gas bubbles is an exothermic process. Meanwhile, a higher gas flow rate would also increase the separation efficiency at lower rates, as higher gas flow rates would increase the mean bubble radius and the higher rising velocities from the larger and faster bubbles could result in lower residence times, hence decreasing its effectiveness (Lu and Zhu, 2007). Other parameters of the bulk liquid medium can also affect the solvent sublation, such as surfactant concentration and ionic strength of the solution. As surfactants tend to decrease the surface tension of the solution, which decreases the size of the generated bubbles, the larger interfacial area of the bubbles would significantly offset the decreased mass transfer rate from the adsorption of surfactants on the bubbles, hence increasing its adsorption efficiency. However, excess amounts of surfactants would compete for adsorption sites on the bubbles, as well as trapping the hydrophobic molecules (molecules of interest) from the self-aggregation of surfactant molecules, decreasing the efficiency of the sublation process (Clark, et al., 1989). The effects of ionic strength in the liquid medium on the sublation process vary case by case, but usually an increase in ionic strength would decrease the efficiency from the competition between the non-surface-active components and the ions on the bubble surface. Nonetheless, high concentrations of inorganic salts would decrease the aqueous solubility of hydrophobic compounds like lipids, since the salt ions would saturate on the water molecules, leaving little to no free water molecules to interact with the hydrophobic compound, which is also known as the "salting out" effect. This phenomenon increases the sublation efficiency and brings us to the next section where the "salting out" effect is used in conjunction with the solvent sublation technique.

7.8 LIQUID BIPHASIC FLOTATION AS AN EMERGING EXTRACTION TECHNIQUE

In order to combat the issues of toxicity or high costs, an emerging technique in the form of LBF has slowly caught the attention of some scholars. As the name implies, this technique combines an aqueous two-phase system with the aforementioned solvent sublation, in other words, a flotation or bubbling system. Since C1 to C4 alcohols (methanol to butanol) are miscible in water, salts whose limiting concentrations are exceeded will be required to generate a two-phase system since salts have greater affinity toward water, hence the alcohol will not form hydrogen bonds with the water since it is already saturated with salt ions, and stay afloat at the top phase instead of dissolving in the water; this phenomenon is otherwise known as the "salting out" effect mentioned in the previous section. The mechanism of the LBF system is that the surface-active biocomponents will attach to the surface of the air bubbles that are rising from the very bottom of the salt solution in the bottom phase. The bubbles will disassociate once they approach the hydrophilic alcohol phase, leaving and accumulating the adsorbed compounds on the top or alcohol phase (Phong et al., 2017).

To date, there have been no works to study the lipid recovery from wet diatom biomass using LBF systems. However, there have been numerous studies regarding protein recovery from diatom biomass. Phong et al. extracted proteins using an alcohol/salt LBF system from ultrasonication-pretreated wet *Chlorella sorokiniana*, and they managed to recover 88.9% of the total proteins in the third recycling run, the optimized biphasic solvent was 60% (v/v) 2-propanol/$(NH_4)_2SO_4$. However, the extractive performance of the alcohol started to sharply deteriorate starting from the fourth recycling run, thereby indicating the benefits of solvent recycling in this LBF technique and the need to verify the life cycle of the solvent to fully utilize it before discarding (Phong et al., 2017). Another study involving protein extraction from osmotic shock pretreated *C. vulgaris* FSP-E used undiluted 1-propanol/$MgSO_4$ as the solvent and achieved 93.0% protein recovery and 64.9% separation efficiency on the top phase of the protein concentration. The authors also performed a scale-up of their optimized experiment which involved five times the initial volume of the solvent used, at 1 L of alcohol and salt solution respectively, and still managed 91.2% protein recovery and higher separation efficiency than the previous run at 82.5%, which demonstrated the feasibility and potential for further scale-up (Krishna Koyande et al., 2020). Moreover, different approaches were also considered in terms of the solvent in the LBF system, whereby sugaring-out was studied instead of the usual salting-out. An experiment using ultrasonication with acetonitrile/glucose LBF system on *C. vulgaris* FSP-E yielded 93.3% of total proteins and 86.4% separation efficiency, a large-scale study by five times to 1.5 L total solvent also obtained comparable results with their small-scale counterpart by having 92.4% and 85.3% yield and separation efficiency, respectively, which again proved the scale-up potential in LBF systems (Sankaran, Manickam, et al., 2018).

Apart from proteins, this technique is also effective in extracting high-value biocompounds for the food and pharmaceutical industries, such as a carotenoid and natural pigment known as astaxanthin which is an antioxidant that gives sea creatures their pinkish or reddish hue. Conventional astaxanthin extraction from *Haematococcus pluvialis* species often involved concentration hydrochloric acid, dimethyl sulfoxide, dichloromethane, and other solvents that are toxic for human ingestion, and emerging technologies like supramolecular solvent extraction, as well as the aforementioned SC-CO_2 extraction and IL extraction methods often incurred high equipment and operating costs. On the other hand, the 2-propanol/$(NH_4)_2SO_4$ LBF system obtained very high recovery and separation efficiencies of 95.1% and 99.8%, respectively, while using recyclable, environmentally friendly, and food-grade solvents. However, the scale-up from 30 mL to 300 mL solvent volume slightly compromised the recovery yield to 78.4% while maintaining similar separation efficiency (Khoo et al., 2019). Another natural pigment in the form of C-phycocyanin (CPC) was also subjected to LBF extraction using polyethylene glycol (PEG) 4000/K_2HPO_4, which resulted in a maximum CPC recovery of 90.4%, hence providing a viable alternative to the conventionally long and expensive CPC purification methods that would otherwise have difficulties in scaling-up (Chew et al., 2019).

Several advantages of implementing LBF extraction include environmental friendliness from the absence of very acidic or basic solvents, hydrocarbon-based solvents, and the feasibility of multiple recycle processes or a high degree of alcohol recovery to reduce solvent waste, high product yield and separation efficiency from the previous studies mentioned, mild conditions toward the biomass and reduced product degradation, and simple operation and low costs of solvents and equipment. However, the research into LBF techniques, particularly on diatom product extraction, is still in its infancy in terms of commercialization and has the potential to be further optimized to discover more ground-breaking discoveries and results, as well as the lack of studies into LBF extraction on microalgal lipids which will be addressed in this work. Table 7.3 summarizes some of the works carried out on microalgal product recovery by LBF systems.

TABLE 7.3
Summary of Optimized LBF Extraction Specifications on Various Diatom

Diatom strain	Type and concentration of biphasic solvents [volume ratio]	Air flotation properties, FR–rate, FT–time	Biomass loading	Product, recovery	Separation efficiency	References
Chlorella sorokiniana	60% (v/v) 2-propanol/250 g/L $(NH_4)_2SO_4$ [1:1]	FR: 4 mm^3/min FT: 10 min	20 g/L	Protein, 88.86%	97.44%	(Phong et al., 2017)
Chlorella sorokiniana CY1	100% acetonitrile/ 200 g/L glucose [1.25:1]	FR: 10 mL/min FT: 5 min	0.25 g/300 mL	Protein, 81%	73%	(Chia et al., 2019)
Chlorella vulgaris FSP-E	100% 1-propanol/ 150 g/L $MgSO_4$ [9:1]	FR: 75 mL/min FT: 5 min	0.25% w/w	Protein, 92.98%	64.91%	(Krishna Koyande et al., 2020)
Chlorella vulgaris FSP-E	100% acetonitrile/ 200 g/L glucose [1:1]	FR: 100 mL/min FT: 5 min (5 s on, 10 s off pulse mode)	0.6% w/w	Protein, 93.33%	86.38%	(Sankaran, Manickam, et al., 2018)
Chlorella vulgaris FSP-E	100% acetonitrile / 300 g/L glucose [1:1]	FR: 150 mL/min FT: 15 min	0.05 g/300 mL	Protein, 69.67%	74.00%	(Koyande et al., 2019)
Haematococcus pluvialis	100% 2-propanol/ 350 g/L $(NH_4)_2SO_4$ [1:1]	FR: (N/A) FT: 15 min	10 mg/30 mL	Astaxanthin, 95.11%	99.84%	(Khoo et al., 2019)
Spirulina platensis	250 g/L PEG 4000/250 g/L KH_2PO_4 [0.85:1]	FR: (N/A) FT: 7 min	0.625% w/w	C-phycocyanin, 90.4%		(Chew et al., 2019)
Spirulina platensis	250 g/L PEG 4000/250 g/L KH_2PO_4 [1:1]	FR: 100 mL/min FT: 7 min	0.45 g/300 mL	C-phycocyanin, 95.10 %		(Chia et al., 2020)
Burkholderia cepacia	80% 1-propanol/ 300 g/L $(NH_4)_2SO_4$ [1:10]	FR: 200 mL/min FT: 30 min	40% w/w	Lipase, 95.73%	92.29%	(Sankaran, Show, Lee, et al., 2018)

7.9 INCORPORATION OF ELECTROLYSIS PRETREATMENT WITH LBF EXTRACTION

In light of the electrolysis pretreatment on the diatom cell wall and diatom product extraction using the LBF technique, there have been several recent studies that involved both the pretreatment and extraction technique, which is also known as LBEF, to enhance the extraction efficiency. Sankaran et al. extracted protein from wet *C. sorokiniana* CY-1 using LBEF with 60% (v/v) propanol/250 g/L K_2HPO_4 with a 1:1 volume ratio, 0.1 g diatom loading in 200 mL solvent, air flotation rate and time of 150 mL/min and 10 min, respectively, electrolysis of 20 V, and achieved 23.41% protein recovery from total protein content and 173.09% separation efficiency. A comparison between LBF with and without electrolysis pretreatment was also carried out and the electrolysis provided an extra 2.7% of protein recovery (Sankaran, Show, Cheng, et al., 2018). In another study, protein was also extracted from *C. vulgaris* FSP-E via a sugaring-out-assisted LBEF method using 0.05 g biomass loading in 300 mL solvent, 15 V DC supply in electrolysis, 150 mL/min flotation rate with 15 min flotation time, 100% acetonitrile/300 g/L glucose solution with a volume ratio of 1:1, resulting in protein recovery and separation efficiency of 69.67% and 74.00%, respectively, which were both higher than that of the LBF system at 48.78% and 61.58%, respectively. A five-fold scale-up of the optimized system was also tested which resulted in 73.29% protein recovery and 52.87% separation efficiency. The high protein yield and favorable results in scale-up tests hinted at its potential for commercialization (Koyande et al., 2019).

Besides, LBEF has also been applied to other biomass types, such as *Hylocereus polyrhizus*, commonly known as red dragon fruit, for the extraction of betacyanins for its health-promoting and antioxidant features. The optimized operating conditions of 1 g fresh extract in 30 mL solvent, 100% ethanol/200 g/L K_2HPO_4 with volume ratio 1:2, 20–30 mL/min flotation rate, 15 min flotation time, and 3 V electrolysis voltage obtained an alcohol-rich top-phase betacyanin concentration of 99.01% and separation efficiency of 98.38% from the peel (Leong et al., 2019). The studies proved that electrolysis pretreatment is an excellent pairing with the LBF extraction technique due to the synergistic effect of increasing the biomass membrane permeability by the electrical treatment to improve the extraction efficiency of the LBF system, particularly by facilitating the degree of penetration of the bubbles into the intracellular matrix to liberate the desired biocontents, moreover neither the pretreatment nor the extraction methods involve high temperatures which can prevent thermal degradation or oxidation of more sensitive compounds.

7.10 CONCLUSIONS AND PERSPECTIVES

Diatom lipids are among of the most promising biofuel feedstocks due to their potential productivity and sustainability benefits compared to that from terrestrial crops, as well as containing the ideal fatty acid composition for superior biofuel quality. The proposed ELBNF method of extracting lipids from *C. vulgaris* for biofuel production is expected to be very effective, given the past success in LBF and LBEF, and the favorable conditions for nanobubbles in the biphasic solution of salt and alcohol. Several factors that govern the lipid recovery yield of the ELBNF method are also discussed, such as electrode type, electrolysis voltage, position of electrodes, type and concentration of alcohol and salt solution, volume ratio of alcohol to salt solution, air flotation rate and time, biomass loading, and recycling of phase components. The use of non-toxic alcohols and salts as well as the ability to recycle the alcohol could serve as attractions for potential commercialization for diatom lipid extraction.

However, one concerning factor that may deter this method from widespread commercialization is the high electricity consumption from electrolysis in scaled-up models. Therefore, life cycle assessment could be done to identify the feasibility and energy balance of the large-scale process, coupled with multidisciplinary research to encourage optimization in this technique from different points of view. One of these would be via nanotechnology with the development of the electrode materials to allow greater current density and create electricity savings, such as the study by Li et al. that developed a highly graphitic mesoporous electrode with an MgO-template, which achieved cellulose electrolysis rates comparable to the more expensive Pt/C electrode (Li et al., 2020). Renewable energy could also be integrated into the proposed ELBNF applications so that the goal of energy sustainability will be more aligned, such as sunlight-driven electrolysis. Lastly, the existing utilization of electrolysis as a diatom flocculation method could provide opportunities for cost optimization as the need for expensive standalone flocculation processes could potentially be eliminated.

REFERENCES

Abdul Razack, S., S. Duraiarasan, and V. Mani. (2016). Biosynthesis of silver nanoparticle and its application in cell wall disruption to release carbohydrate and lipid from *C. vulgaris* for biofuel production. *Biotechnology Reports (Amsterdam)* 11: 70–76.

Aguirre, A. M., and A. Bassi. (2014). Investigation of high pressure steaming (HPS) as a thermal treatment for lipid extraction from *Chlorella vulgaris*. *Bioresources Technology* 164: 136–142.

Aguoru, C.U., and P.O. Okibe. (2015). Content and composition of lipid production by *Chlorella vulgaris* for biodiesel production. *Advances in Life Science and Technology* 36: 96–100.

Almutairi, Adel Waslallah, and Hussein Elsayed Toulibah. (2017). Effect of salinity and pH on fatty acid profile of the green algae *Tetraselmis suecica*. *Journal of Petroleum & Environmental Biotechnology* 8(3).

Anto, Susaimanickam, Arivalagan Pugazhendhi, and Thangavel Mathimani. (2019). Lipid enhancement through nutrient starvation in *Chlorella* sp. and its fatty acid profiling for appropriate bioenergy feedstock. *Biocatalysis and Agricultural Biotechnology* 20: 101179.

Araujo, Glacio S., Leonardo J.B.L. Matos, Jader O. Fernandes, Samuel J.M. Cartaxo, Luciana R.B. Gonçalves, Fabiano A.N. Fernandes, and Wladimir R.L. Farias. (2013). Extraction of lipids from microalgae by ultrasound application: prospection of the optimal extraction method. *Ultrasonics Sonochemistry* 20: 95–98.

Bligh, E. Graham, and W. Justin Dyer. (1959). A rapid method of total lipid extraction and purification. *Canadian Journal of Biochemistry and Physiology* 37: 911–917.

Cheirsilp, Benjamas, Sirasit Srinuanpan, and Yohanis Irenius Mandik. (2020). Efficient harvesting of microalgal biomass and direct conversion of microalgal lipids into biodiesel. In: *Microalgae Cultivation for Biofuels Production*, Abu Yousuf (ed.) (pp. 83–96). Academic Press.

Chen, Wei, Yongmei Liu, Lirong Song, Milton Sommerfeld, and Qiang Hu. (2020). Automated accelerated solvent extraction method for total lipid analysis of microalgae. *Algal Research* 51: 102080.

Chew, K. W., S. R. Chia, R. Krishnamoorthy, Y. Tao, D. T. Chu, and P. L. Show. (2019). Liquid biphasic flotation for the purification of C-phycocyanin from *Spirulina platensis* microalga. *Bioresources Technology* 288: 121519.

Chia, Shir Reen, Kit Wayne Chew, Hui Yi Leong, Sivakumar Manickam, Pau Loke Show, and The Hong Phong Nguyen. (2020). Sonoprocessing-assisted solvent extraction for the recovery of pigment-protein complex from *Spirulina platensis*. *Chemical Engineering Journal* 398: 125613.

Chia, Shir Reen, Kit Wayne Chew, Pau Loke Show, Manickam Sivakumar, Tau Chuan Ling, and Yang Tao. (2019). Isolation of protein from *Chlorella sorokiniana* CY1 using liquid biphasic flotation assisted with sonication through sugaring-out effect. *Journal of Oceanology and Limnology* 37: 898–908.

Choi, Sun-A, You-Kwan Oh, Min-Ji Jeong, Seung Wook Kim, Jin-Suk Lee, and Ji-Yeon Park. (2014). Effects of ionic liquid mixtures on lipid extraction from *Chlorella vulgaris*. *Renewable Energy* 65: 169–174.

Clark, BJ, William Wesley Eckenfelder, and John Med Morriss. (1989). *Industrial Water Pollution Control*. McGraw-Hill.

Couto, Ricardo Miguel, Pedro Calado Simões, Alberto Reis, Teresa Lopes Da Silva, Vítor Hugo Martins, and Yolanda Sánchez-Vicente. (2010). Supercritical fluid extraction of lipids from the heterotrophic microalga *Crypthecodinium cohnii*. *Engineering in Life Sciences* 10(2): 158–164.

Coward, Thea, Jonathan G.M. Lee, and Gary S. Caldwell. (2014). Harvesting microalgae by CTAB-aided foam flotation increases lipid recovery and improves fatty acid methyl ester characteristics. *Biomass and Bioenergy* 67: 354–362.

Day, Creina, and Garth Day. (2017). Climate change, fossil fuel prices and depletion: The rationale for a falling export tax. *Economic Modelling* 63: 153–160.

de Jesus, Sérgio S., Gabriela F. Ferreira, Larissa S. Moreira, Maria Regina Wolf Maciel, and Rubens Maciel Filho. (2019). Comparison of several methods for effective lipid extraction from wet microalgae using green solvents. *Renewable Energy* 143: 130–141.

Fidalgo, J.P., A. Cid, E. Torres, A. Sukenik, and C. Herrero. (1998). Effects of nitrogen source and growth phase on proximate biochemical composition, lipid classes and fatty acid profile of the marine microalga *Isochrysis galbana*. *Aquaculture* 166: 105–116.

Folch, Jordi, Mark Lees, and G.H. Sloane Stanley. (1957). A simple method for the isolation and purification of total lipides from animal tissues. *Journal of Biological Chemistry*, 226: 497–509.

Ge, Shijian, Pascale Champagne, William C. Plaxton, Gustavo B. Leite, and Francesca Marazzi. (2016). Microalgal cultivation with waste streams and metabolic constraints to triacylglycerides accumulation for biofuel production. *Biofuels, Bioproducts and Biorefining* 11: 325–343.

Halim, R., M. K. Danquah, and P. A. Webley. (2012). Extraction of oil from microalgae for biodiesel production: A review. *Biotechnology Advances* 30: 709–732.

Han, Song-Fang, Wenbiao Jin, Qian Yang, Abd El-Fatah Abomohra, Xu Zhou, Renjie Tu, Chuan Chen, Guo-Jun Xie, and Qilin Wang. (2019). Application of pulse electric field pretreatment for enhancing lipid extraction from *Chlorella pyrenoidosa* grown in wastewater. *Renewable Energy* 133: 233–239.

Hattab, M. Al, and A. Ghaly. (2015). Microalgae oil extraction pre-treatment methods: Critical review and comparative analysis. *Journal of Fundamentals of Renewable Energy and Applications* 5: 1–26.

Hossain, S. M. Zakir, Shaikh A. Razzak, Abdulla F. Al-Shater, Muhammad Moniruzzaman, and Mohammad M. Hossain. (2020). Recent advances in enzymatic conversion of microalgal lipids into biodiesel. *Energy & Fuels* 34: 6735–6750.

Hsu, Chang Samuel, and Paul R. Robinson. (2017). *Springer Handbook of Petroleum Technology*. Springer.

Hu, Hanhua, and Kunshan Gao. (2003). Optimization of growth and fatty acid composition of a unicellular marine picoplankton, *Nannochloropsis sp.*, with enriched carbon sources. *Biotechnology Letters* 25: 421–425.

Kaniapan, Sivabalan, Suhaimi Hassan, Hamdan Ya, Kartikeyan Patma Nesan, and Mohammad Azeem. (2021). The utilisation of palm oil and oil palm residues and the related challenges as a sustainable alternative in biofuel, bioenergy, and transportation sector: A review. *Sustainability* 13: 3110.

Kapoore, R. V., T. O. Butler, J. Pandhal, and S. Vaidyanathan. (2018). Microwave-assisted extraction for microalgae: From biofuels to biorefinery. *Biology (Basel)* 7(1): 18.

Karim, Ahasanul, M. Amirul Islam, Zaied Bin Khalid, Che Ku Mohammad Faizal, Md Maksudur Rahman Khan, and Abu Yousuf. (2020). Microalgal cell disruption and lipid extraction techniques for potential biofuel production. In: *Microalgae Cultivation for Biofuels Production*. Elsevier.

Khan, M. I., J. H. Shin, and J. D. Kim. (2018). The promising future of microalgae: current status, challenges, and optimization of a sustainable and renewable industry for biofuels, feed, and other products. *Microbial Cell Factories* 17: 36.

Khoo, K. S., K. W. Chew, C. W. Ooi, H. C. Ong, T. C. Ling, and P. L. Show. (2019). Extraction of natural astaxanthin from *Haematococcus pluvialis* using liquid biphasic flotation system. *Bioresources Technology* 290: 121794.

Khoo, K. S., K. W. Chew, G. Y. Yew, W. H. Leong, Y. H. Chai, P. L. Show, and W. H. Chen. (2020). Recent advances in downstream processing of microalgae lipid recovery for biofuel production. *Bioresources Technology* 304: 122996.

Kim, Young-Hoo, Saerom Park, Min Hoo Kim, Yong-Keun Choi, Yung-Hun Yang, Hyung Joo Kim, Hyungsup Kim, Han-Soo Kim, Kyung-Guen Song, and Sang Hyun Lee. (2013). Ultrasound-assisted extraction of lipids from *Chlorella vulgaris* using [Bmim][$MeSO_4$]. *Biomass and Bioenergy* 56: 99–103.

Koyande, A. K., K. W. Chew, J. W. Lim, S. Y. Lee, M. K. Lam, and P. L. Show. (2019). Optimization of protein extraction from *Chlorella vulgaris* via novel sugaring-out assisted liquid biphasic electric flotation system. *Engineering in Life Sciences* 19: 968–977.

Krishna Koyande, Apurav, Vera Tanzil, Haridharan Murraly Dharan, Manivarman Subramaniam, Ryann Noel Robert, Phei-Li Lau, Ianatul Khoiroh, and Pau Loke Show. (2020). Integration of osmotic shock assisted liquid biphasic system for protein extraction from microalgae *Chlorella vulgaris*. *Biochemical Engineering Journal* 157: 107532.

Kumar, Vinod, Manisha Nanda, H. C. Joshi, Ajay Singh, Sonal Sharma, and Monu Verma. (2018). Production of biodiesel and bioethanol using algal biomass harvested from fresh water river. *Renewable Energy* 116: 606–612.

Lai, YenJung Sean, Federica De Francesco, Alyssa Aguinaga, Prathap Parameswaran, and Bruce E. Rittmann. (2016). Improving lipid recovery from *Scenedesmus* wet biomass by surfactant-assisted disruption. *Green Chemistry* 18: 1319–1326.

Lai, YenJung Sean, Yun Zhou, Rebecca Martarella, Zhaocheng Wang, and Bruce E. Rittmann. (2016). Synergistic integration of C12–C16 cationic surfactants for flocculation and lipid extraction from *Chlorella* biomass. *ACS Sustainable Chemistry & Engineering* 5: 752–757.

Lee, J. Y., C. Yoo, S. Y. Jun, C. Y. Ahn, and H. M. Oh. (2010). Comparison of several methods for effective lipid extraction from microalgae. *Bioresources Technology* 101 Suppl 1: S75–77.

Leong, Hui Yi, Chien Wei Ooi, Chung Lim Law, Advina Lizah Julkifle, Tomohisa Katsuda, and Pau Loke Show. (2019). Integration process for betacyanins extraction from peel and flesh of *Hylocereus polyrhizus* using liquid biphasic electric flotation system and antioxidant activity evaluation. *Separation and Purification Technology* 209: 193–201.

Li, Yaorong, Masahiro Nagao, Kazuyo Kobayashi, Yongcheng Jin, and Takashi Hibino. (2020). A cellulose electrolysis cell with metal-free carbon electrodes. *Catalysts* 10(1): 106.

Liu, J., G. Pei, J. Diao, Z. Chen, L. Liu, L. Chen, and W. Zhang. (2017). Screening and transcriptomic analysis of *Crypthecodinium cohnii* mutants with high growth and lipid content using the acetyl-CoA carboxylase inhibitor sethoxydim. *Applied Microbiology and Biotechnology* 101: 6179–6191.

Lu, Yujuan, and Xihai Zhu. (2007). Solvent sublation: Theory and application. *Separation and Purification Methods* 30: 157–189.

Martín, L. A., C. A. Popovich, M. C. Damiani, and P. I. Leonardi. (2020). A practical tool for selecting microalgal species for biodiesel production. *Journal of Renewable and Sustainable Energy* 12(6).

Mathimani, Thangavel, Lakshmanan Uma, and Dharmar Prabaharan. (2018). Formulation of low-cost seawater medium for high cell density and high lipid content of *Chlorella vulgaris* BDUG 91771 using central composite design in biodiesel perspective. *Journal of Cleaner Production* 198: 575–586.

Obeid, Sara, Nicholas Beaufils, Séverine Camy, Hosni Takache, Ali Ismail, and Pierre-Yves Pontalier. (2018). Supercritical carbon dioxide extraction and fractionation of lipids from freeze-dried microalgae Nannochloropsis oculata and Chlorella vulgaris', *Algal Research*, 34: 49–56.

Phong, W. N., P. L. Show, W. H. Teh, T. X. Teh, H. M. Y. Lim, N. S. B. Nazri, C. H. Tan, J. S. Chang, and T. C. Ling. (2017). Proteins recovery from wet microalgae using liquid biphasic flotation (LBF). *Bioresources Technology* 244: 1329–1336.

Piligaev, A. V., K. N. Sorokina, M. V. Shashkov, and V. N. Parmon. (2018). Screening and comparative metabolic profiling of high lipid content microalgae strains for application in wastewater treatment. *Bioresources Technology* 250: 538–547.

Rashid, Naim, Muhammad Saif Ur Rehman, Madeha Sadiq, Tariq Mahmood, and Jong-In Han. (2014). Current status, issues and developments in microalgae derived biodiesel production. *Renewable and Sustainable Energy Reviews* 40: 760–778.

Ritchie, Hannah. (2019). Who has contributed most to global CO_2 emissions? *Our World in Data.* Available at: https://ourworldindata. org/contributed-most-global-co2

Rokicka, Magdalena, Marcin Zieliński, Magda Dudek, and Marcin Dębowski. (2020). Effects of ultrasonic and microwave pretreatment on lipid extraction of microalgae and methane production from the residual extracted biomass. *BioEnergy Research* 14: 752–760.

Sadeghin, Bahareh, Mohammad-Hossein Sarrafzadeh, Jian Jin, Catherine Dupre, Makoto Watanabe, Jack Legrand, and Dominique Grizeau. (2018). Variation of fatty acids composition in the hydrocarbon producer *Botryococcus braunii* BOT 22. *Biomass and Bioenergy* 119: 456–461.

Safi, Carl, Bachar Zebib, Othmane Merah, Pierre-Yves Pontalier, and Carlos Vaca-Garcia. (2014). Morphology, composition, production, processing and applications of *Chlorella vulgaris*: A review. *Renewable and Sustainable Energy Reviews* 35: 265–278.

Saha, Sampa, Arun Sharma, Srijita Purkayastha, Kalpana Pandey, and Shaifali Dhingra. (2018). 14 Bio-plastics and biofuel: Is it the way in future development for end users? *Plastics to Energy: Fuel, Chemicals, and Sustainability Implications* 365–376.

Sajjadi, Baharak, Wei-Yin Chen, Abdul Aziz Abdul Raman, and Shaliza Ibrahim. (2018). Microalgae lipid and biomass for biofuel production: A comprehensive review on lipid enhancement strategies and their effects on fatty acid composition. *Renewable and Sustainable Energy Reviews* 97: 200–232.

Sakarika, M., and M. Kornaros. (2016). Effect of pH on growth and lipid accumulation kinetics of the microalga *Chlorella vulgaris* grown heterotrophically under sulfur limitation. *Bioresources Technology* 219: 694–701.

Sankaran, R., S. Manickam, Y. J. Yap, T. C. Ling, J. S. Chang, and P. L. Show. (2018). Extraction of proteins from microalgae using integrated method of sugaring-out assisted liquid biphasic flotation (LBF) and ultrasound. *Ultrasonics Sonochemistry* 48: 231–239.

Sankaran, R., P. L. Show, Y. S. Cheng, Y. Tao, X. Ao, T. D. P. Nguyen, and D. Van Quyen. (2018). Integration process for protein extraction from microalgae using liquid biphasic electric flotation (LBEF) system. *Molecular Biotechnology* 60: 749–761.

Sankaran, R., P. L. Show, S. Y. Lee, Y. J. Yap, and T. C. Ling. (2018). Integration process of fermentation and liquid biphasic flotation for lipase separation from *Burkholderia cepacia*. *Bioresources Technology* 250: 306–316.

Sarayloo, E., S. Simsek, Y. S. Unlu, G. Cevahir, C. Erkey, and I. H. Kavakli. (2018). Enhancement of the lipid productivity and fatty acid methyl ester profile of *Chlorella vulgaris* by two rounds of mutagenesis. *Bioresources Technology* 250: 764–769.

Seo, Jung Yoon, Minjeong G. Kim, Kyubock Lee, Young-Chul Lee, Jeong-Geol Na, Sang Goo Jeon, Seung Bin Park, and You-Kwan Oh. (2016). Multifunctional nanoparticle applications to microalgal biorefinery. In: *Nanotechnology for Bioenergy and Biofuel Production*. Springer, Cham: Switzerland.

Sobianowska, Katarzyna, Władysław Walkowiak, and Cezary Kozłowski. (2009). Principles and applications of solvent sublation – a review. *Ars Separatoria Acta* 7: 23–38.

Speight, James G. (1990). *Fuel Science and Technology Handbook.* United States. CRC Press.

Stansell, Graham Robert, Vincent Myles Gray, and Stuart David Sym. (2011). Microalgal fatty acid composition: implications for biodiesel quality. *Journal of Applied Phycology* 24: 791–801.

Sun, X., Y. Cao, H. Xu, Y. Liu, J. Sun, D. Qiao, and Y. Cao. (2014). Effect of nitrogen-starvation, light intensity and iron on triacylglyceride/carbohydrate production and fatty acid profile of *Neochloris oleoabundans* HK-129 by a two-stage process. *Bioresources Technology* 155: 204–12.

Tiong, Irene Kai Ru, Yeong Yik Sung, Malinna Jusoh, Mohd Effendy Abdul Wahid, and Thilahgavani Nagappan. (2020). *Chlorella vulgaris*: a perspective on its potential for combining high biomass with high value bioproducts. *Applied Phycology* 1: 2–11.

van Gerpen, J., B. Shanks, R. Pruszko, D. Clements, and G. Knothe. (2004). *Biodiesel Production Technology.* U.S. Department of Energy. Oak Ridge, TN.

Vasistha, S., A. Khanra, M. Clifford, and M. P. Rai. (2021). Current advances in microalgae harvesting and lipid extraction processes for improved biodiesel production: A review. *Renewable and Sustainable Energy Reviews* 137: 110498.

Wahidin, S., A. Idris, and S. R. Shaleh. (2013). The influence of light intensity and photoperiod on the growth and lipid content of microalgae *Nannochloropsis* sp. *Bioresources Technology* 129: 7–11.

Yang, Y. H., L. Du, M. Hosokawa, K. Miyashita, Y. Kokubun, H. Arai, and H. Taroda. (2017). Fatty acid and lipid class composition of the microalga *Phaeodactylum tricornutum*. *Journal of Oleo Science* 66: 363–368.

Yang, Yi-Hung, Worasaung Klinthong, and Chung-Sung Tan. (2015). Optimization of continuous lipid extraction from *Chlorella vulgaris* by CO_2-expanded methanol for biodiesel production. *Bioresource Technology* 198: 550–556.

Zheng, H., J. Yin, Z. Gao, H. Huang, X. Ji, and C. Dou. (2011). Disruption of *Chlorella vulgaris* cells for the release of biodiesel-producing lipids: a comparison of grinding, ultrasonication, bead milling, enzymatic lysis, and microwaves. *Applied Biochemistry Biotechnology* 164: 1215–1224.

Zouboulis, A.I., K.A. Matis, and G.A. Stalidis. (1990). Parameters influencing flotation in removal of metal ions. *International Journal of Environmental Studies* 35: 183–96.

8 Antioxidative Compounds from Diatoms

Heli Siti Halimatul Munawaroh[1], *Riska Nur Pratiwi*[1], *and Nur Akmalia Hidayati*[2]
[1]Universitas Pendidikan, Bandung, Indonesia
[2]Research Center for Applied Microbiology, The National Research and Innovation Agency (BRIN), Jakarta, Indonesia

8.1 INTRODUCTION

Oxidative damage caused by reactive oxygen species (ROS) on lipids, proteins, and nucleic acids can trigger various chronic diseases, such as coronary heart disease, atherosclerosis, cancer, and aging (Adwas et al., 2019). In recent years, there has been an increased interest in the use of antioxidant nutritional supplements, such as vitamins, minerals, and other dietary constituents that have shown a protective effect in preventing or reducing the severity of the disease (Tan et al., 2018; Karthikeyan, 2013).

Microalgae are one of the potential food constituents because they have sustainable antioxidant reserves for the future (Foo et al., 2015), and so they have emerged as a significant source of food, chemicals, and energy due to the presence of various chemical compounds that accumulate in their cells. Many species of microalgae, such as *Chlorella vulgaris*, *Spirulina* (Arthrospira) *platensis*, and *Haematococcus pluvialis*, can be cultured on a large scale and grow rapidly under photosynthetic conditions. Diatoms are one of the most beneficial microalgae from the Bacillariophyceae class that are available abundantly in the oceans, which contribute 45% of marine primary productivity and 20–25% of total global carbon fixation due to their higher growth rate and competitive characteristics compared to other microalgae groups. Diatoms are unicellular eukaryotic microscopic plants which constitute about 200 genera and more than 100,000 species. Diatoms have a siliceous cell wall (frustula), which is synthesized intracellularly by polymerization of silicic acid monomers, whereas the cell wall is a key component of diatoms that have many industrial applications (Pekkoh et al., 2022; Chasapis et al., 2022).

Diatoms are microorganisms that are rich in primary and secondary metabolites with beneficial biological activities, such as high value-added compounds, especially lipids, fatty acids, and sterols. In addition, diatom extracts may also contain polyphenols, pigments, proteins, and polysaccharides (Tiwari et al., 2021). Diatom's wide distribution, abundance, and diversity of chemical compounds make them ideal for various applications such as anticancer and antioxidant agents, as healthy food and drug sources, as biomaterials synthesis agents and in nanotechnology (drug delivery, molecular separation, biomimetics, photonics, structural materials, electronics, optics, and biosensing). Recently, they have also been used for fuel synthesis, sewage bioremediation, heavy metals and polluted water treatment, and biomineralization (Hussein and Abdullah, 2020).

According to Foo et al., extracting a group of active compounds is more beneficial than a single compound because the interaction in the combinational group will show a synergistic and protagonist effect that contributes to an increased antioxidant capacity. Diatoms contain not only carotenoids

 DOI: 10.1201/9781003436553-8

but also phenolic compounds. Preparation of antioxidant-rich extracts that consist of carotenoids and phenolic compounds from diatoms can be economically feasible, when the amount of biomass can provide higher antioxidant activity due to co-extraction of other active compounds (Foo et al., 2015). However, considering all the richness in biological elements of diatoms, the current literature lacks sufficient information on the therapeutic and applied aspects of biological macromolecules of diatoms (Tiwari et al., 2021).

8.2 OXIDATIVE STRESS AND ROLE OF ANTIOXIDANTS

8.2.1 Oxidative Stress

Many natural biological processes in our body, such as respiration, food digestion, alcohol, and drug metabolism, and fat-to-energy conversion produce harmful compounds called free radicals (Sharifi-Rad et al., 2020). Free radicals are defined as chemical species that are able to stand alone and contain one or more unpaired electrons. There are several free radicals, including reactive oxygen species (ROS), reactive nitrogen species (RNS), reactive sulfur species (RSS), reactive carbonyl species (RCS), and reactive selenium species (RSeS). These reactive species continue to be formed from endogenous and exogenous processes that take place in the body (Ferdous et al., 2021). ROS and RNS are free radicals that associate with oxygen atoms (O) or their equivalent and are highly reactive compared to O_2, by oxidizing (removing electrons from) other atoms, or sometimes reducing (donating electrons to) other atoms (Adwas et al., 2019; Kurutas, 2016).

Mitochondria are the main organelles responsible for generating energy through oxidative phosphorylation to produce adenosine triphosphate (ATP), a molecule that is essential for cellular actions. The electron transport chain consumes up to 90% of the total oxygen (O_2) taken up by the cell. During this process, ROS are generated as a by-product for the partial reduction of four electrons of O_2 to produce a water molecule, which is the final electron acceptor in the ATP generation process. Nearly 0.1–0.5% of the inhaled O_2 is converted to superoxide (O_2^-) during normal physiological conditions (Tan et al., 2018). In addition, the endoplasmic reticulum and peroxisomes are other sources of ROS (García-Sánchez, et al., 2020). External causes, such as radiation, cigarette smoke, and pollutants, can also indirectly generate ROS (Goh et al., 2010).

ROS/RNS usually exhibit four species, namely superoxide anion (O_2^-), hydrogen peroxide (H_2O_2), hydroxyl radical (OH), and singlet oxygen ($^1O_2$). There are also other biologically important free radicals, lipid hydroperoxides (ROOH), lipid peroxyl radicals (ROO), and alkoxyl lipid radicals (RO), which are associated with membrane lipids; nitric oxide (NO), nitrogen dioxide (NO_2), and peroxynitrite ($ONOO^-$), which are reactive nitrogen species; and the thiol radical species (RSS), which has an unpaired electron on a sulfur atom (Rahal et al., 2014).

The most important free radicals in many disease states are oxygen derivatives, especially superoxide anions and hydroxyl radicals (Manisha et al., 2017). The formation of radicals in the body occurs through several mechanisms, involving both endogenous and environmental factors. The superoxide anion is produced by the addition of one electron to oxygen, and there are several mechanisms by which superoxide can be produced in vivo. Some molecules such as flavin nucleotides and thiol compounds are oxidized in the presence of oxygen to give superoxide, and this reaction is greatly accelerated in the presence of a transition metal such as iron or copper. The electron transport chain in the inner mitochondrial membrane reduces oxygen to water. During this process, free radical intermediates are generated, which are generally tightly bound to transport chain components (Kurutas, 2016).

ROS play an important role in several biological processes, such as the destruction of intracellular bacteria by phagocytes (mainly by granulocytes and macrophages), and are involved in several cellular signaling processes, known as redox signaling (Sharifi-Rad et al., 2020). In addition, ROS are also required to maintain cell homeostasis, which may involve signal transduction, gene expression, receptor activation, pathogen recognition, ensuring cell viability, proliferation, migration, and differentiation (Pisoschi et al., 2021). Under a normal, healthy condition, the formation and oxidation of ROS occur in a controlled manner. In contrast, ROS production is increased under high stress conditions or in disease states (Tan et al., 2018).

Oxidative stress can lead to structural modification of cellular proteins and changes in their function, leading to cellular dysfunction and disruption of vital cellular processes. Protein damage involves site-specific amino acid modification, peptide chain fragmentation, aggregation of cross-linked reaction products, changes in electrical charge, enzymatic inactivation, and proteolysis susceptibility (Sharifi-Rad et al., 2020). Therefore, the inhibition of oxidative damage with antioxidant supplementation becomes an attractive therapeutic strategy to reduce the risk of diseases caused by oxidative stress (Adwas et al., 2019).

8.2.1.1 The Role of Antioxidants

Naturally, organisms possess several antioxidant defenses to minimize radical damage and to protect against excessive production of free radicals by maintaining a balance between oxidants and antioxidants (Lakshmegowda et al., 2020). Antioxidants are compounds that play an important role in eliminating free radicals or inhibiting other oxidation reactions by oxidizing themselves. The mechanisms carried out by antioxidant defenses consist of; (1) blocking the production of free radicals; (2) scavenging oxidants; (3) converting toxic free radicals into less toxic substances; (4) blocking the production of secondary toxic metabolites and inflammatory mediators; (5) blocking secondary oxidant chain propagation; (6) repairing injured molecules; and (7) initiation and enhancement of the endogenous antioxidant defense system. All of these defense mechanisms work hand in hand to protect the body from oxidative stress (Adwas et al., 2019). The antioxidant system in the human body consists of non-enzymatic and enzymatic antioxidants that work synergistically, and together protect the cells and organ systems of the body from free radical damage. Antioxidants can be obtained endogenously or exogenously as part of the diet or as a dietary supplement (Kurutas, 2016).

8.2.2 Enzymatic Antioxidants

Several enzymes inhibit the formation of free radicals, some of which act directly in scavenging ROS (primary enzymes), whereas secondary enzymes act indirectly by supporting other endogenous antioxidants (Aslani and Ghobadi, 2016).

8.2.2.1 Primary Enzymes

Primary antioxidants are chain-breaking antioxidants that have slow initiation or interfere with propagation, acting by capturing radical species, converting them into more stable radicals or non-radical species (Pisoschi et al., 2021). Primary enzymes act directly on major ROS arising from incomplete reduction of O_2, such as O_2^- and H_2O_2. The most efficient enzymatic antioxidants contain superoxide dismutase (SOD), catalase (CAT), and glutathione peroxidase (GPX) (García-Sánchez, et al., 2020). SOD scavenges the former, whereas CAT and GPX remove the latter (Singh et al., 2017).

8.2.2.1.1 Superoxide Dismutase (SOD)

SOD (EC 1.15.1.1) is one of the most potent intracellular enzymatic antioxidants. SOD is a metalloenzyme that catalyzes the conversion of superoxide anions into H_2O_2 and molecular oxygen (reaction 8.1). In turn, H_2O_2 can be removed by other enzymatic antioxidant agents, such as CAT or GPX (Sharifi-Rad et al., 2020; Kurutas, 2016).

$$2O_2^{\cdot -} + 2H^+ \xrightarrow{SOD} H_2O_2 + O_2 \qquad (8.1)$$

SOD can be divided into four groups, with different metal cofactors. Copper–zinc SOD is the most abundant in the chloroplast, cytosol, and extracellular space. Iron SOD is found in plant cytosol and in microbial cells, whereas manganese SOD is in mitochondria. SOD also competes for superoxide anion with NO. Therefore, SOD also indirectly reduces the formation of another damaging ROS, peroxy-nitrite ($ONOO^-$) (reaction 8.2), and increases the biological availability of NO, an important modulator for endothelial function (Alzahrani et al., 2017).

$$O_2^{\cdot -} + NO^{\cdot} \longrightarrow ONOO^- \quad (8.2)$$

8.2.2.1.2 Catalase (CAT)

CAT (E.C. 1.11.1.6) is a tetrameric ferriheme oxidoreductase, which catalyzes the conversion of H_2O_2 to water and oxygen (reaction 8.3) (Grigoras, 2017).

$$2H_2O_2 \xrightarrow{CAT} 2H_2O + O_2 \quad (8.3)$$

CATs are mainly located in peroxisomes, where the highest activity is found in the liver and red blood cells, but some CATs are found in all tissues. CAT has one of the highest turnover rates for all enzymes: one CAT molecule can convert about six million hydrogen peroxide molecules into water and oxygen every minute (Kurutas, 2016). Recent advances demonstrated the ability of CAT to scavenge ONOO– (Sharifi-Rad et al., 2020).

8.2.2.1.3 Glutathione Peroxidase (GPX)

GPX (EC 1.11.1.19) is a selenium-dependent oxidoreductase, which employs H_2O_2 or an organic hydroperoxide as an oxidant, and the tripeptide GSH as an electron donor, in a typical class I peroxidase catalytic cycle (reactions 8.4 and 8.5) (Cardoso et al., 2017).

$$H_2O_2 + 2GSH \xrightarrow{GPX} 2H_2O + GSSG \quad (8.4)$$

$$ROOH + 2GSH \xrightarrow{GPX} ROH + H_2O + GSSG \quad (8.5)$$

Other peroxides, including lipid hydroperoxides, can also act as substrates for this enzyme, thereby playing a role in repairing damage caused by lipid peroxidation. There are two forms of this enzyme, one that is selenium-dependent (GPx, EC1.11.1.19) and the other that is selenium-independent (glutathione-S-transferase, GST, EC 2.5.1.18).

8.2.2.2 Secondary Enzymes

All of the enzymatic activities described above depend on continuous regeneration of the reduced form. This is usually accomplished by several NADPH-dependent reductases (such as glutathione reductase E.C. 1.8.1.7 and thioredoxin reductase E.C. 1.8.1.9). However, reduction of NADPH, in turn, is required by this reductase for its sustained action. Thus, the enzymes responsible for the constant production of NADPH can be considered as secondary antioxidants, as their malfunctioning can affect the overall ROS balance. The main source of NADPH metabolism is the pentose phosphate pathway, through the first two enzymatic activities: glucose-6-phosphate dehydrogenase (EC 1.1.1.49) and 6-phosphogluconate dehydrogenase (EC 1.1.1.44). However, another contribution comes from the malate enzyme (EC 1.1.1.40) and a recently identified folate-dependent mechanism (Sharifi-Rad et al., 2020). Secondary antioxidants may show a synergistic effect in combination with primary antioxidants. In this case, secondary antioxidants stabilize primary antioxidants by creating an acidic environment, regenerating primary antioxidants by donating hydrogen, chelating metal ions, or quenching molecular oxygen (Pisoschi et al., 2021).

8.2.3 Non-Enzymatic Antioxidants

Some chemical molecules with low molecular weight also act directly as antioxidants. In this case, the mechanism is not considered as a catalytic reaction, since it always requires the regeneration of antioxidants or their supply from food. Non-enzymatic antioxidants can therefore be divided into endogenous (if eukaryotic cells are able to synthesize them) and exogenous (if antioxidants need to be ingested through food) (Sharifi-Rad et al., 2020).

8.2.3.1 Endogenous Non-Enzymatic Antioxidants

8.2.3.1.1 Thiol Antioxidant

Both protein-bound and nonprotein-thiol-bound thiols act as cellular reducing agents and protective agents against major inorganic pollutants, via the –SH group. Therefore, thiols are often the first line of defense against oxidative stress (Adwas et al., 2019).

8.2.3.1.1.1 Glutathione The most abundant thiol antioxidant is tripeptide glutathione/γ-glutamyl-cysteinyl-glycine (GSH), which is a multifunctional intracellular antioxidant, that is known to be the main thiol-disulfide redox buffer of cells and which plays a central role in maintaining the redox state of cells (Adwas et al., 2019). GSH is abundant in the cytosol, nucleus, peroxisomes, and mitochondria. Despite its ubiquity, the liver is the main site for its synthesis (Amir Aslani and Ghobadi, 2016). The reduced form of glutathione is GSH (glutathione), while the oxidized form is GSSG (glutathione disulfide). The antioxidant capacity of thiol compounds is due to the existence of a sulfur atom, thereby it can accommodate the loss of one electron. The protective role of glutathione against oxidative stress is that it can act as an electron donor to inactivate the free radicals, serve as a co-factor for several detoxification enzymes, participate in amino acid transport across plasma membranes, scavenge hydroxyl radicals and singlet oxygen directly, and regenerate vitamins C and E back to their active forms (Kurutas, 2016; Smerilli et al., 2019).

8.2.3.1.1.2 Thioredoxin Another thiol antioxidant is the thioredoxin system (TRX), which is a protein with oxidoreductase activity and is universal in both mammalian and prokaryotic cells. It also contains disulfide and has two redox-active cysteine in a conserved active site (Cys-Gly-Pro-Cys). Thioredoxin contains two adjacent –SH groups in their reduced form which are converted to disulfide units in oxidized TRX when subjected to redox reactions with many proteins involved. The level of thioredoxin is much less than that of GSH, however, TRX and GSH may have overlapping functions in the activation and regulation of transcription factors (Balsera and Buchanan, 2019).

8.2.3.1.1.3 α-Lipoic Acid (1,2-Dithione-3-Pentanoic Acid) The third thiol antioxidant is the natural compound lipoic acid (ALA), which is a sulfur-containing antioxidant with metal-chelating and antiglycation abilities. In contrast to many antioxidants, which are only active in the lipid or water phase, lipoic acid is active in both lipid and water phases. Lipoic acid (LA) stabilizes lysosomes against oxidative stress, by chelating intralysosomal iron and preventing intralysosomal Fenton radical reactions (Zhang and McCullough, 2016). As an antioxidant, LA directly inhibits free radicals, chelates the transition metal ions (e.g., iron and copper), increases cytosolic glutathione and vitamin C levels, and serves as a potential therapeutic treatment for neurodegenerative disorders (Sharifi-Rad et al., 2020).

The possible mechanisms for the protective effect of LA against oxidative stress are as follows: (a) LA is easily digested, absorbed, and rapidly reduced to dihydrolipoic acid by NADH or NADPH; (b) meanwhile, DHLA is a strong antioxidant to scavenge excess oxidants and recycle antioxidants, such as vitamins E and C, and glutathione; (c) DHLA chelates metals to prevent the formation of free radicals, so as to reduce oxidant attack on bio-macromolecules; (d) LA is a key co-factor of pyruvate dehydrogenase and ketoglutarate dehydrogenase-sensitive enzymes which function against oxidative stress; (e) sufficient LA supplementation can stimulate enzyme activity, thereby increasing and improving oxidative phosphorylation and mitochondrial respiration; and (f) LA enhances antioxidant defense by inducing phase 2 enzymes, such as glutathione synthetase, to increase antioxidant GSH (Kurutas, 2016; Rochette et al., 2015).

8.2.3.1.1.4 N-acetylcysteine (NAC) N-acetyl-L-cysteine (NAC) is a thiol-containing antioxidant that has been used to reduce conditions of oxidative stress. Its antioxidant action is ascribed to its ability to stimulate GSH synthesis, thereby maintaining intracellular GSH levels and scavenging ROS/RNS. NAC is rapidly deacetylated to cysteine and thus it can increase the GSH levels by providing a substrate for the rate-limiting step in GSH synthesis. The thiol groups present in NAC act to reduce free radicals and provide chelating sites for metals (Mokhtari et al., 2017; Kurutas, 2016).

8.2.3.1.2 Melatonin (N-Acetyl-5-Methoxytryptamine)

Melatonin is the main secretory product of the pineal gland in the brain which is renowned for its functional versatility. Melatonin has been reported as a direct free radical scavenger and an indirect antioxidant, as well as an important immunomodulatory agent. It is shown that the radical scavenging ability of melatonin acts via an electron donor to remove various ROS/RNS, including the highly toxic hydroxyl radical. Furthermore, melatonin stimulates a number of antioxidant enzymes including SOD, GPX, glutathione reductase (GR), and CAT. Melatonin also inhibits pro-oxidant enzymes, such as nitric oxide synthase, xanthine oxidase, and lipoxygenase. Overall, there is evidence to show that melatonin acts to stabilize the cell membranes, thus possibly it might alleviate the oxidative damage (Kurutas, 2016; Tarocco et al., 2019).

8.2.3.1.3 Coenzyme Q10 or Ubiquinone

Coenzyme Q10, or ubiquinone (2,3-dimethoxy-5-methyl-6-polyisoprene parabenzoquinone), is an isoprenoid antioxidant present in cell membranes. It is one of the few fat-soluble antioxidants, ensuring the protection of lipoproteins and lipids from radical chain reactions, peroxidation, and oxidative damage. In its active form (quinol), coenzyme Q10 could scavenge some ROS or regenerate other oxidized antioxidants (including vitamins C and E). In turn, the quinone form can be reduced again by several NAD(P)H-dependent enzymatic systems (Sharifi-Rad et al., 2020).

8.2.3.2 Exogenous Non-Enzymatic Antioxidants

Exogenous antioxidants need to be added continuously through the diet because the synthetic pathway is only present in microbial or plant cells. Intake of these exogenous antioxidants may delay the uptake of endogenous antioxidants, keeping the antioxidant potential in the cells unchanged (Lee et al., 2017).

8.2.3.2.1 Vitamin C (Ascorbic Acid)

Vitamin C should be supplemented by food (or supplements), especially from tomatoes, pineapples, watermelons, and all citrus fruits, since human do not express the L-gulonolactone oxidase enzyme (E.C. 1.1.3.8). Vitamin C effectively quenches ROS, either directly or jointly regenerating vitamin E, GSH, and oxidized carotenoids, which serves as a reducing agent to reduce ROS, such as superoxide anions, $^1O_2$, H_2O_2, organic peroxides, OH, or hypochlorous acid (HClO) (Aslani and Ghobadi, 2016). In combination with vitamin E, it helps regeneration of α-tocopherol from α-tocopherol radicals in membranes and lipoproteins, and increases glutathione levels in cells, thus playing an important role in protecting protein thiol groups against oxidation (Manisha et al. 2017).

8.2.3.2.2 Vitamin E

Vitamin E is a fat-soluble vitamin, found mostly in some vegetable oils, nuts, broccoli, and fish. It has eight different forms, which have been reported as α-, β-, γ-, and δ-tocopherol, and α-, β-, γ-, and δ-tocotrienol. Meanwhile, α-tocopherol has the highest antioxidant activity, especially in cell membranes (Salehi et al., 2020).

8.2.3.2.3 Carotenoids

Carotenoids are a group of fat-soluble antioxidants belonging to the tetraterpene class that are widely distributed among plants (Fiedor and Burda, 2014). Carotenoids are classified as strong ROS scavengers. In a physical mechanism, the electron-rich structure of carotenoids absorbs excess singlet oxygen energy and quenches peroxides, thereby preventing the formation of ROS. The conjugated double-bond structure in carotenoids is responsible for this ability (Smerilli et al., 2019).

8.2.3.2.4 Polyphenols

Polyphenols are classified as plant secondary metabolites, whereas the synthesis of polyphenols only occurs in higher plants. Some biological functions of polyphenols include anti-inflammatory, antioxidant, and antimicrobial. The most widely recognized classes of polyphenols are flavonoids, phenolic acids, stilbenes (including resveratrol), tannins, coumarins, curcuminoids, and lignans. For example, one of the most studied polyphenols is curcumin. This phytochemical turned out to have potent activity as a scavenger of superoxide anions, H_2O_2, lipid peroxides, OH, and some RNS. Curcumin can also increase cellular GSH levels (Sharifi-Rad et al., 2020).

8.2.3.2.5 Epigallocatechin-3-Gallate (EGCG)

EGCG is a well-known antioxidant mostly found in green tea. The green tea-derived catechins include catechin, epicatechin, epigallocatechin, epicatechin gallate, and epigallocatechin gallate. Catechins can react with superoxide and OH anions, and are involved in chelating metal ions (Imam et al., 2017).

8.2.3.2.6 Flavonoids

Flavonoids are regarded as having strong antioxidant properties. They are comprised of scavenging peroxyl radicals, and act as effective inhibitors of lipid peroxidation and redox active metals chelating agents (such as iron and copper). Therefore, they prevents the catalytic breakdown reaction of hydrogen peroxide (Fenton) (Kurutas, 2016). Moreover, quercetin is one of the most widely distributed flavonols that is capable of chelating and stabilizing iron molecules. Meanwhile, catechol and galloyl derivatives are also widely known metal chelators (Sharifi-Rad et al., 2020).

8.2.3.2.7 Organosulfur Compounds

Organosulfur compounds have also been suggested as powerful antioxidants. The most extensively studied compounds are probably some of the sulfur-containing metabolites present in garlic, especially S-allyl-mercapto cysteine, S-allyl cysteine, diallyl sulfide, and diallyl trisulfide. Their antioxidant actions include ROS scavenging and inhibiting lipid peroxidation (Kimura et al., 2017).

8.2.4 Structure of Antioxidative Molecules in Diatoms

Bioactive compounds are additional nutritional components that are biologically present in small amounts in plants, animal products, and lipid-rich foods. Due to their properties, various clinical and experimental trials have been carried out to estimate the health effects and identify these compounds from different plant and animal sources. They are the most common bioactive compounds available in plants, including phenolic compounds, fatty acids, lipids, carbohydrates, pigments, and others (Kuppusamy et al., 2017). The antioxidant compounds contained in diatoms are shown in Table 8.1.

TABLE 8.1
The Antioxidant Compounds Contained in Diatoms

Metabolite compound	Diatom species	Antioxidative compounds	References
Pigment	*Pseudo-nitszchia*	Chlorophyll a, c_2, c_1	Stonik and Stonik, 2015
Carotenoid	*Pleurosigma angulatum, Chaetoceros simplex, Odentella aurita, Chaetoceros calcitrans, Cylindrotheca closterium, Phaeodactylum tricornutum*	Fucoxanthin, β-carotene, lycopene, β-cryptoxanthin, diadinoxanthin, diatoxanthin, antheraxanthin, zeaxanthin, lutein, astaxanthin, violaxanthin, and canthaxanthin	(Peng et al., 2011; Xia et al., 2013; Tiwari et al., 2021; Khaw et al., 2021; Azizan et al., 2018
Protein	*Skeletonema costatum*	10-Hydroxy-phaeophorbide A and phaeophorbide A (product of chlorophyll breakdown, known as chlorophyll proteins)	Lauritano et al., 2016; Tiwari et al., 2021
	Nitzschia laevis	Hydrolysates	Alzahrani, Perera, and Hemar, 2017; Tiwari et al., 2021
Polysaccharide	*Phaeodactylum tricornutum, Odentella aurita, Synedra acus, Chaetoceros muelleri, Nitzschia sigmoidea, Cymatopleura solea, Pinnularia* sp., *Melosira varians, Saccharina latissima, Skeletonema costatum*	Chrysolaminarin	Baoyan et al., 2017; Xia et al., 2014; Tiwari et al., 2021; Gügi et al., 2015
	Cylindrotheca closterium, Thalassiosira pseudonana, Cylindrotheca fusiformis	Exopolysaccharide (EPS)	Gügi et al., 2015
	Navicula sp.	Sulfated polysaccharide (SPs)	Fimbres-Olivarria et al., 2018
	Didymosphenia geminate	Crude polysaccharides	Figueroa et al., 2019; Tiwari et al., 2021
Polyphenol	*Phaeodactylum tricornutum, Nitzschia laevis, Phaeodactylum* sp., *Nitzschia palea, Chaetoceros calcitrans, Skeletonema marinoi*	Gallic acid, protocatechuic acid, catechin, vanillic acid, epicatechin, syringic acid, chlorogenic acid, gentisic acid, caffeic acid, coumaric acid, ferulic acid, rutin, myricetin, quercetin, p-hydroxybenzoic acid, sinapic acid, rosmaric acid, and cinnamic acid	Rico et al., 2013; Foo et al., 2015; Lakshmegowda et al., 2020; Smerilli et al., 2019; Alzahrani, Perera, and Hemar, 2018
Fatty acid	*Odentella aurita, Phaeodactylum tricornutum, Stauroneis* sp., *Anomoeoneis* sp., *Rhopalodia* sp., *Skeletonema costatum, Synedra acus, Fistulifera* sp.	Saturated fatty acid (SFA), monounsaturated fatty acid (MUFA), and polyunsaturated fatty acid (PUFA)	Haimeur et al., 2012; Sharma et al., 2021; Parkes et al., 2021; Pekkoh et al., 2022; Stonik and Stonik, 2015

8.2.5 Pigment Antioxidants

Diatoms are the main source of natural pigments consisting of light-harvesting chlorophyll a, chlorophyll c, fucoxanthin (Fx), as well as photo-protectives such as β-carotene, diadinoxanthin (Ddx), diatoxanthin (Dtx), zeaxanthin (Zx), antheraxanthin (Ax), and violaxanthin (Vx) (Figure 8.1). In diatoms, there are only two forms of chlorophyll: chlorophyll a (chl a) which plays a central role in the photochemical energy conversion of most photosynthetic organisms and chlorophyll c (chl c) which participates effectively in photosynthesis as an accessory pigment. There are several chlorophylls c in diatoms, but the most abundant are chlorophyll c_1 and chlorophyll c_2. The different structure of chlorophyll c brings about a change in the absorption spectrum to produce a stronger (blue) absorption band compared with the weaker red absorption band. The ratio of band I (at ~630 nm) to band II (at ~580 nm) was >1 for chlorophyll c_1-like chromophores, ~1 for chlorophyll c_2-like chromophores, and <1 for chlorophyll c_3-like chromophores (Kuczynska, et al., 2015). Chlorophyll shows high antioxidant activity by having a tetrapyrrole structure substituted with a centrally bonded magnesium atom (Scrob, et al., 2019). Fucoxanthin, diadinoxanthin, and diatoxanthin are mainly diatom-specific carotenoids of great importance in the food, cosmetic, and pharmaceutical industries due to their many beneficial properties such as anti-oxidant, anti-cancer, anti-inflammatory, anti-obesity, and neuroprotective effects (Stonik and Stonik, 2015; Tiwari et al., 2021).

Fucoxanthin is an allelic carotenoid with an allelic carbon chain, an acyclic keto group, and a hydroxyl group on one of the β-ionone rings. It can be extracted from edible brown seaweed diatoms, where they attach to the thylakoid membrane to transmit energy excitation to the photosynthetic electron transport chain via chlorophyll a (Tiwari et al., 2021). Fucoxanthin is found in Bacillariophyta diatoms, including *Chaetoseros* sp., *Cylindrotheca closterium*, *Phaeodactylum tricornutum*, and *Odontella aurita* (Hussein and Abdullah, 2020). Diatom *P. tricornutum*, a potential commercial source of fucoxanthin which contains at least 10 times more fucoxanthin per gram dry weight than brown algae (Stonik and Stonik, 2015).

Meanwhile, fucoxanthin is the most abundant pigment in diatoms, accounting for 1–2.5% of their dry cell weight. It exhibits significant anti-oxidant, anti-obesity, anti-inflammatory, antidiabetic, and antihypertensive activities. Studies have shown that fucoxanthin is effective at preventing DNA damage, intracellular ROS species formation, and H_2O_2-induced apoptosis. Since fucoxanthin enhances the action of catalase, it showed higher hydroxyl radical scavenging activity than fucoxanthinol, α-tocopherol, and halocynthiaxanthin. The antioxidant activity of fucoxanthin is mainly due to the fact that it contains an allenic bond at the C-7 site, in combination with the 5,6-monoepoxide carbonyl, and acetyl region which is responsible for the higher anti-oxidation activity compared with other pigments. It also consists of six oxygen atoms which are more sensitive to radicals under anoxic conditions, which is why fucoxanthin exerts more anti-oxidant activity under these conditions (Hussein and Abdullah, 2020; Tiwari et al., 2021).

Diadinoxanthin and diatoxanthin contain allene bonds and acetyl and epoxy groups, thereby expressing high bioactivity, which act as health-promoting substances. Zeaxanthin, antheraxanthin, and violaxanthin exhibit strong antioxidant properties, and important marine metabolites exhibit health-promoting activity and beneficial effects (Athanasakoglou and Kampranis, 2019).

8.2.6 Polysaccharide Antioxidants

Diatoms produce many storage compounds in the form of carbohydrates, which aid in cell motility and act as phototrophic biofilms. The main polysaccharides in diatoms are intracellular food storage polymers, extracellular mucilaginous matrices, or exopolysaccharides (EPS), and the cell wall or frustule, particularly sulfated polysaccharides. Diatomaceous polysaccharides are known to have different health effects such as anti-inflammatory, anti-cancer, immunostimulant, and antioxidant properties which are effective for scavenging ROS (Fimbres-Olivarria et al., 2018).

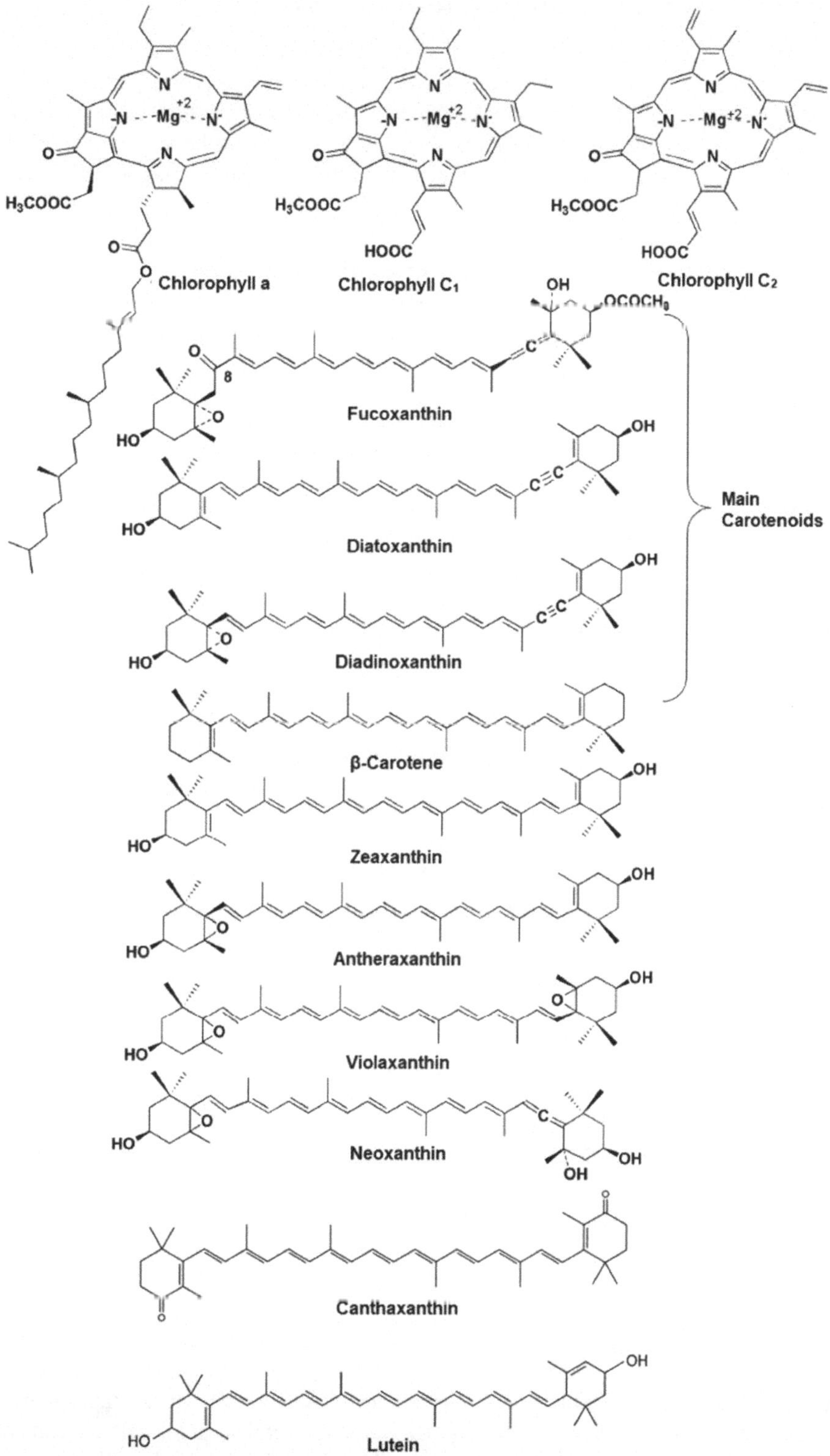

FIGURE 8.1 Diatom pigment (modified from Stonik and Stonik, 2015).

8.2.7 Food Storage Polysaccharides: Chrysolaminarin

Chrysolaminarin is the most abundant carbohydrate storage in marine diatoms, which is known to inhibit the proliferation of colon tumor cells. The structure consists of a β-glucan backbone in an 11:1 ratio of (β-1,3)-glucose and (β-1,6)-glucose, respectively. Several environmental factors such as temperature, light exposure, nutrients, and salinity can affect the amount of carbohydrates present in algae. The higher the amount of nutrients in the medium, the higher the accumulation of chrysolaminerin, reaching up to 33% of dry weight. Chrysolaminarin extracted from marine diatoms *Odontella aurita* and *Synedra acus* was also shown to have strong hydroxyl radical scavenging activity in concomitant with the activity to inhibit colon tumor cell formation, cancer, infection, inflammation, and influenza (Tiwari et al., 2021; Xia et al., 2013).

8.2.8 Exopolysaccharides (EPS)

Diatoms produce extracellular mucus which has a high carbohydrate content. This matrix is known as EPS, which may include other extracellular polymeric substances. The diverse components of EPS determine the morphology of EPS, which varies from very stiff fibers to mucilaginous capsules. Due to its complex and heterogeneous composition, EPS is difficult to extract and purify. The most-studied EPS is in the soluble form. In addition, EPS of diatoms composed by specialized metabolites, such as methyl-pentoses in the form of rhamnose, seems to be involved in their adhesion properties and play a part in diatoms' medical applications as bioderived adhesives to seal injured tissues (Gügi et al., 2015).

8.2.9 Sulfated Polysaccharides (SPs)

SPs are found in the cell walls of diatoms (frustules). Sulfur compounds are involved in frustule formation, and their amount and configuration vary according to environmental conditions. SPs have been the most studied diatom polymers regarding their biological activity as their sulfate group (–SO3H) gives them antimicrobial, antiviral, and antioxidant characteristics. Previous studies have suggested that SPs and β-glucans are effective bioactive compounds in the ROS scavenging mechanism. Thus they act as a protective system against oxidative stress (Tiwari et al., 2021).

8.2.10 Polyphenol Antioxidants

Polyphenols (also known as polyhydroxyphenols) are characterized by multiple structural units of phenol. The amount and structural characteristics of phenols contribute to the uniqueness of polyphenolic compounds in terms of chemistry, physics, and biology. Generally, polyphenols are secondary metabolites produced by plants, macroalgae, and microalgae which are thought to protect against ultraviolet radiation and pathogen invasion (Tan et al., 2018). The antioxidant activity of polyphenols on diatoms *P. tricornutum*, *Nitzschia laevis*, *Phaeodactylum* sp., and *Nitzschia palea* has shown that DPPH bleaching is one of the most common methods to determine the radical scavenging activity which may act to chelate or bind metal ions, donating hydrogen atoms, and generating radical intermediates to stabilize reactive oxygen species. The hexane extract of *N. palea* (NPHE) was found to inhibit oxidative DNA and protein damage; this suggests that polyphenols and other volatile compounds in diatoms are responsible for the prevention of oxidative damage associated with several stress conditions such as cancer, diabetes, and neurological diseases (Tiwari et al., 2021). Polyphenols derived from *P. tricornutum* are gallic acid, protocatechuic acid, catechin, vanillic acid, epicatechin, syringic acid, chlorogenic acid, gentisic acid, caffeic acid, coumaric acid, ferulic acid, rutin, myricetin, and quercetin. The most dominant phenolic compounds are myricetin and catechin, which have been confirmed to have radical scavenging activity (Rico et al., 2013). Meanwhile, polyphenols derived from *Chaetoceros calcitrans* are classified as gallic acid, protocatechuic acid, chlorogenic acid, gentisic acid, p-hydroxybenzoic acid, caffeic acid, vanillic acid, syringic acid, sinapic acid, ferulic acid, rosmaric acid, and cinnamic acid (Foo et al., 2015).

8.2.11 Miscellaneous Antioxidants

Diatom *H. ostrearia* produces blue pigment and is soluble in water, which is reported to have antioxidant and antiviral activities against herpes simplex virus type 1 (HSV-1). Trigonelin is a product of niacin metabolism in *P. tricornutum* which has distinct therapeutic potential for diabetes, hyperlipidemia, cardiovascular disease, cancer, and central nervous system diseases. Domoic acid synthesized by *Nitzschia pungens* has been used in anticancer treatment due to its cytotoxic and antioxidant activities (Tiwari et al., 2021).

Diatomaceous lipids consists of various group of fatty acids including saturated fatty acids (SFAs), monounsaturated fatty acids (MUFAs), and polyunsaturated fatty acids (PUFAs) which play important roles in human metabolism. Fatty acid antioxidants derived from the diatom *Anomoeoneis* sp. AARL D039 and *Rhopalodia* sp. AARL D020 have been suggested to have the potential to protect cells from oxidative stress by scavenging reactive oxygen species (ROS) or free radicals [e.g., 2,2-diphenyl-1-picrylhydrazyl (DPPH) and 2,2'-azino-bis(3)-ethylbenzthiazoline-6-sulfonic acid (ABTS)]. Fatty acid neutralizes free radicals by interacting with oxygen molecules during the destabilization process to prevent cellular damage, thereby slowing down the release of ROS. Moreover, EPA (eicosapentaenoic acid) is synthesized abundantly by several marine microalgae diatoms, such as *Odentella aurita*, *Phaeodactylum tricornutum*, *Stauroneis* sp., (Parkes et al., 2021; Haimeur et al., 2012; Sharma et al., 2021).

8.2.12 Biosynthetic Pathways of Selected Bioactive Compounds in Diatoms

8.2.12.1 Biosynthesis of Fucoxanthin

The biosynthesis of fucoxanthin in diatoms is a complex process, including the formation of β-carotene in the early stages of the transformation of gerenylgeranyl diphosphate through phytoene, β-carotene and lycopene through the act of phytoene synthase, and phytoene desaturase and β-carotene desaturase. Lycopene forms β-carotene on the action of lycopene–cyclase. The latter pigment is converted to zeaxanthin through catalysis by β-carotene hydroxylase. The next step includes the role of zeaxanthin epoxidase which leads to the synthesis of antheroxanthin and violaxanthin. All the corresponding genes encoding this enzyme were found in *P. tricornutum*. In addition, several putative genes which encode violaxanthin de-epoxidase (VDE gene) and zeaxanthin epoxidase (ZEP gene) that participate in the last step of xanthine biosynthesis in diatoms were detected. It has been suggested that the diversity of VDE and ZEP genes represent early eukaryotic invasion to higher plants. Neoxanthin, isolated from *P. tricornutum*, is a branch point in the biosynthesis of diadinotoxin and fucoxanthin in diatoms. Diadinoxanthin is formed in a single reaction along with the formation of an acetylene bond from the alenat double bond, whereas the formation of fucoxanthin by acetylation of the hydroxyl group at C3′ takes place as shown in Figure 8.2 (Stonik and Stonik, 2015).

8.2.12.2 Protein Biosynthesis in Diatoms

N-glycosylation is a major co- and post-translational protein modification in eukaryotes that occurs in both ER and Golgi apparatus (Figure 8.3). In this process, a lipid-associated oligosaccharide consisting of three glucose (Glc), nine mannose (Man), and two GlcNAc residues (Glc3Man9GlcNAc2) is first assembled by the stepwise addition of monosaccharides to dolicholpyrophosphate on the cytosolic side, which then continue in the ER lumen. These oligosaccharide precursors are then transferred by the oligosaccharyl transferase (OST) complex to the asparagine residue, consisting of the Asn-XSer/Thr consensus sequence. Other sequences such as Asn-X-Cys and Asn-X-Val

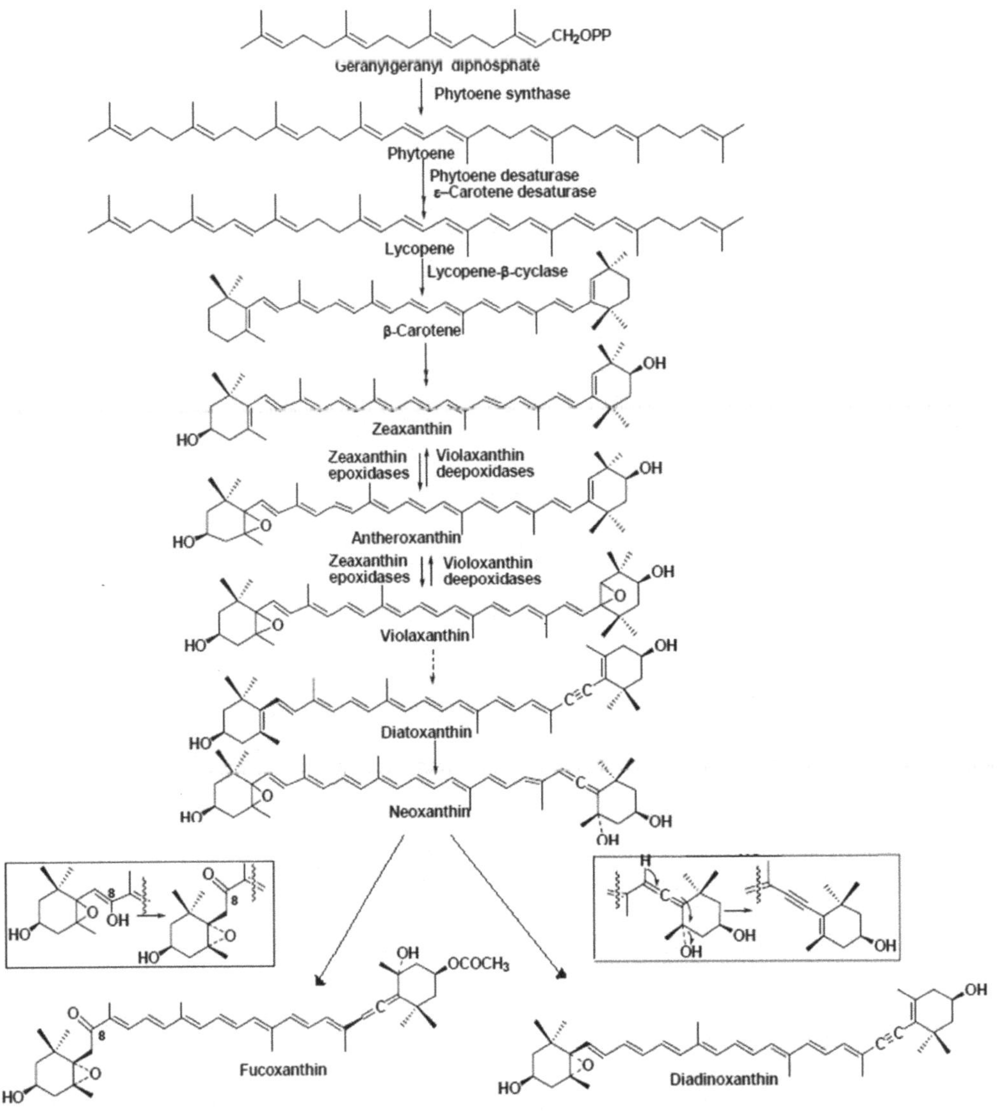

FIGURE 8.2 Proposed hypothetical pathway of fucoxanthin biosynthesis in *P. tricornutum* (modified from Stonik and Stonik, 2015).

were glycosylated in endogenous or recombinant proteins produced in mammalian or plant cells. Glycoproteins are deglucosylated by α-glucosidase I and II and then re-glucosylated by uridine diphosphate (UDP)-glucose glycoprotein glucosyl transferase (UGGT) to ensure proper folding of the nascent protein through its interaction with ER-resident chaperones, such as calnexin and calreticulin. These ER events are conserved in eukaryotes as they are essential for efficient protein folding. Bioinformatics analysis showed that most of the gene-encoding enzymes involved in the biosynthesis of dolicholpyrophosphate-linked oligosaccharides, related to asparagine-linked glycosylation (ALG), were putatively found in the genomes of diatoms (*P. tricornutum*, *Thalassiosira pseudonana*, *Fragilariopsis cylindrus*, and *Aureococcus anophagereffens*). The only exception is ALG 10, an (1,2)-glucosyl transferase that is responsible for the transfer of the Glc terminal residue in the triglucosyl extension of the N-glycan precursor, for which no homology was

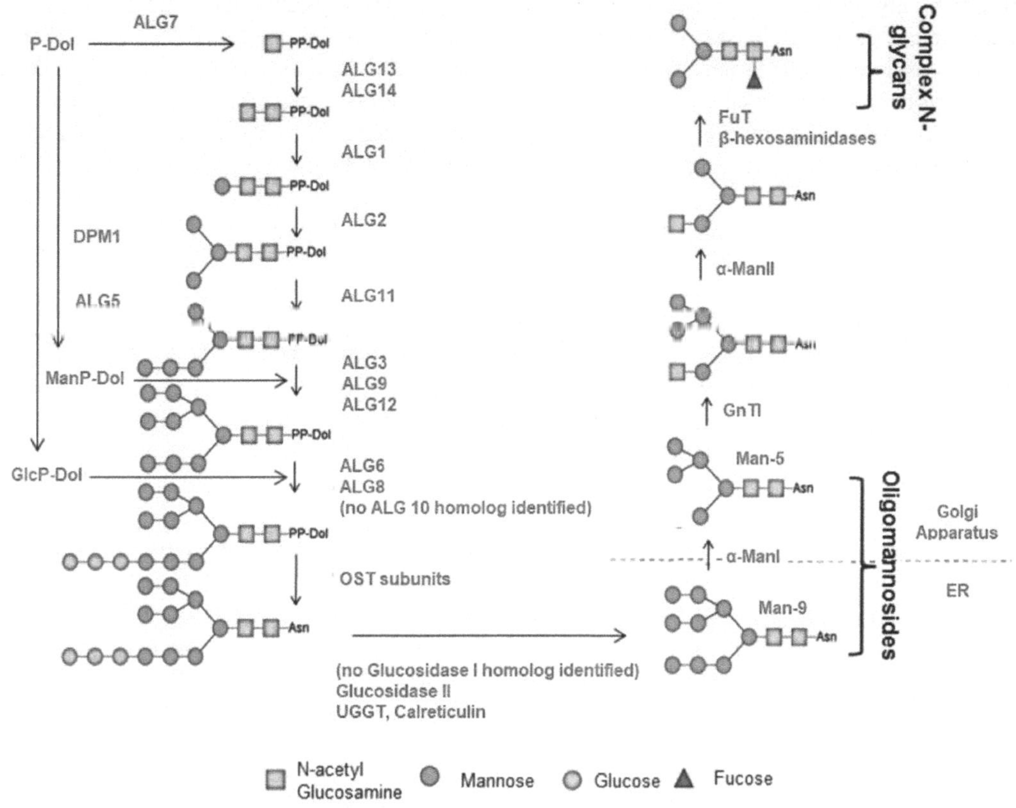

FIGURE 8.3 Proposed N-glycosylation pathway in *Phaeodactylum tricornutum*. The predicted sequences in the *P. tricornutum* genome are shown in bold. The ALG10 and glucosidase I genes have not been identified so far. RE: endoplasmic reticulum; DPM1: mannosyl dolichol-phosphate transferase; ALG: asparagine-associated glycosylation; PP-Dol: pyrophosphate dolichol; P-Dol: dolichol phosphate; OST: oligosaccharyl transferase; Asn: asparagine; UGGT: UDP-glucose glycoprotein glucosyl transferase; GnT: N-acetylglucosamine transferase; α-Man: α-mannosidase; FuT: fucosyl transferase; Man-5–Man-9: oligomannoside containing 5–9 mannose residues (modified from Gügi et al., 2015).

found. In addition to the ALG gene, other gene-encoding oligosaccharyl transferase subunits have also been identified in the diatom genome, particularly in *P. tricornutum*, where α-glucosidase II (but not α-glucosidase I) as well as resident and chaperone ER-UGGT such as calreticulin are also predicted. This protein is a key element for the protein synthesis which occurs in the ED. Large oligomannosides, such as Man9GlcNA, were found in the *P. tricornutum* glycoprotein, suggesting that the oligosaccharide precursor synthesis and the quality control of secreted proteins may occur in a similar manner as observed in other eukaryotes. However, in the ER, α-glucosidase I appear to abolish the terminal of (1,2)-glucosyl transferase which is transferred by ALG10. The absence of the ALG 10 and α-glucosidase I genes in *P. tricornutum* indicates that the N-glycan precursor is not completely glucosylated in diatoms to $Glc_3Man_9GlcNAc_2$ (Gügi et al., 2015).

In contrast to ER events, the evolutionary adaptation of N-glycan processing in the Golgi apparatus has given rise to a variety of organism-specific complex structures. First, α-mannosidase (α-Man I) degraded the oligosaccharide precursors to oligomannosides ranging from $Man_9GlcNAc_2$ to $Man_5GlcNAc_2$ (Man-9 to Man-5). N-acetylglucosaminyl transferase I (GnT I) then transfers the first N-acetylglucosaminyl (GlcNAc) residue to Man-5 and initiates the synthesis of a wide variety of structurally distinct complex-type N-glycans. This step continues with the removal of the two mannosyl residues, which leads to the decoration of the N-glycans by the repertoire action of specific glycosyl transferases such as α-fucosyl transferases (FuT). Therefore, mature proteins released from the secretory pathway by carrying the organism-specific N-glycan complexes that allow protein to acquire a range of glycan-mediated biological functions. The investigation for candidate genes encoding Golgi glycosidase and glycosyl transferases in the diatom genome involved in N-glycan processing has led to the identification of candidate α-Man I, GnT I, and FuT [putative (1,3)-FuT]. The predicted GnT I gene in the *P. tricornutum* genome has been shown to encode a functional active enzyme which is capable of restoring the maturation of N-linked glycans to complex-type N-glycans in the CHO Lec_1 mutant affected by its endogenous GnT I. In addition, structural analysis of N-linked glycans to proteins secreted by *P. tricornutum* diatoms showed that these oligosaccharides are processed via a GnT I-dependent pathway to partially fucosylated $Man_3GlcNAc_2$. These cleaved and fucosylated N-linked glycans likely resulted from pruning of two mannose residues from Man-5 by α-Man II and subsequent transfer of (1,3)-fucose residues. Then, the terminal GlcNAc introduced by Golgi *P. tricornutum* GnT I might be eliminated by α-hexosaminidases, as previously described in land plants and insects. Two putative α-hexosaminidases have been identified in *P. tricornutum* (Gügi et al. 2015).

In addition, diatoms may have O-glycan-containing glycoproteins. For example, Swift and Wheeler demonstrated that several silica shell-associated proteins, called frustule-associated components (FACs), could be easily extracted from cell walls by EDTA-based treatment. This fraction may contain glycoproteins. In *Cylindrotheca fusiformis*, the soluble protein EDTA, called frustulin, contains a glycan moiety consisting of rhamnose (25%), galactose (20%), xylose (20%), glucose (2%), and mannose (1%). Although first described as proteoglycans, glycoproteins have also been isolated in urea-extracted *Craspedosaurus australis* and exhibited a xylose-rich composition (41%), with lower amounts of galactose, rhamnose, and mannose of about 14%, 12%, and 10%, respectively (Gügi et al., 2015).

8.2.12.3 Biosynthesis of Ribose Nucleotides in Diatoms

Monosaccharides represent the building blocks for glycans and polysaccharides. They are normally synthesized and converted to nucleotide sugars via cytosolic interconversion metabolism (KEGG map 00520). In turn, sugar nucleotides—which are universal sugar donors—are involved in the formation of polysaccharides, glycoproteins, proteoglycans, and glycolipids (Figure 8.4). This type of metabolism is highly conserved in prokaryotes and eukaryotes and involves a set of phosphorylases, epimerases and reductases, as well as amino fructose-6-phosphate transferases that allow the synthesis of amino sugars. Nucleotide sugars can also be generated from the salvage pathways involving the hydrolysis of glycans to free sugars, their phosphorylation, and finally their nucleotidylation. The identification of genes encoding cytosolic enzymes in the genomes of diatoms, such as *P. tricornutum*, *T. pseudonana*, *Fragilariopsis cylindrus*, and *Aureococcus anophagefferens* generates several putative candidates for the synthesis of UDP sugars, such as UDP-galactose (UDP-galactose), UDP-galacturonic acid (UDP-GalA), UDP-glucuronic acid (UDP-GlcA), UDP-xylose (UDP-Xyl), and UDP-rhamnose (UDP-Rha), which are directly derived from UDP-Glc. In addition, gene predictions also included enzymes required for the synthesis of the sugar guanine diphosphate (GDP) derived from Man-6P. Aminosugar like GlcNAc biosynthesis is likely to occur by C2 amination of fructose-6P (Fru-6P) as reported in other organisms. Other gene predictions based on the diatom genome include genes encoding several sugar phosphorylases of the rescue pathway. However, neither UDP-galacturonate decarboxylases, nor UDP-arabinose 4-epimerase required for L-arabinose biosynthesis were discovered in the genome. Apart from arabinose, predictions of nucleotide sugar metabolism were based on the sugar composition of polysaccharides and glycans which are isolated from diatoms (Gügi et al., 2015). Figure 8.4 shows the prediction of nucleotide sugar metabolism in diatoms.

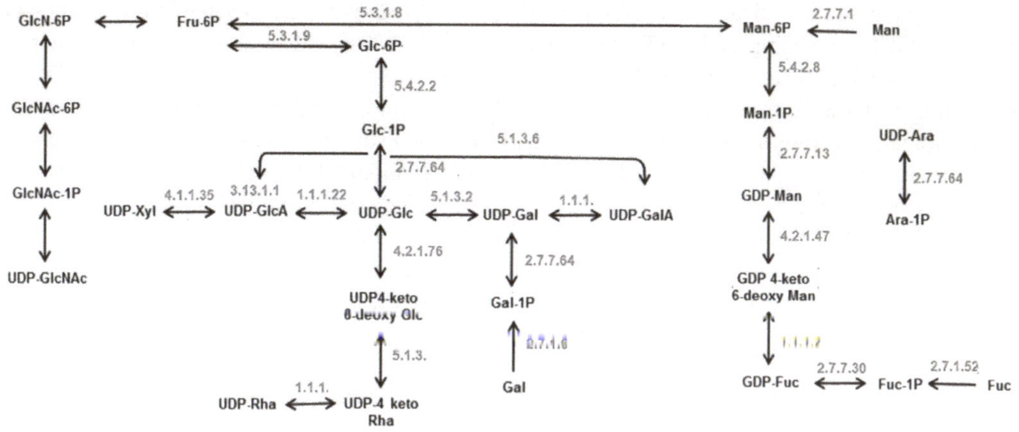

FIGURE 8.4 Prediction of nucleotide sugar metabolism in diatoms based on the comparative genomic analysis in *Phaeodactylum tricornutum*, *Thalassiosira pseudonana*, *Fragilariopsis cylindrus*, and *Aureococcus anophagefferens* (modified from Gügi et al., 2015).

8.3 IN VITRO EVALUATIONS OF ANTIOXIDANT ACTIVITY OF DIATOMS

Reactive oxygen species (ROS) cause oxidative stress in mammalian cells, which leads to the development of various diseases. The free radicals 1,1-diphenyl-2-picrylhydrazyl (DPPH) and 2,2'-azino-bis (3-ethylbenzthiazolin-6-sulfonic acid) (ABTS) are harmful molecules that are detrimental to human health, causing disorders such as cancer and hypertension. The DPPH and ABTS radical scavenger assays are widely used to assess the biological potential of natural extracts due to their simplicity and methodological stability (Pekkoh et al., 2022). The DPPH assay works by a mechanism in which antioxidants act to inhibit lipid oxidation, which then scavenges DPPH radicals and determines the free radical scavenging capacity. The ABTS scavenging activity assay works through a single electron transfer mechanism, which is proved as a reliable method to test food products containing hydrophilic and highly pigmented antioxidants (Foo et al., 2015).

Hydroxyl radicals are believed to be the most dangerous free radicals in ROS, since they may cause severe damage to adjacent biomolecules. Superoxide radicals are considered to be one of the important free radicals in vivo, which contribute to the oxidative stress of vital cells by reducing metal transitions and reacting with hydroxyl radicals. Therefore, analyses of the hydroxyl and superoxide radical scavenging activities were beneficial to determine the antioxidant potential (Xia et al., 2013; Alzahrani et al., 2017).

The total phenolic content can be determined by Folin-Ciocalteu assay, which is an electron transfer-based test that provides a reduction capacity that is expressed (Alzahrani, et al., 2017). In addition, the iron chelation assay is carried out to measure the ability of compounds to chelate or bond with metal ions. The ferric-reducing anti-oxidant power (FRAP) assay is used to evaluate the reducing ability of compounds that depend on the donation of hydrogen atoms to inactivate free radical formation in living systems. The deoxyribose assay is performed to determine the constant rate of the reaction between the anti-oxidant molecules which present in the extract and the hydroxyl radicals. Lastly, NBT assay is used to measure antioxidants through superoxide radical scavenging activity (Lakshmegowda et al., 2020).

The DPPH and ABTS radical scavenger assays were carried out to evaluate the antioxidant properties of two diatom lipid contents, *Anomoeoneis* sp. AARL D039 (Ano) and *Rhopalodia* sp. AARL D020 (Rho), with Ano, showed a lower half maximum inhibitory concentration (IC50) than Rho (Pekkoh et al., 2022). The lower the IC50 value, the more potent the extract to be used in pharmaceutical, nutraceutical, and cosmetic applications. Meanwhile, IC50 values for Ano indicate that the DPPH scavenging and ABTS free radicals were 3.65 mg/mL and 4.83 mg/mL, whereas IC50 values for Rho were 4.49 mg/mL and 11.26 mg/mL. The reference compound used for DPPH scavenging and ABTS free radicals is Trolox, with IC50 values of 7.68 g/mL and 0.30 mg/mL. When compared, the antioxidant activities of fatty acids in Ano and Rho were found to be lower than that of Trolox. This might be because Trolox is a single compound that acts as a radical scavenger, while fatty acids contain both active and inactive forms. The results showed that the lipid extract of Ano and Rho had the ability to reduce the unstable DPPH free radicals by converting them to a more stable form and reducing the ABTS free radicals through oxidation process. The overall results suggested that Ano- and Rho-derived lipids can be used as a supplement or directly as food because of their bioactive potential (Pekkoh et al., 2022).

A study conducted by Lakshmegowda et al. (2020) showed that total phenolic and flavonoid contents in *Nitzschia palea* hexane extract (NPHE) to be equivalent with 346.28 ± 34.74 mg/g gallic acid and 17.22 ± 0.70 mg/g catechins, respectively. Antioxidant activity was tested by DPPH radical scavenging, iron ion chelating, nitric oxide and ABTS radical scavenging activity, deoxyribose, and FRAP assay. The free radical scavenging activity of DPPH from NPHE showed an IC50 value equivalent to 194.75 ± 6.6 g/ml BHT (butylated hydroxytoluene). The iron ion chelating property of the hexane fraction was shown to be 139.36 ± 6.97 g/mL as compared with standard EDTA equivalent. The results of the nitric oxide test showed a remarkable reduction of nitric oxide by the hexane fraction of 303.76 ± 6.93 g/mL. The ABTS radical scavenging activity by NPHE showed excellent result of 26.52 ± 1.2 g/mL equivalent to butylated hydroxyanisole (BHA) standard, whereas the deoxyribose assay showed an IC50 value of 246.81 ± 24.51 g/mL. Overall, NPHE was capable of inhibiting the formation of NBT blue colored in superoxide radical scavenging activity, in terms of an IC50 value of 102.42 ± 10.39 g/ml. NPHE showed dose-dependent inhibition of lipid peroxidation with a corresponding IC50 value of 269.97 ± 7185 g/mL. The FRAP assay was used to determine the reduction capacity of Fe^{3+} ions to Fe^{2+} ions, and the calculated reduction value was 11.19 ± 0.25 mM/mg (Lakshmegowda et al., 2020).

Another study conducted by Smerilli et al. (2019) showed that *Skeletonema marinoi* is rich in phenolic compounds which are the most widespread antioxidants in photosynthetic organisms. Phenolic compounds act directly against radical species or indirectly through inhibition of pro-oxidant enzymes such as lipoxygenase or through metal chelation, preventing the Haber-Weiss and Fenton reactions, which are important sources of radical species. The antioxidant compounds in *Skeletonema marinoi*, such as ascorbic, phenolic, and flavonoid acids displayed ABTS scavenging activity under low light conditions (Smerilli et al., 2019).

Meanwhile, a study conducted by Fimbres-Olivarria et al. (2018) into *Navicula* sp. has shown that polysaccharides in the form of WSPN (sulfated polysaccharide from *Navicula* sp. in white wavelength), BSPN (sulfated polysaccharide from *Navicula* sp. in blue wavelength), and RSPN (sulfated polysaccharide from *Navicula* sp. in red wavelength) showed DPPH scavenging activity at various concentrations (Fimbres-Olivarria et al., 2018). The free radical scavenging rate of DPPH varied from 14.0% to 48.7% when the concentration was increased from 25 to 200 g/mL. BSPN showed a clear increase of antioxidant activity when the concentration was increased from 25 to 200 g/mL, whereas the WSPN and RSPN samples remained nearly constant across the evaluated concentrations. The highest scavenging activity was observed at 200 g/mL for the three samples, which was higher for WSPN (48.7 ± 3.1%) in relation to BSPN (35.5 ± 3.2%) and RSPN (27.2 ± 2.1%). IC50 was used to compare the radical scavenging activity of different extracts. This value represents the sample concentration required to scavenge 50% of free radicals. WSPN presents the lowest IC50 value (238 g/mL) compared to BSPN (326 g/mL) and RSPN (3066 g/mL). Based on the ABTS method test, WSPN showed higher activity than BSPN and RSPN (68.17±4.68 mol Trolox/g). BSPN and RSPN showed the same values (53.85±0.45 and 53.69±2.89 mol Trolox/g). In both methods (DPPH and ABTS), the scavenging activity of WSPN was higher than that of BSPN and

RSPN; this could be attributed to the lower MW (17 kDa), sulfate content (0.40% w/w), and amount of rhamnose (35% w/w).

It is widely known that polysaccharides with low Mw tend to have the strongest reducing power, which is due to the polysaccharide chains having a higher content of reducing ends in the terminal reductive hydroxyl groups to remove and accept free radicals. In addition, the content of rhamnose in the polysaccharide is the most significant determining factor related to antioxidant properties, and the higher the sulfate content, the greater the antioxidant ability. This suggests that sulfated polysaccharides from microalgae are attractive natural antioxidants that could be successfully applied in different functional products (Fimbres-Olivarria et al., 2018).

A study conducted by Alzahrani et al. (2018) showed the determination of the total phenolic content/reducing capacity of protein extract and *Nitzschia laevis* hydrolysate by Folin-Ciocalteu reagent. *N. laevis* protein extract showed a reduction capacity of 0.56/0.01 mg GAE/100 g, which was lower than that of *Spirulina* protein extract (0.60/0.05 mg GAE/100 g, $P < 0.05$). Gallic acid equivalent (GAE) is a general reference for measuring a number of phenolic compounds contained in a material. The hydrolysis process increased the reduction capacity of the extract, which could be due to the complex fragmentation and the production of smaller bioactive peptides by the enzymes. For *Nitzschia*, alcalase hydrolysate after 90 minutes, while Flavorzyme hydrolysate after 30 min yielded the highest reducing capacities of 2.40 ± 0.02 and 1.96 ± 0.04 mg GAE/100 g, respectively. The trypsin hydrolysate after 30 minutes showed a considerable effect on the reduction capacity (Alzahrani et al., 2018).

Protein extract and hydrolysate from *Nitzschia* showed high ABTS radical scavenging activity. The hydrolysate produced using the three enzymes increased antioxidant activity and showed higher ABTS scavenging activity compared to the unhydrolyzed protein, especially at a longer exposure ($P < 0.05$). The trypsin hydrolysate generally showed the lowest ABTS value compared to Alcalase and Flavorzyme, especially at the hydrolysis times of 30, 60 and 120 minutes. The higher ABTS radical quenching activity was observed mostly with Alcalase hydrolysate, especially at 60 min of hydrolysis. It can be concluded that the peptides produced by the three proteases act as antioxidants through electron transfer reactions (Alzahrani et al., 2018).

The DPPH radical scavenging assay indicated all protein extracts and hydrolysates that showed antioxidant activity against DPPH. The longer hydrolysis time showed better DPPH scavenging activity for the hydrolysate compared to the control, except for trypsin. Protein hydrolysate obtained by Alcalase showed the highest DPPH activity, which was hydrolyzed by Flavorzyme giving DPPH values similar to Alcalase. Trypsin hydrolysate showed the least DPPH activity ($P < 0.05$). The antioxidant activity of this hydrolysate can be caused by structural modification of proteins by certain enzymes after a certain hydrolysis time. Structural modification is achieved by disrupting the tertiary structure of the protein, generating fragments with highly active sites (i.e. peptides, amino acids), thereby chelating free radicals (Alzahrani et al., 2018).

In a study by Foo et al. (2015), the main antioxidant compounds (carotenoids and phenolic compounds) in *Chaetoceros calcitrans* diatoms were extracted using six different solvents: methanol, ethanol, acetone (ACE), 9:1 v/v acetone and water (AW), 9:1 v/v acetone and chloroform (AC), and 8:1:1 v/v/v acetone, chloroform and methanol (ACM). In terms of antioxidant activity, it was observed that the methanol extract showed the highest DPPH scavenging activity of 1,435 mg TE/g DW. It was significantly different ($P < 0.05$) from ethanol extract (0.844 mg TE/g DW), AW (0.665 mg TE/g DW), ACM (0.173 mg TE/g DW), and AC (0.067 mg TE/g DW). The ABTS assay showed a trend of antioxidant efficacy which was similar to the DPPH scavenger assay. It was found that the methanol extract showed the highest antioxidant activity (10,593 mg TE/g DW), followed by ethanol (7,210 mg TE/g DW), AW (4,159 mg TE/g DW), acetone (3,046 mg TE/g DW), extract ACM (2,528 mg TE/g DW), and AC (1,211 mg TE/g DW). In addition to potent antioxidant efficacy through primary antioxidant mechanisms, all extracts also showed positive effects through secondary antioxidant mechanisms. Following a similar trend to the radical scavenging activity assay, the methanol extract consistently showed the highest iron chelating activity (18.52 mg Na-EDTA/g DW) which was significantly different ($P < 0.05$) from ethanol (14.17 mg Na-EDTA/g DW), AW (13.79 mg Na-EDTA/g DW), ACM (4.25 mg Na-EDTA/g DW), acetone (2.64 mg Na-EDTA/g DW), and AC (0.29 mg Na-EDTA/g DW). Overall, all extracts exhibited free radical scavenging activity but at different levels (Foo et al., 2015).

From the correlated study, the compounds responsible for the scavenging activity of DPPH were phenolics ($r = 0.918$, $P < 0.05$) and carotenoids ($r = 0.762$, $P < 0.05$). Phenolic compounds (gallic acid and protocatechuic acid) have hydroxyl groups that loosely bonded to their aromatic rings, thus they can easily donate hydrogen atoms or electrons to reactive radicals, making them important radical scavengers. This study indicated that methanol is the recommended solvent for the production of antioxidant-rich extracts of *C. calcitrans*. Methanol extract consistently showed the highest extraction yield [(22.71 ± 0.96) g/100 g DW], total carotenoids [(4.46 ± 0.36) mg/g DW], fucoxanthin content [(2.08 ± 0.03) mg FX/g DW], and total phenolic [(2.49 ± 0.08) mg/g DW] with good free radical scavenging and iron chelating activity (Foo et al., 2015).

In the study conducted by Xia et al. (2014), a novel chrysolaminarin, named CL2, with a molecular mass of 7.75 kDa, was purified from the marine diatom *Odontella aurita*, using DEAE-52 cellulose anion exchange chromatography and Sephadex G-200 gel filtration chromatography. Monosaccharides and structural analysis revealed that CL2 is a glucan consisting mainly of glucose, linked by –D-(1→3) (main chain) and –D-(1→6) (side chain) glycosidic bonds. Antioxidant activity assay revealed that CL2 showed a stronger hydroxyl radical scavenging activity with increased concentration. At 1 mg/mL, CL2 showed a hydroxyl radical scavenging activity of 17.72% ± 1.98%, which gradually increased to 83.54% ± 6.71% at 10 mg/mL. However, the scavenging activity of CL2 was much lower than that of ascorbic acid, as it reached 99.38% ± 2.089% at 1 mg/mL. Furthermore, for the assessment of reducing power, namely the Fe^{3+}–Fe^{2+} transformation of CL2 which was investigated using the potassium ferricyanide reduction method, it was shown that the reducing power of CL2 was weak. At 2 mg/mL, CL2 showed a reducing power of 0.046 ± 0.015 abs, which gradually increased to 0.554 ± 0.139 abs at 100 mg/mL; while the reducing power of ascorbic acid reached 2,508 ± 0.13 abs at 2 mg/mL. In addition, the DPPH radical scavenging assay showed that CL2 scavenging activity increased slowly with increasing concentration. At 100 mg/mL CL2 concentration, the scavenging activity reached 42.455% ± 4.671%. Ascorbic acid showed much higher scavenging ability of 93.774% ± 2.089% at 2 mg/mL. In conclusion, chrysolaminarin CL2 from *Odontella aurita* can be explored as a natural antioxidant agent for applications in the aquaculture, food, and pharmaceutical fields (Xia et al. 2013; Xia et al., 2014).

REFERENCES

Adwas, Almokhtar A., Ata Sedik Ibrahim Elsayed, Azab Elsayed Azab, and Fawzia Amhimmid Quwaydir. (2019). Oxidative stress and antioxidant mechanisms in human body. *Journal of Applied Biotechnology & Bioengineering* 6(1): 43–47. https://doi.org/10.15406/jabb.2019.06.00173

Alzahrani, Mona Ahmed J., Conrad O. Perera, and Yacine Hemar. (2018). Production of bioactive proteins and peptides from the diatom *Nitzschia laevis* and comparison of their in vitro antioxidant activities with those from *Spirulina platensis* and *Chlorella vulgaris*. *International Journal of Food Science and Technology* 53(3): 676–682. https://doi.org/10.1111/ijfs.13642

Alzahrani, Mona Ahmed J., Conrad O Perera, and Yacine Hemar. (2017). Production of bioactive proteins and peptides from the diatom *Nitzschia laevis* and comparison of their in vitro antioxidant activities with those from *Spirulina platensis* and *Chlorella vulgaris*. *International Journal of Food Science & Technology* 53(3): 1–7. https://doi.org/10.1111/ijfs.13642

Amir Aslani, Banafsheh, and Sirous Ghobadi. (2016). Studies on oxidants and antioxidants with a brief glance at their relevance to the immune system. *Life Sciences* 146: 163–173. https://doi.org/10.1016/j.lfs.2016.01.014

Athanasakoglou, Anastasia, and Sotirios C. Kampranis. (2019). Diatom isoprenoids: advances and biotechnological potential. *Biotechnology Advances* 37(8). https://doi.org/10.1016/j.biotechadv.2019.107417

Azizan, Awanis, Muhammad Safwan Ahamad Bustamam, M. Maulidiani, Khozirah Shaari, Intan Safinar Ismail, Norio Nagao, and Faridah Abas. (2018). Metabolite profiling of the microalgal diatom *Chaetoceros calcitrans* and correlation with antioxidant and nitric oxide inhibitory activities via 1H NMR-based metabolomics. *Marine Drugs* 16(5): 1V. https://doi.org/10.3390/md16050154

Balsera, Monica, and Bob B. Buchanan. (2019). Evolution of the thioredoxin system as a step enabling adaptation to oxidative stress. *Free Radical Biology and Medicine* 140: 28–35. https://doi.org/10.1016/j.freeradbiomed.2019.03.003

Baoyan, Gao, Chen Ailing, Zhang Wenyuan, Li Aifen, and Zhang Chengwu. (2017). Co-production of lipids, eicosapentaenoic acid, fucoxanthin, and chrysolaminarin by *Phaeodactylum tricornutum* cultured in a flat-plate photobioreactor under varying nitrogen conditions. *Journal of the Ocean University of China* 16(5): 916–924. https://doi.org/10.1007/s11802-017-3174-2

Cardoso, B. R., D. J. Hare, A. I. Bush, and B. R. Roberts. (2017). Glutathione peroxidase 4: a new player in neurodegeneration? *Molecular Psychiatry* 22(3): 328–335. https://doi.org/10.1038/mp.2016.196

Chasapis, Christos T., Massimiliano Peana, and Vlasoula Bekiari. (2022). Structural identification of metalloproteomes in marine diatoms, an efficient algae model in toxic metals bioremediation. *Molecules* 27(2): 1–10. https://doi.org/10.3390/molecules27020378

Ferdous, Umme Tamanna, Zetty Norhana, and Balia Yusof. (2021). Medicinal prospects of antioxidants from algal sources in cancer therapy. *Frontiers in Pharmacology* 12: 1–23. https://doi.org/10.3389/fphar.2021.593116

Fiedor, Joanna, and Květoslava Burda. (2014). Potential role of carotenoids as antioxidants in human health and disease. *Nutrients* 6(2): 466–488. https://doi.org/10.3390/nu6020466

Figueroa, Fabián A, Roberto Abdala-díaz, Víctor Hernández, Pablo Pedreros, Mario Aranda, Jaime R. Cabrera-pardo, Claudia Pérez, José Becerra, and Roberto Urrutia. (2019). Invasive diatom *Didymosphenia geminata* as a source of polysaccharides with antioxidant and immunomodulatory effects on macrophage cell lines. *Journal of Applied Phycology* 32: 93–102.

Fimbres-Olivarria, Diana, Elizabeth Carvajal-Millan, Jose Antonio Lopez-Elias, Karla Guadalupe Martinez-Robinson, Anselmo Miranda-Baeza, Luis Rafael Martinez-Cordova, Fernando Enriquez-Ocaña, and Jose Eduardo Valdez-Holguin. (2018). Chemical characterization and antioxidant activity of sulfated polysaccharides from *Navicula* sp." *Food Hydrocolloids* 75: 229–236. https://doi.org/10.1016/j.foodhyd.2017.08.002

Foo, Su Chern, Fatimah Md Yusoff, Maznah Ismail, Mahiran Basri, Nicholas Mun Hoe Khong, Kim Wei Chan, and Sook Kun Yau. (2015). Efficient solvent extraction of antioxidant-rich extract from a tropical diatom, *Chaetoceros calcitrans* (Paulsen) Takano 1968. *Asian Pacific Journal of Tropical Biomedicine* 5(10): 834–840. https://doi.org/10.1016/j.apjtb.2015.06.003

García-Sánchez, Andrés, Alejandra Guillermina Miranda-Díaz, and Ernesto Germán Cardona-Muñoz. (2020). The role of oxidative stress in physiopathology and pharmacological treatment with pro- and antioxidant properties in chronic diseases. *Oxidative Medicine and Cellular Longevity* 1–16. https://doi.org/10.1155/2020/2082145

Goh, Su-Hua, Fatimah Md Yusoff, and Su Peng Loh. (2010). A comparison of the antioxidant properties and total phenolic content in a diatom, *Chaetoceros* sp. and a green microalga, *Nannochloropsis* sp. *Journal of Agricultural Science* 2(3). https://doi.org/10.5539/jas.v2n3p123

Grigoras, Anca Giorgiana. (2017). Catalase immobilization—a review. *Biochemical Engineering Journal* 117: 1–20. https://doi.org/10.1016/j.bej.2016.10.021

Gügi, Bruno, Tinaïg Le Costaouec, Carole Burel, Patrice Lerouge, William Helbert, and Muriel Bardor. (2015). Diatom-specific oligosaccharide and polysaccharide structures help to unravel biosynthetic capabilities in diatoms. *Marine Drugs* 13(9): 5993–6018. https://doi.org/10.3390/md13095993

Haimeur, Adil, Lionel Ulmann, Virginie Mimouni, Frédérique Guéno, Fabienne Pineau-Vincent, Nadia Meskini, and Gérard Tremblin. (2012). The role of *Odontella aurita*, a marine diatom rich in EPA, as a dietary supplement in dyslipidemia, platelet function and oxidative stress in high-fat fed rats. *Lipids in Health and Disease* 11: 1–13. https://doi.org/10.1186/1476-511X-11-147

Hussein, Hanaa Ali, and Mohd Azmuddin Abdullah. (2020). Anticancer compounds derived from marine diatoms. *Marine Drugs* 18(7). https://doi.org/10.3390/md18070356

Imam, Mustapha Umar, Shenshen Zhang, Jifei Ma, Hao Wang, and Fudi Wang. (2017). Antioxidants mediate both iron homeostasis and oxidative stress. *Nutrients* 9(7): 1–19. https://doi.org/10.3390/nu9070671

Karthikeyan, P. (2013). In vitro antioxidant activity of marine diatoms. *IOSR Journal of Environmental Science, Toxicology and Food Technology* 5(2): 32–37. https://doi.org/10.9790/2402-0523237

Khaw, Yam Sim, Fatimah Yusoff, Hui Teng Tan, Nur Amirah, Izyan Noor, Muhammad Farhan Nazarudin, Noor Azmi Shaharuddin, and Abdul Rahman Omar. (2021). The critical studies of fucoxanthin research trends from 1928 to June 2021: a bibliometric review. *Marine Drugs* 19(606): 1–25.

Kimura, Shunsuke, Yen Chen Tung, Min Hsiung Pan, Nan Wei Su, Ying Jang Lai, and Kuan Chen Cheng. (2017). Black garlic: a critical review of its production, bioactivity, and application. *Journal of Food and Drug Analysis* 25(1): 62–70. https://doi.org/10.1016/j.jfda.2016.11.003

Kuczynska, Paulina, Malgorzata Jemiola-rzeminska, and Kazimierz Strzalka. (2015). Photosynthetic pigments in diatoms. *Marine Drugs* 13: 5847–5881. https://doi.org/10.3390/md13095847

Kuppusamy, Palaniselvam, Ilavenil Soundharrajan, Srisesharam Srigopalram, Mashitah M. Yusoff, Gaanty Pragas Maniam, Natanamurugaraj Govindan, and Ki Choon Choi. (2017). Potential pharmaceutical and biomedical applications of diatoms microalgae – an overview." *Indian Journal of Geo-Marine Sciences* 46(4): 663–667.

Kurutas, Ergul Belge. (2016). The importance of antioxidants which play the role in cellular response against oxidative/nitrosative stress: current state. *Nutrition Journal* 15(1): 1–22. https://doi.org/10.1186/s12937-016-0186-5.

Lakshmegowda, Sudarshan Bilikallahalli, Sanjay Konasur Rajesh, Hemanth Kumar Kandikattu, Ilaiyaraja Nallamuthu, and Farhath Khanum. (2020). In vitro and in vivo studies on hexane fraction of *Nitzschia palea*, a freshwater diatom for oxidative damage protective and anti-inflammatory response. *Revista Brasileira de Farmacognosia* 30(2): 189–201. https://doi.org/10.1007/s43450-020-00008-6.

Lauritano, Chiara, Jeanette H. Andersen, Espen Hansen, Marte Albrigtsen, Giovanna Romano, and Adrianna Ianora. (2016). Bioactivity screening of microalgae for antioxidant, anti-inflammatory, antibacterial activities. *Frontiers in Marine Science* 3: 1–12. https://doi.org/10.3389/fmars.2016.00068

Lee, Sean Qiu En, Tsu Soo Tan, Makoto Kawamukai, and Ee Sin Chen. (2017). Cellular factories for coenzyme Q10 production. *Microbial Cell Factories* 16(1): 1–16. https://doi.org/10.1186/s12934-017-0646-4

Manisha, Whidul Hasan, Richa Rajak, and Deepali Jat. (2017). Oxidative stress and antioxidant: an overview. *International Journal of Advanced Research and Review* 2(9): 110–119. https://doi.org/10.1016/j.cnd.2009.09.001

Mokhtari, Vida, Parvaneh Afsharian, Maryam Shahhoseini, Seyed Mehdi Kalantar, and Ashraf Moini. (2017). A review on various uses of N-acetyl cysteine citation: Mokhtari V, Afsharian P, Shahhoseini M, Kalantar SM, Moini A. A review on various uses of N-acetyl." *Cell Journal (Yakhteh)* 19(191): 11–17.

Parkes, Rachel, Lorraine Archer, Dónal McGee, Thomas J. Smyth, Eoin Gillespie, and Nicolas Touzet. (2021). Differential responses in EPA and fucoxanthin production by the marine diatom *Stauroneis* sp. under varying cultivation conditions. *Biotechnology Progress* 37(6): 2. https://doi.org/10.1002/btpr.3197

Pekkoh, Jeeraporn, Kittiya Phinyo, Theera Thurakit, Sureeporn Lomakool, Kritsana Duangjan, Khomsan Ruangrit, Chayakorn Pumas, et al. (2022). Lipid profile, antioxidant and antihypertensive activity, and computational molecular docking of diatom fatty acids as ACE inhibitors. *Antioxidants* 11(2): 1–16. https://doi.org/10.3390/antiox11020186

Peng, Juan, Jian-ping Yuan, Chou-fei Wu, and Jiang-hai Wang. (2011). Fucoxanthin, a marine carotenoid present in brown seaweeds and diatoms: metabolism and bioactivities relevant to human health. *Marine Drugs* 9: 1806–1828. https://doi.org/10.3390/md9101806

Pisoschi, Aurelia Magdalena, Aneta Pop, Florin Iordache, Loredana Stanca, Gabriel Predoi, and Andreea Iren Serban. (2021). Oxidative stress mitigation by antioxidants – an overview on their chemistry and influences on health status. *European Journal of Medicinal Chemistry* 209: 112891. https://doi.org/10.1016/j.ejmech.2020.112891

Rahal, Anu, Amit Kumar, Vivek Singh, Brijesh Yadav, Ruchi Tiwari, Sandip Chakraborty, and Kuldeep Dhama. (2014). Oxidative stress, prooxidants, and antioxidants: the interplay. *BioMed Research International* 1–19. https://doi.org/10.1155/2014/761264

Rico, Milagros, Aroa López, J. Magdalena Santana-Casiano, Aridane G. González, and Melchor González-Dávila. (2013). Variability of the phenolic profile in the diatom *Phaeodactylum tricornutum* growing under copper and iron stress. *Limnology and Oceanography* 58(1): 144–152. https://doi.org/10.4319/lo.2013.58.1.0144

Rochette, Luc, Steliana Ghibu, Adriana Muresan, and Catherine Vergely. (2015). Alpha-lipoic acid: molecular mechanisms and therapeutic potential in diabetes. *Canadian Journal of Physiology and Pharmacology* 93(12): 1021–1027. https://doi.org/10.1139/cjpp-2014-0353

Salehi, Bahare, Antonio Rescigno, Tinuccia Dettori, Daniela Calina, Anca Oana Docea, Laxman Singh, Fatma Cebeci, et al. (2020). Avocado–soybean unsaponifiables: a panoply of potentialities to be exploited. *Biomolecules* 10(1): 1–20. https://doi.org/10.3390/biom10010130

Scrob, Teodora, Anamaria Hosu, and Claudia Cimpoiu. (2019). The influence of in vitro gastrointestinal digestion of *Brassica oleracea* florets on the antioxidant activity and chlorophyll, carotenoid and phenolic content. *Antioxidants* 8(212): 1–11.

Sharifi-Rad, Mehdi, Nanjangud V. Anil Kumar, Paolo Zucca, Elena Maria Varoni, Luciana Dini, Elisa Panzarini, Jovana Rajkovic, et al. (2020). Lifestyle, oxidative stress, and antioxidants: back and forth in the pathophysiology of chronic diseases. *Frontiers in Physiology* 11: 1–21. https://doi.org/10.3389/fphys.2020.00694

Sharma, Nikunj, Daris Pazhukkunnel Simon, Aracely Maribel Diaz-Garza, Elisa Fantino, Anis Messaabi, Fatma Meddeb-Mouelhi, Hugo Germain, and Isabel Desgagné-Penix. (2021). Diatoms biotechnology: various industrial applications for a greener tomorrow. *Frontiers in Marine Science* 8 (February). https://doi.org/10.3389/fmars.2021.636613

Singh, Yogendra Pratap, Ram N. Patel, Yogendra Singh, Ray J. Butcher, Pradeep Kumar Vishakarma, and R. K. Bhubon Singh. (2017). Structure and antioxidant superoxide dismutase activity of copper(II) hydrazone complexes. *Polyhedron* 122: 1–15. https://doi.org/10.1016/j.poly.2016.11.013

Smerilli, Arianna, Sergio Balzano, Maira Maselli, Martina Blasio, Ida Orefice, Christian Galasso, Clementina Sansone, and Christophe Brunet. (2019). Antioxidant and photoprotection networking in the coastal diatom *Skeletonema marinoi*. *Antioxidants* 8(6). https://doi.org/10.3390/antiox8060154

Stonik, Valentin Stonik, and Inna Stonik. (2015). Low-molecular-weight metabolites from diatoms: structures, biological roles and biosynthesis. *Marine Drugs* 13(6): 3672–3709. https://doi.org/10.3390/md13063672

Tan, Bee Ling, Mohd Esa Norhaizan, Winnie Pui Liew, and Heshu Sulaiman Rahman. (2018). Antioxidant and oxidative stress: a mutual interplay in age-related diseases. *Frontiers in Pharmacology* 9: 1–28. https://doi.org/10.3389/fphar.2018.01162

Tarocco, Anna, Natascia Caroccia, Giampaolo Morciano, Mariusz R. Wieckowski, Gina Ancora, Giampaolo Garani, and Paolo Pinton. (2019). Melatonin as a master regulator of cell death and inflammation: molecular mechanisms and clinical implications for newborn care. *Cell Death and Disease* 10(4): 1–12. https://doi.org/10.1038/s41419-019-1556-7

Tiwari, Archana, Elda M. Melchor-Martínez, Abhishek Saxena, Neha Kapoor, Kawal Jeet Singh, Sara Saldarriaga-Hernández, Roberto Parra-Saldívar, and Hafiz M.N. Iqbal. (2021). Therapeutic attributes and applied aspects of biological macromolecules (polypeptides, fucoxanthin, sterols, fatty acids, polysaccharides, and polyphenols) from diatoms — a review. *International Journal of Biological Macromolecules* 171: 398–413. https://doi.org/10.1016/j.ijbiomac.2020.12.219

Xia, Song, Baoyan Gao, Aifen Li, Jihai Xiong, Ziqiang Ao, and Chengwu Zhang. (2014). Preliminary characterization, antioxidant properties and production of chrysolaminarin from marine diatom *Odontella aurita*. *Marine Drugs* 12(9): 4883–4897. https://doi.org/10.3390/md12094883

Xia, Song, Ke Wang, Linglin Wan, Aifen Li, Qiang Hu, and Chengwu Zhang. (2013). Production, characterization, and antioxidant activity of fucoxanthin from the marine diatom *Odontella aurita*. *Marine Drugs* 11(7): 2667–2681. https://doi.org/10.3390/md11072667

Zhang, Jun, and Peter A. McCullough. (2016). Lipoic acid in the prevention of acute kidney injury. *Nephron* 134(3): 133–140. https://doi.org/10.1159/000448666

9 Wastewater Remediation Potential of Diatoms

Azalea Dyah Maysarah Satya[1], *Awalina Satya*[2], *Tjandra Chrismadha*[2], *Ika Atman Satya*[3], *Sara Kazemi Yazdi*[1], *and Pau Loke Show*[1]
[1]Department of Chemical and Environmental Engineering, Faculty of Science and Engineering, University of Nottingham Malaysia, Selangor, Malaysia
[2]Research Center for Limnology and Water Resources-National Research and Innovation Agency, Bogor, Indonesia
[3]Research Center for Computation, National Research and Innovation Agency, (BRIN), Komplek Cisitu, Bandung, Indonesia

9.1 INTRODUCTION

Wastewater as an untapped resource has long been a reliable resource in alleviating water scarcity, which is still experienced by 500 million people in areas with use of water exceeding locally renewable water resources (UNESCO, 2017). It is to note that of household wastewater flows amongst 42 countries (representative of 18% of the global population), only 56% were safely treated in 2020 (United Nations, 2021). And overall, in a global scale, the production of wastewater is estimated to reach 360 km^3 $year^{-1}$, of which 41 km^3 $year^{-1}$ is treated by wastewater treatment plants and then reused, 149 km^3 $year^{-1}$ is treated and discharged, and 170 km^3 $year^{-1}$ did not even go to a wastewater treatment plant but instead was discharged directly into the environment, as can be seen in Figure 9.1 (Jones et al., 2021). These discharged wastewaters more often than not still contain considerable amount of nutrients, contaminants, and harmful chemicals that can pose a risk to the environment, such as through eutrophication and, in some cases, the outflow of these nutrients is not monitored and causes harm far from its source (Grmela et al., 2014; Agwanda and Iqbal, 2019; Choi et al., 2019). Human activities pose a significant threat to water quality and the ecological environment, which makes these areas inhabitable for many organisms and is thus related to a loss of biodiversity if left untreated, which in turn results in degradation of water quality (Cunningham and Gharipour, 2018). Water consumption will only increase over time, which idealizes the valorizing of wastewater to avoid further damage to the environment which could have devastating consequences with a 40% water deficit by 2030 as projected from current water utilization practices globally (Liyanage and Yamada, 2017; Al-Jabri et al., 2021). Therefore, remediating wastewater is of importance not only for the environment, but also the future of humankind.

Diatoms are extremely diverse microcellular algae, consisting of more than 100,000 different species through extrapolation of an eclectic sample of genera and species complexes (Mann and Vanormelingen, 2013). They are the most successful group of photosynthetic eukaryotic microbes that inhabit almost all aquatic ecosystems, contributing ~20% to annual global carbon fixation, which in turn drives the global biogeochemical cycle as well as underpinning major aquatic food

 DOI: 10.1201/9781003436553-9

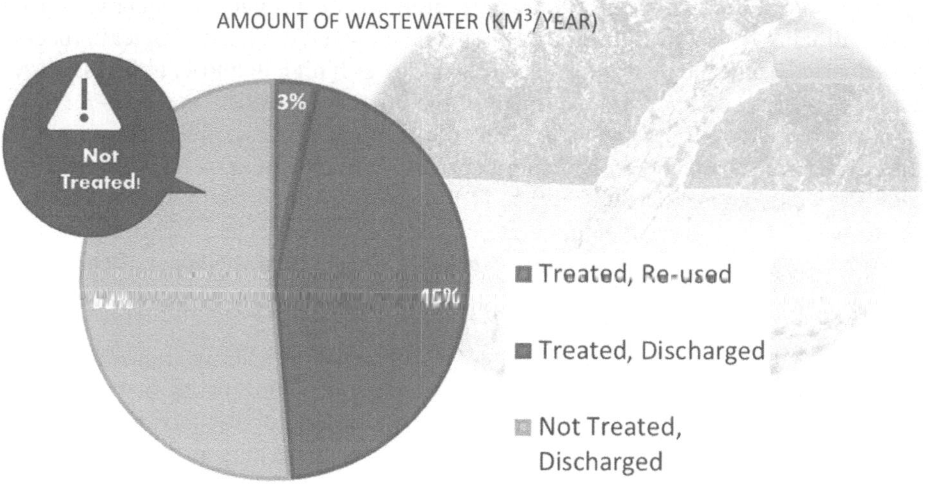

FIGURE 9.1 Global wastewater (km^3/year) as reported by Jones et al. (2021).

webs (José et al., 2019). Their distribution is worldwide, with a diversity of lifestyles and survival strategies which is likely attributed to their exceptional adaptability to high dynamic aquatic environments (Tréguer et al., 2018).

Unlike any other microalgae, diatoms are adorned with porous silica on the nano- to microscale and have been found to exceed synthetic approaches due to fundamentally different formation correlating processes between their genes and proteins involved (Hildebrand et al., 2006). Diatoms have also been known to have distinct interactions with marine bacteria for over 100 years, playing a large role in nutrient cycling between oxidized and reduction states, impacting bioavailability which in turn affects trophic levels (Amin et al., 2012). With diatoms' hierarchical structure, it is noteworthy that this organism inspired many technologies due to the high surface area and porosity, prominently in catalytic science, medicinal science, photoelectricity for imaging, and even metamaterial absorbents (A. Li et al., 2019; Gholami, et al., 2020b). Diatoms are also known to be exceedingly robust and therefore highly useful for a rapid response with their capacity to react to freezing sea by excreting antifreeze proteins (AFPs) consisting of ice-binding proteins which is prevalent in the Antarctic marine diatom *Chaetoceros neogracile* (Kim et al., 2017; Janech et al., 2006).

In this chapter, the possibility of diatoms acting as a bioremediating component in wastewater treatment is explored with recent findings and prospects for future research interested in utilizing diatoms as a bioremediatory agent in the environment. As sustainability is the goal of industries nowadays, biorefineries of diatoms were also included to further interest in developing diatoms to their full potential as bioremediatory agents.

9.2 PHYTOREMEDIATION

Wastewater's value is often neglected as a sustainable source of water, energy, nutrients, and other recoverable by-products. It, however, plays a critical role in balancing the protection of natural resources and environmental sustainability (UNESCO, 2017). Any liquid water discharged from domestic, commercial, industry, or agriculture uses that can contain a wide range of potential contaminants and concentrations is considered wastewater. Traditional wastewater treatment has four primary procedures. These procedures are preliminary, primary, secondary, and tertiary procedures utilizing biological, physical, and chemical processes. The biological process commonly used is activated sludge, however its effluents are still high in nutrients. Thus, wastewater remediation is still needed and pre-treatment of the effluent is needed as there's a limit to the amount of costly sources, for example, by using the chemical process, precipitate excess ammonium (NH_3) and phosphate (PO_4^{3-}) by using the struvite precipitation technique which requires a ratio of 1:1:1 ratio of magnesium (Mg) to NH_3 and PO_4^{3}, which then requires a lot of water in order to readjust that ratio (Ngatiman et al., 2021; Siciliano et al., 2020). Aside from the limited resources, wastewater remediation, especially for physical methods, often requires the construction of a reactor which in turn takes up space in arable land that could have been used for farming.

As wastewater treatment in turn requires a lot of water and additional chemicals to reach its full functionality, the use of biologically available and nature-based materials has been highlighted since the late 19th century (Frankland, 1888). Essentially, this process of treating wastewater is to use a living agent. Natural techniques or biotreatments were first applied in treating spills and anthropogenic compounds in *Beneficial Stimulation of Bacterial Activity in Groundwater Containing Petroleum Products* in 1975, which inspired similar research for years (Litchfield, 2005). Bioremediation for wastewater is often researched on highly complex plants through rhizofiltration, phytoextraction, phytoimmobilization, phytodegradation, and rhizodegradation (Arthur et al., 2005). Phytoremediation is believed to be a promising approach for heavily polluted land. However, the use of complex macrophytes has practical implications as it requires a longer time to cultivate and more space to implement. The use of macrophytes also can be detrimental to the water–food–energy nexus as can reach up to 80% of the food that we eat (*International Year of Plant Health–Final Report*, 2021). Furthermore, cultivation of macrophytes is a long process and the biomass obtained often cannot be utilized due to its lack of valuable bioproducts. Their response to stressors is also not as positive, with favorable optimum conditions in a small gap of 6–9 for pH and 15–38°C for temperature for municipal wastewater (Shah et al., 2014).

Culturing microalgae began at the end of the 19th century, which in turn became the greatest discovery of botany and science in general. The beginning of research started with creating a method that could appropriate survival and maintenance to later develop knowledge of the photosynthetic mechanisms of organisms. Their biomass is rich in high-value biocompounds such as lipid, carbohydrates, proteins, and many more bioactive compounds. Their cultivation also did not require a large space and they can be utilized by many kinds of wastewater, ranging from municipal sewage water to highly nutrient-dense palm oil milling effluent (POME) (Hariz et al., 2019; Zhu et al., 2017; Chawla et al., 2022). However, despite the fast biomass production and tons of high-quality bioproducts, microalgae often need extra pretreatment before they can be openly and massively cultivated in wastewater which often has varying and extreme concentrations that may hinder their growth.

Diatoms have many advantages in comparison to other microalgae. They have a ubiquitous presence, which allows continuously varying species through seasonal variations for use of the optimal available organism. They also grow rapidly, some are even reported to grow two times their biomass in a few hours, and can easily be regulated by the presence of silicate. Just like microalgae, almost all its biomass can be put into profitable use. They also serve an impactful ecological role when it comes to their survivability in the wild which gives them the capability to produce and function in a more profitable system for simultaneous assimilation of carbon dioxide and removal of nutrients, while producing valuable bioproducts and fuels in the long term (Wang and Seibert, 2017; Agusti et al., 2015).

9.3 DIATOMS' VALUED SIGNATURES AND THEIR ECOLOGICAL ROLE

9.3.1 Diatom Rigidity

Diatoms are microscopic (~1–500 μm in length) single-celled microalgae renowned for their silica cell walls with intricate morphologies. These morphologies show the formation of nano- and micrometer-sized pores which gives their structure a remarkable power by providing a considerable amount of surface area as well as rigid cell walls (frustules) to survive harsh conditions. The linking structure that holds these cells together can either be stiff or flexible, according to the species. However, the exterior itself is rigid and exhibits a similar property to many micro/nanofluidic devices, which has inspired the formation of hierarchical nanomaterials (Pawolski et al., 2018; Gao et al., 2018; Liu et al., 2019; Mishra et al., 2017).

There are two types of frustules, depending on the shape of the diatom. The most common is the centric diatom, which has radial symmetry on its shells, while the pennate diatom has bilateral symmetry. Centric diatom's frustule consisted of two overlapping thecae, where the larger one is called the epitheca and the small one is called the hypotheca. Each one has a valve and accompanying series of girdle bands. When a cross-section is made, rows of pores called interstriae can be seen with ribs between them called striae. In the cross-section of the diatom, there are three distinct layers with different distributions of foramens or opening holes across the surface perpendicularly connected by cross walls in a rectangular or honeycomb pattern (Round et al., 2000). Simplified individual cross-sections can be seen in Figure 9.2 where *Coscinodiscus* sp.'s frustule was examined (Aitken et al., 2016). The structure of frustules exhibits a 3D-hierarchical pore system which has been of interest in research for its excellent adsorption capacity in efficient wastewater treatment due to its unique structure and high surface area that allows functionalization and more adsorption sites with maximum adsorption capacity ranging from 925.93 mg/g to 1281.39 mg/g (J. Li et al., 2021; Wu et al., 2022).

Using three-point bending and nanoindentation experiments, a group of researchers investigated the mechanical properties and fracture behavior of *Coscinodiscus* sp.'s frustule (Aitken et al., 2016). The experiments showed that the elastic properties of the biosilica are like those in a girdle band with an average modulus of 36.4 ± 8.3 GPa. The discovery illustrated an unprecedently high specific strength compared with that of all natural biomaterials with similar honeycomb sandwich

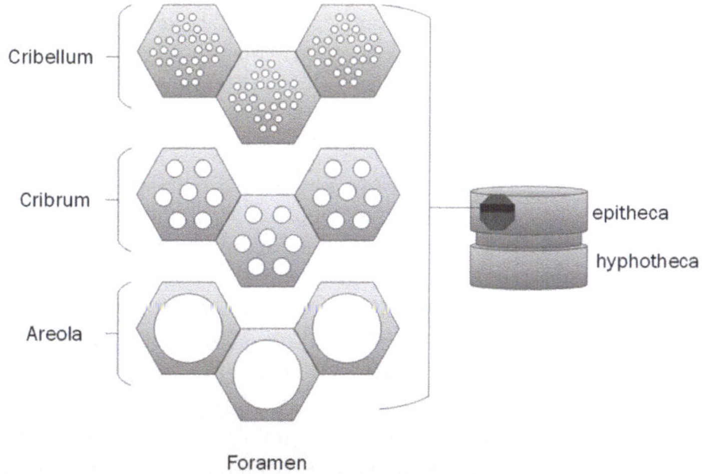

FIGURE 9.2 Structure of a frustule on a diatom (sourced from Aitken et al., 2016).

plate architecture. This result proved the extreme resilience of diatoms, opening a new class of bioinspired, low-density, and high-strength materials.

The basic repetitive structure of diatoms may explain their success in a variety of environments, as seen in the case of *Thalassiosira eccentrica*, a centric diatom, which is dominant in the euryhaline zone where the salinity changes regularly (Bilbao et al., 2021). This diatom was even reported as abundant in the coastal waters of the North Sea (Muylaert and Sabbe, 1996). Diatom's sturdiness also plays a role in their ability to live in extreme temperatures. As such, a study carried out on pennate diatoms, *Bacillariophyceae* class, represented by *Sellaphora disjuncta* and *Archnanthes brevipes* var. *intermedia* was found to have survived in the sulfurous geothermal zone of Comanjilla's microbial matt (Puy-Alquiza et al., 2021). In bioremediation, this sturdiness gives the potential of diatoms to be applied in post-treatment of high-nutrient and high-temperature polluted wastewater to avoid the addition of sterilizer condensate and cooling water such as that of palm oil mill effluent which often still exhibits these features and often requires a longer time to have a permissible content as it has a high amount of oil and grease (4000 mg/L) and COD (50,000 mg/L) with a very high BOD (25,000 mg/L) which can be detrimental to aquatic life when this waste is discharged in large quantities (Kamyab, 2018). The use of diatoms in wastewater as a bioremediatory has good potential due to its rigidity that allows it to live in the harshest conditions where other macro- and microphytes would not survive. Also, owing to frustule's porous material, it could perhaps be used as a material for composites (Hamm et al., 2003), filler for polymers (Conradi, 2013), or even an additive in metal manufacturing.

9.3.2 Biological Signature and Its Response to Extreme Conditions

Diatoms' abundance is known to be dependent on silicates which secure its relatively competitive success when it comes to other microalgae in growth studies through its ability to sequester other nutrients and an advanced carbon concentration metabolism compared to other microalgae (Schoefs et al., 2017). In contrast to land plants, a study of the *Ochrophyta* phylum of diatoms on its Calvin-Benson-Bassham cycle showed a regulation process based on transcriptional regulation which is very different to land plants' enzymatic level. This was believed to be an evolutionary adaptation by diatoms for when CO_2 values are limited. The limitations of CO_2 come from the effect of environmental pH values which dissolve inorganic carbon as bicarbonate (HCO_3^{2-}) instead. Thus, diatoms evolved a carbon concentration mechanism (CCM). The CCM works by taking up dissolved inorganic carbon by enabling direct uptake of HCO_3^{2} through a plasma-membrane type solute carrier (SLC) 4 family which serves as its transporter. The other more reliant form of CCM that is present in diatoms is by passively diffusing CO_2 formed by an external carbonic anhydrase. This carbonic anyhydrases play a significant role by providing mobilization for the dissolved inorganic carbon which then progressively moves to fixation (Matsuda et al., 2017). Due to the ecological importance of diatoms, their biochemistry, and physiology, they are believed to share the mechanism. However, through genome sequencing data, it was revealed that the core set of biochemical reactions of diatoms is closer to that of land plants with unexpected pathways such as the urea cycle (Allen et al., 2011). This unique metabolism was believed to be the key to diatoms' overall amazing adaptability to diverse environmental conditions, but that was not all. A study of *Phaedocatylum tricornutum* in varying light intensity shifts in a CO_2-starved condition indicated that diatoms rearranged their carbon metabolism toward production of phosphophenolpyruvate (PEP) which was later converted differently depending on the light intensity. When light was low, PEP was used for respiration, while when light was high, lipids and proteins were slightly accumulated (Heydarizadeh et al., 2017).

In addition to the biological signature of diatoms' flexibility, they exhibit the ability to cover themselves completely with a silica wall throughout their division phase in asexual reproduction. During this process, they still undergo reproduction which forces newly formed thecae to be formed within the confines of existing thecae (Not et al., 2012). This caused a rigid frustule which imposes two evolutionary constraints on diatoms. Those two evolutionary constraints present during mitotic division at which new valves form back-to-back inside parental cells to produce two or more daughter cells. The cells then journeyed onto the vast ocean as solitary or uniseriate filaments which might exhibit a similar look to multicellular beings. As hypotheca must fit inside the epitheca, the size of the daughter that inherits the parental hypotheca must use it as their epitheca, thus their average cell size decreases (Kooistra et al., 2007). Aside from having a frustule as a protector, its material (silica) acts as an effective pH buffering material which shifts HCO_3^{2-} into CO_2 which can be transferred in cell fluids easily to be metabolized (Milligan and Morel, 2001). Thus, diatom survivability is no doubt an asset when it comes to bioremediating certain effluents in extreme conditions with minimum preparation and thus reducing costs of bioremediation with the addition of obtaining refined materials such as silicates, pigments, and biofuel, which will be discussed briefly in the following sections and as previously explained in earlier chapters.

9.3.3 Diatoms and Carbon Capture

Carbon emissions in the form of atmospheric greenhouse gases (GHGs) into the atmosphere pose great risks, one of which is accelerating the process of global warming, with melting of ice caps that could release harmful gases or substances that have long been frozen within the permafrost (Malavin et al., 2018). The sinking of atmospheric greenhouse gas is one of the Earth's mechanisms for fighting climate change, which is known as the carbon concentrating mechanism and which is illustrated in the following reference (Karlusich et al., 2021). The biomass of diatoms in the sea is believed to be of equal to that of the tropical rain forests, and is estimated to contribute ~40% of primary production in oceans all over the world with an estimated fixing of 25% of atmospheric CO_2 (Seth et al., 2021). Diatoms' ability to fix carbon is reported to be higher than for any other classes of algae examined under both field and laboratory conditions (Samantaray et al., 2020). Amongst all microalgae, diatoms have a significantly efficient CCM which has resulted in these microalgae being known as the "biological carbon pump" (Bowler et al., 2009). Diatoms' frustule also provides them with high surface-area-to-volume ratios which enable rapid uptake of nutrients and CO_2 which enables diatoms' ability to mitigate pollutants and CO_2 in advancement of wastewater treatment processes even in non-productive land. In countries with a high density of people and wastewater generated such as China and the United States this cane result in a real improvement for self-sufficiency in sustainable fuel production (Wang and Seibert, 2017). The aqueous environment opens up more efficient access, as calculated by Weyer et al., who put together a conceptual maximum efficiency of algal production at 354,000 L per hectare per year of bio-oil with the best case real-world production at 40,700–53,000 L per hectare per year of bio-oil (Weyer et al., 2009).

9.4 WASTEWATER POLLUTANTS AND DIATOM TREATMENT POTENTIAL

9.4.1 Microplastics

Microplastics are regarded as an emerging threat of a great scale to the environment, especially marine life. On top of that, the top distributor of microplastics to enter aquatic environments is wastewater treatment plants (WWTPs). The majority of those being removed from the plant were found in sewage sludge. Its small particles (~150 μm) make the impact of this material impact marine life by disrupting hormones and growth. This includes the use of resins, such as Bisphenol A (BPA), which mimics estrogen and thus affects fish (Wang et al., 2019).

Diatoms are known to form biofilms around marine litter as part of their benthic carbon cycle (Gordon et al., 2009). These biofilms are formed by *exopolysaccharides* (EPS) that are mainly produced by bacteria, microalgae and cyanobacteria, protists, fungi, and even yeasts. Marine litter is first colonized by bacteria, followed by microalgae. The surface roughness of marine litter and

EPS produced by diatoms and bacteria varies, bringing diverse changes to the environment (Fazey and Ryan, 2016).

A comparative study of diatoms and bacteria consortia on different substrates showed that diatoms strongly adhere to hydrophobic substrates, e.g., plastics. The diatoms (*Nitzschia* sp.) were found to secrete EPS which gave them a strong anchorage with hydrophobic substrates such as plastics (Khan et al. 2020). The plastics undergo photolysis immediately, releasing particles, ions, and free radicals that result in changes to the genomes of marine life, including diatoms.

Developing a wastewater treatment plant for plastic might not look too promising but a study has shown that three marine diatoms (*Phaeodactylum tricornutum*, *Chaetoceros gracilis*, *Thalassiosira* sp.) could reduce the toxic effects of microplastics through hetero-aggregation and homo-aggregation between microplastics and diatoms. This was found to increase the photosynthetic efficiency of diatoms as they gradually recover to their normal growth (Wang et al., 2020).

9.4.2 Heavy Metals

Heavy metals are a group of metals that have high density, and are usually located in group B of the periodic table. Their electronic structures allow for coordination by accepting a pair of electrons. As heavy metals act as Lewis's acid, they are abundantly found as part of minerals, such as chalcopyrite ($CuFeS_2$) which contains two heavy metals: copper and iron. Naturally, rainwater leaching into these minerals would be rare as they are often found deep underground. To further complicate matters, heavy metals are used in industries as precursors to catalyze the production of paints, cosmetics, herbicides, and even pesticides (Gautam et al., 2014; Rehman et al., 2008; Yona et al., 2018). Dealing with this type of wastewater is often costly and detrimental to the environment due to how slow the process takes to separate these heavy metals from the environment and how persistent they are in soils.

There have been many remediation studies carried out on heavy metals by diatoms, mainly through adsorption. Adsorption in diatoms, for *Nitzschia closterium*, was found to be linear with its silica content. The lower density of the silanol group attached, the weaker the Cd adsorption observed (Ma et al., 2018). The same diatom was reportedly able to remediate other heavy metals such as Cu, Zn, Co, and Mn. This is done through adsorption but there are those that remediate through surface ligand modification such as *Planothidium lanceolatum* for Cd, Cu, and Zn (Marella et al., 2020).

9.4.3 Pesticides

Agricultural pollution is a global environmental concern. It is mainly caused by the application of farming inputs (e.g., fertilizers and pesticides) and practices (runoff). One of the pollutants most notorious in agriculture is pesticides. Due to their high toxicity, they pose a grave threat to the environment that can be escalated into a great environmental imbalance (Khan et al., 2020). Among all pesticides, organophosphorus pesticides are one of the most used to control several pests, especially insects. As it is used in large amounts, aquatic creatures are in dangers, which is where diatoms come into play. Diatoms species are capable of learning to switch between two modes of photosynthesis which allows them to either have an inhibited growth (during pesticide exposure) or be later adapted to grow even during pesticide pollution. Through taxonomic identification, it was found that variants of *Melosira* in French agriculture dominated by herbicides (isoproturon and s-metolachlor) showed that *Nitzschia dissipata* and *Cocconeis placentula*, which are individual species of pennate diatoms, were not affected by the presence of these two herbicides. The results obtained proved that these diatoms could switch their trophic mode, and thus it did not affect their photosynthesis but rather their cell multiplications (Debenest et al., 2009). A study of *Nitzschia palae* strains on atrazine showed that different strains develop different methods of dealing with the presence of herbicides (Esteves et al., 2018). Atrazine interferes directly with electron transfer in the chloroplast and also causes damage to chlorophyll and photodamage by inhibiting photosystem-II and blocking flow from primary quinone to secondary quinone (Suresh Kumar et al., 2014). The tolerant strain was found to mitigate the damage caused by atrazine by significantly increasing the content of carotenoid which is known to increase under oxidative stress and be involved in short-term photoprotection which is in contrast with the intolerant strain *Phaeodactylum tricornutum* used in earlier research (Bai et al. 2015).

Based on Lozano's research, diatoms tend to alter their antioxidant response when exposed to herbicides (Lozano et al., 2014). This shift is what makes diatoms a good bioindicator for the presence of micropollutants in the environment. And at the same time, it indicates that diatoms have their own mechanism in developing immunity toward pesticides, although more research is needed for this, especially on other types of pesticides.

9.4.4 Nitrogen and Phosphate

There have only been a few reports on nutrient remediation in WWTP by diatoms. One of which was Minggat et al. (2021)'s report on nutrient removal using a marine diatom *Chaetoceros muelleri* in WWTP. This diatom is commonly used for aquacultural feed as well as being known for its fast growth and easy maintenance. Their research objective was to perform an evaluation on the potential of *C. muelleri* to be applied for nutrient removal in eutrophic saline wastewaters and for biomass production. The diatoms were cultured under a wide range of temperatures, salinities, photoperiods, and light intensities. The optimum temperature for the biomass production was observed to be 30°C, but the algae could maintain at least 66% of their highest production between 20°C and 35°C. The optimum salinity for the biomass production was 22–25%. Both light intensity and photoperiod affected the algal biomass production, and the minimum light required was 100 μmol $m^{-2} \cdot s^{-1}$ for 6 hours to maintain biomass production alongside nitrogen (N) and phosphorus (P) absorption. N and P absorption in the study could have been largely controlled by the kinetic reaction in the algal metabolism. Throughout all the experiments, the N and P absorption increased with the biomass production, but the ratio of N and P to biomass exponentially decreased with biomass production. These results showed that *C. muelleri* is tolerant to a wide range of environmental conditions, absorbing nutrients and producing organic matter. *C. muelleri* has great potential to be introduced into water treatment processes, especially where the temperature and salinity fluctuate.

Another report by Panneerselvam et al. (2013) investigated the suitability of the marine diatoms, *Chaetoceros curvisetus* and *C. simplex*, in removing macronutrients from different sources. They observed the growth and nitrate-phosphate removal properties under varied nitrate, ammonium, and urea nitrogen sources. Three separate experiments were conducted using modified F/2 medium with 12.35 mg L^{-1} total nitrogen and 1.12 mg L^{-1} total phosphorus to simulate the typical concentration of nitrogen and phosphorus in secondary effluent as growth medium. The maximum cell densities of *C. curvisetus* and *C. simplex* were $7.16 \pm 0.34\ 9 \times 10^4$ cells mL^{-1} in NO_3 and $3.88 \pm 0.32 \times 10^5$ cells mL^{-1} in urea. The maximum chlorophyll a per cell was 1.7 and 4.7 pg for *C. simplex* and *C. curvisetus*, cultured with urea and nitrate. High protein contents of 4.7 pg $cell^{-1}$ in *C. simplex* with urea and 19.7 pg $cell^{-1}$ in *C. curvisetus* with nitrate nitrogen sources were found. Higher cell density and protein content of both species were found from urea and nitrate nitrogen sources ($p<0.05$). This proved that these nitrogen sources were utilized by microalgae and were converted to protein. *C. simplex* and *C. curvisetus* showed maximum removal efficiencies of nitrate by 97.86% and 91.62% and phosphate by 98.5% and 100%, respectively, when urea was used as a nitrogen source rather than ammonia. The results indicated that *C. simplex* was more efficient than *C. curvisetus* and suitable for the removal of macronutrients when cultured with urea and nitrate nitrogen sources.

9.5 VALUABLE BY-PRODUCTS

9.5.1 Biofuel

Unicellular microalgae as biofuel-production organisms have been selected as they are more efficient than vascular plants at converting light and nutrients into valuable products. The economics of algal biofuels is critically dependent on productivity. Productivity is defined as the ability of a species to grow rapidly to high density and accumulate fuel precursor molecules. For a given class of algae, there will be differences in measured productivity values when comparing individual species. These differences can emerge from intrinsic factors related to their cell biology or from growth conditions at the time of sampling, which can involve both environmental and ecological variables. Comparing productivity in controlled culture frequently is not reliable since conditions are never exactly the same in two culture systems. Therefore, algal biofuels will be better developed on a regional basis since a wide variety of environments occurs. To obtain a good portion of production for the whole of the year, it is desirable to have strains that are consistently productive under various environmental conditions than strains that only have higher productivity under optimal conditions for a shorter period of time. Another consideration is the algal class' productivity must be on the basis of its abundance in nature in a variety of environments (Kumar et al., 2021).

Diatoms show characteristics which lead them to be amenable to large-scale biofuel cultivation. The US Aquatic Species Program in 1998 recommended a list of 50 microalgal strains, obtained from a pool of over 3000 candidates, which held the most promise as biofuel-production organisms (Sheehan et al., 1998). More than 50% of those strains were chosen based on criteria such as high growth rates and lipid yields, tolerance of harsh environmental conditions, and performance in large-scale cultures. Diatoms are excellent lipid accumulators, as illustrated in Figure 9.3. Commonly, diatoms show that a substantial portion of the cells' volume is occupied by lipid droplets that accumulate rapidly, especially under Si limitation conditions.

The oceanic diatoms are among the most productive classes of microalgae. They are responsible for a significant portion (40%) of marine primary productivity and most of the CO_2 export (drawdown and sinking) from the atmosphere, of the order of 50% (Brennan and Owende, 2010). Diatoms are efficient at nutrient utilization. The main contributing factors to diatom dominance are that they possess a large nutrient storage vacuole. The capability for storing in a vacuole has important consequences in terms of productivity. The efficiency for utilizing nutrients is determined by the surface-area-to-volume ratio, which for smaller cells tends to be more efficient for this purpose. For this reason, to compensate for a larger surface area nutrients are stored inside the cell in a vacuole, which in essence reduces the cytoplasmic volume and increases the surface area from which nutrients can be utilized by the cytoplasm (de Castro Araújo and Garcia, 2005). Supposing that extracellular nutrients are present at levels that reduce diffusive limitation to the cell surface, the gains in productivity in cells which have vacuoles are quite considerable. Therefore, in nutrient-lacking conditions, diatoms will reserve excess nutrients, depositing enough nitrate to allow for several cell divisions, and averting the growth of other phytoplankton (Sheehan et al., 1998).

Diatoms also demonstrate a greater carbon fixing ability than other co-existing groups of microalgae. This has not only been repeatedly confirmed in laboratory cultures under optimal conditions but also in environmental samples from different locations. In a comparative study, diatoms grew faster under low light conditions than coexisting algae. The photosynthetic efficiency for diatoms is higher than for green algae. For example, *P. tricornutum* compared to *Chlorella vulgaris* revealed that under nonfluctuating light conditions, efficiencies of transforming light into biomass were equivalent, but under fluctuating light conditions, the diatom was nearly twice as efficient as *C. vulgaris*. Efficient carbon fixation ability along with nutrient utilization could increase the interest in diatoms for co-processes such as CO_2 sequestration and wastewater remediation (Sharma et al., 2021).

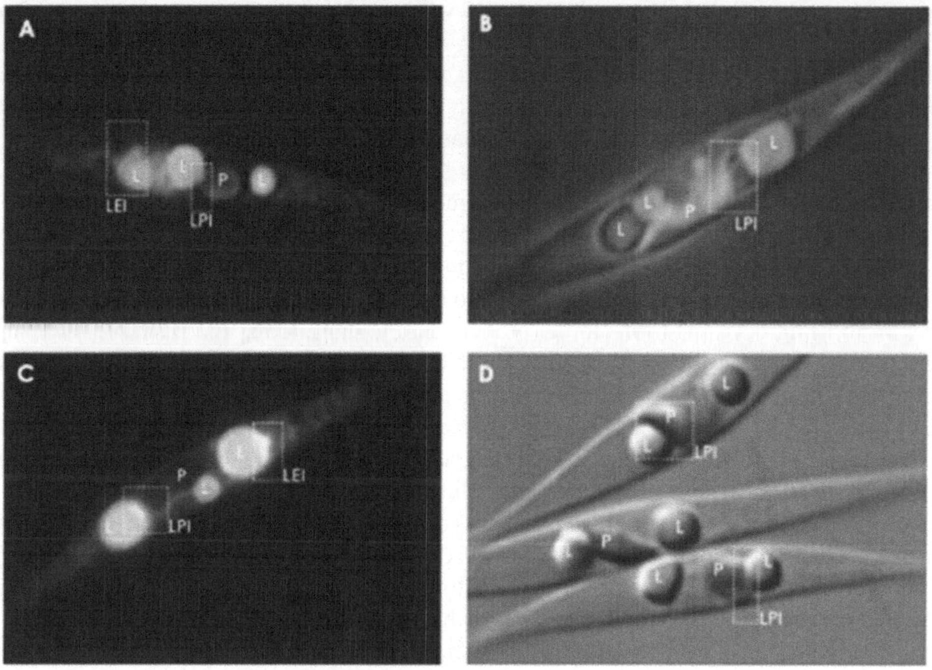

FIGURE 9.3 Micrographs of nitrogen-starved *P. tricornutum* cells, illustrating potential interconnectivity between lipid droplets and various cellular compartments (modified from Leyland et al., 2020).

Diatoms naturally bloom under nutrient-replete conditions, resulting in a maximal growth rate to the highest density. This is an ideal characteristic for a biofuel production system. Diatom blooms also tend to crash suddenly, which is a frequent occurrence for all bloom species. Collapse can result from several factors including nutrient exhaustion, grazing, sedimentation, or viral lysis, and remains a fundamentally poorly understood process in all microalgae. Many reports have shown that once chlorophyll levels begin to decrease (suggesting the beginning of the bloom termination), the triacylglycerol (TAG) content increases. Turning that into a biofuel production scenario, TAG accumulation leads to cell death and thus crashing would not necessarily be detrimental to production (Akita and Kamo, 2015).

The diatom ability for sequestering nutrients could be beneficial for cultivation efforts. More rapid nutrient uptake from growth medium causes a reduction in the nutrient availability for competing organisms. The greater size of the vacuole means a higher storage capacity. For example, Chlorophytes, including *Chlorella* species, *Chlamydomonas reinhardtii*, and *Dunaliella salina*, have small vacuoles (3–14% of their protoplast volume) in contrast with diatoms at 45–90% of their protoplast volume. The chlorophytes should have greater dependence on external nutrient conditions including consistency of supply. In terms of lipid productivity, cells with larger vacuoles may take longer than cells with smaller ones to become nutrient limited, due to their enhanced nutrient storage capability (Silva et al., 2015).

For decades, diatoms have been cultivated in large-scale outdoor systems for purposes other than biofuel production, such as aquaculture. Widely ranging values for algal biomass productivity are reported in the literature and are due in large part to differences in cultivation parameters. The cultivation parameters are light source and its intensity, depth of culture, mixing rate, temperature, nutrients, and the availability of supplementary carbon (sugars or CO_2). To limit some of the variability in comparison, only values obtained from microalgae grown in outdoor ponds, raceways, or open photobioreactors illuminated with natural light are discussed here. Diatom biomass productivity values range from 5 to 25 g m^{-2} day^{-1}, and are in line with values obtained for other types of microalgae including the Eustigmatophyte *Nannochloropsis salina*, the cyanobacterium *Spirulina platensis*, and the Chlorophyte *Chlorella* sp., whose biomass productivity values range from 10 to 25 g m^{-2} day^{-1} (Cerón García et al., 2005; Chauton et al., 2013; Pruvost et al., 2011).

Diatoms are found in a wide variety of environments. They not only can live in diverse environments such as marine, freshwater, and soil, but also specific extreme or variable environments of salinity, temperature (ice or hot springs), or pH. The most intensively studied diatoms are marine, and all of the species with sequenced genomes fall into that category. This does not exclude freshwater diatoms from being developed for biofuel production, but considering the issue of water supply, saline-tolerant species may be the most appropriate for development. Strains of *P. tricornutum* were shown to have near-linear growth rates (divisions per day) in salinities ranging from 5 to 70 psu (Gerhardt et al., 2008). Numerous species grow under acidic conditions (pH 3.3–4.4) and diatoms have been found to grow at pHs consistently maintained between 1 and 2.5 (Souza-Egipsy et al., 2011). Some species also have been found in alkaline lakes, such as those in East Africa, where the pH can reach 10.6 (Hecky Kilham, 1973). Diatoms can grow at 60–65°C in hot springs (Brianskaia et al., 2006) and at 40°C in thermal muds, which are a 5 g L^{-1} colloidal aluminum silicate suspension mixed with thermomineral water, rich in sodium chloride and sulfates (Tolomio et al., 1999). Psychrophilic diatoms grow on polar ice and in hypersaline brine channels at –1 to 2°C (Choi et al., 2008), with photosynthetic activities optimized between 4 and 7°C (Neori Holm-Hansen, 1982). Diatoms can form biomasses in ice with chlorophyll concentrations that are matched only in hypertrophic waters or in extremely eutrophicated sediments (Thomas Dieckmann, 2002). The *Fragilariopsis* genus may be the most prevalent biomass producer among polar plankton (Mock Hoch, 2005; Gersonde, 1997), serving as the base of the entire polar food web. The range of different environments where diatoms can dominate sheds light on their diversity; in terms of biofuels production, growth in extreme conditions, such as salinity or pH, can be utilized to keep pond contaminants and predators at low levels.

Diatoms exhibit a trophic flexibility. Most species are autotrophic, but mixotrophy and obligate heterotrophy also occur. *Cyclotella cryptica*, a candidate biofuel production strain, is capable of slow growth (10% of autotrophic) on glucose only in the dark, likely due to its glucose transport mechanism being induced in the absence of light (Hellebust, 2008). The mixotrophs and obligate heterotrophs are generally able to use simple sugars as carbon sources, but some species are able to break down complex carbohydrates. There are two forms of mixotrophs: those that can grow strictly auto- and heterotrophically, and those whose growth is only enhanced in the presence of an extracellular carbon source (Cerón-García et al., 2013).

Diatoms are excellent lipid accumulators, which partly may be due to extra buoyancy needed to counteract the effect of their relatively heavy silica cell walls. In terms of total lipids, Shifrin and Chisholm (Shifrin and Chisholm, 2008) reported values of 12–45% of dry cell weight (DCW) for diatoms under conditions of N limitation. A comparative analysis of several species in different algal classes by Hu et al. (Hu et al., 2008) showed that on a DCW basis, chlorophytes had an average total lipid content of 25.5% under normal growth and 45.7% under nutrient stress. Regarding diatoms, lipid content was 22.7% DCW during normal growth and 44.6% during stress (Hu et al., 2008). A literature survey by Griffiths and Harrison (Griffiths and Harrison, 2009) of 55 microalgal species in various classes showed that diatoms as a class performed above average at 27.4% (19% better than average) for total lipid content under nutrient-replete conditions and 36.1% (13% better than average) under N-deficient conditions. Diatoms also accumulate lipid as a result of Si limitation, and the average lipid content for diatoms limited for Si relative to N-deficient conditions was 41%, which was 28% higher compared with all species and 13% higher when compared within the diatoms (Griffiths and Harrison, 2009). In terms of neutral lipid (TAG), Si-deficient cultures of *C. cryptica* were reported to contain 64% on an ash-free dry weight (AFDW) basis, compared with 32% in Si-replete cultures, and *Cylindrotheca fusiformis* contained 57% in Si-deficient cultures and 17–20% in Si-replete cultures (Sheehan et al., 1998).

A wide variety of TAG compositions have been identified in diatoms, with fatty acid (FA) side chains ranging from 14:0/16:0/16:1 to 16:1/20:5/22:6; however, there were 10 predominant species, which were enriched in C14 and C16 (Yu et al., 2009). Diatom FAs are among the most highly enriched in C14 chain lengths compared with other classes of algae (Hu et al., 2008), especially chlorophytes, which appear to lack this particular length (Chen et al., 2008; Guckert and Cooksey, 1990; Thompson, 1996). Reports range from 4 to 32% C14:0 of the total FA composition for diatoms (Chen et al., 2008; Hu et al., 2008; Mansour et al., 2005). C16 FAs are highly represented in diatoms and C16:0 and C16:1 combined can represent 65–70% of the total FA, with C16:1 being predominant. C18 lengths are poorly represented in diatoms, but can be a predominant form in chlorophytes, especially polyunsaturated forms (Hu et al., 2008; Thompson, 1996). In general, shorter chain lengths are desirable to improve cold-flow properties of biofuels (Radakovits et al., 2011) and saturated FAs are more desirable because they increase the ignition quality (cetane number) of the fuel (Knothe, 2008).

A critical point to make regarding the measurement of total lipid or TAG in diatoms is that DCW measurements will underestimate the actual amount of lipid in terms of percent cellular carbon when compared with a similar measurement in other classes of algae because of the contribution of the noncarbonaceous silica cell wall. Comparison of eight different diatom species indicated ash contents from 49% to 59% (Reyes et al., 2009); further, diatoms had an average of twice the ash content of a variety of nonsilicified algae and four times the content of two *Chlamydomonas* species examined (Shifrin and Chisholm, 2008). The silica content of three diatom species was measured at 22–47% of the dry weight (Lewin and Guillard, 1963). Factoring this into DCW measurements, this suggests that the lipid content in diatoms on a per cell carbon basis is substantially higher than usually reported. Illustrating this point is a comparison of Si-induced lipid accumulation in *C. cryptica*, in which total lipid based on DCW was 42% (Shifrin and Chisholm, 2008) compared with 54.1% when calculated on an AFDW basis (Roessler, 2008). The data suggest that diatoms are at least on a par with, if not superior to, other classes of algae in terms of their ability to accumulate lipids.

9.5.2 Pigments

There are two types of photosynthetic pigments in diatoms with different functions. These are chlorophylls and carotenoids. Chlorophylls function as light harvesters, trapping light energy of blue and red of the electromagnetic spectrum for photosynthesis. There are only two forms that occur in diatoms which are Chl a (the one that plays a central role in energy conversion during photosynthesis) and Chl c which is found to be in abundance in diatoms (Kuczynska et al., 2015). Aside from chlorophyll, there are carotenoids. The most abundant carotenoids, based on their chemical structures, can be divided into two categories: (1) the carotenes (β-carotene); and (2) the xanthophylls, such as fucoxanthin. These two categories have reportedly been extracted from the model species *Phaedocarylum tricornutum*. This group of diatoms is known as a bio-factory for synthesizing high amounts of fucoxanthin, which is an important source of omega-3 fatty acid. Mixotrophic conditions of cultivating this diatom have been achieved which expressed hexose uptake protein (HUP1) from *Chlorella kesslery*, a chlorophyte, under a nitrate reductase promoter. According to the research, this mixotrophic culture enables *P. tricornutum* cells to take up glucose, which leads to increased biomass in low light intensities (Krämer et al., 2022; Butler et al., 2020).

The production of carotenoids in diatoms, *Phaeodactylum tricornutum*, was reported to be affected by light intensity or photon flux densities qualitatively (Heydarizadeh et al., 2017). This study was observed at a low CO_2 level as it wished to observe the PEP/pyruvate incident of the peculiar metabolic pathway that diatoms undergo. This amount correlates along with chlorophyl a, which was extracted and quantified with methods as stated in previous research. The amount of both pigments was reported to be increased during their investigation which reflects the need to harvest lighter when the number of cells was increasing, and more thylakoid membranes were developed. It was also found that under the lowest light, cells were dividing at a lower rate, resulting in a plateau phase with lower total cell numbers. At the end of the study period, the pigment level grown under medium light was somewhat similar to that of high light intensity. The transcriptome of this diatom was now known to be affected by light as well at which acclimation to medium light has induced alteration in cell biology, physiology, and biochemical acclimation. The use of the diatom *Phaeodactylum tricornutum* grown in a palm oil mill effluent as medium has been done with the addition of urea. This resulted in high productivity of fucoxanthin under light irradiance. The research also showed that urea enhanced their cell density and fucoxanthin productivity in that wastewater (Nur et al., 2019). This shows that diatoms, no matter the conditions, will always try to produce pigment as they tend to adjust their physiological processes to new lighting conditions, which makes wastewater remediation by diatoms a valuable one.

9.5.3 Silica

Silica (SiO_2*H_2O) is valued in the industry due to its shared characteristics with carbon materials. Its chemical structure allows the formation of a chain-like structure with palpability that leads to a sturdy yet functional material that can co-host other materials or substances, much like carbon. Silica is only slightly more metallic than carbon as it loses electrons more easily, thus making it an important component in electronics (Yoshida et al., 2018). Therefore, researching for other sources of silica would be beneficial in the future. The formation of inorganic minerals by organisms (biomineralization) is widespread in nature. Depositing silica can also happen to land plants such as rice and cucumber but the amount is often insignificant to reinforce their tissues as part of a systematic response to pathogenic attack which often only happens at extreme environmental conditions (Ehrlich et al. 2021). Specialized organisms for biomineralizing are considered to have potential economic use for nanocomposites and metal nanoparticles including templating those benefits most from unique silica formations (Ehrlich and Witkowski 2015) which even inspired the formation of 2D hexagonal packing of vertically oriented silicate channels (Yeh et al., 2021). For diatoms especially, silica precipitates out of its silica deposition vesicles (SDVs) under strict biological control that give rise to species-specific nanoscale architectural motifs as they are exocytosed and append on the existing cell wall and, in addition, extracellular silicifications by which diatoms take silica sources from their surroundings to strengthen their cells (Mayzel et al., 2021).

There are many methods of obtaining silica from diatoms. In obtaining biosilica of diatoms as templating, a recent procedure has been to calcinate the diatom frustule of *Archanthidium eutrophilum* at 800°C which exhibits a broad spectrum of light from UV to visible where its inner chamber possesses multiple reflections of light that enhance the photocatalytic activity of diatom shaped titania nanostructures (Chetia et al., 2017). In this procedure, the lab-cultivated diatom was cleaned with distilled water after 30 days then kept in 4% glutaraldehyde solutions for 12 h. The diatom was then cleaned by distilled water and then immersed in 2 M HCl (35%) for another 12 h. The diatom templates were then dehydrated and preserved in ethanol which was then gradually exchanged by isopropanol as 5% of the precursor. Preparing nanoporous silica from *Cyclotella* sp., a freshwater diatom with drum-shaped circular valve of 3–5 μm in diameter, has also been identified by either centrifugating the diatom pellet with 37% aqueous HCl and centrifuging at 3000 rpm for 10 minutes in water bath for 15 minutes at 60°C (Mazumder et al., 2014) or by sono-Fenton process by first preparing a suspension of Fenton agents (Fe^{2+} and H_2O_2) and introducing it into 80 mL of algae suspension with the pH adjusted before sonication with a constant frequency of 40 kHz (Gholami et al., 2020b, 2020a).

The use of natural silica nanoporous biomaterials from marine diatoms as a drug-delivery application has been explored (Gnanamoorthy, Anandhan, and Prabu, 2014). The diatom in this study was identified as *Coscinodiscus concinnus*, a centric diatom, that exhibits a radially symmetrical shape with numerous discoid plastids that denote its phenotypic plasticity (Hevia-Orube et al., 2016). This phenotypic plasticity is what drives the interest in finding the right mix to develop the shape needed for certain applications. As such, in drug-delivery applications, fine particles with a large surface area are preferred as an in vitro oral drug-delivery model (Rojas et al., 2018) The diatomic shells produced were homogeneous in size distribution with a radius of 200 ±15 μm and a 3D quasi-regular construction. The surface featured a mounded topography with about 2 μm wide porous dome organized in a pentagonal packing which can be seen in an FE-SEM from the study (Figure 9.4). The model drug in this research is streptomycin which was found in the bone structure at 33.3 ± 2%. The release profile found to be biphasic which involves an initial burst release (6 h) from the diatom foramen (silica surface) followed by sustained drug release (up to 7 d) from the cribrum (inside pores) and cribellum (structure). The obtaining of silica frustules was done by removing organic contents using a treatment of 30% aqueous hydrogen peroxide at pH 2.5 in this study. This study is the first of its kind to use laboratory-cultured diatom silica frustules as a biocarrier for drug-delivery applications. These lab-cultured diatoms opened future investigations into the purification of frustules for biomedical usage such as *Nitzschia* sp. as diatom microshells for loading of curcumin to later be functionalized with ferrofluid to exhibit magnetically responsive siliceous frustules for chemotherapy uses (Javalkote et al., 2015) and even stimulus-responsive diatom biosilica microcapsules from diatomaceous earth for antibiotics by grafting copolymers of oligo(ethylene glycol) methacrylate (Vasani et al., 2015).

The diatom frustules also serve as a greener approach for nanoparticle production (Pytlik et al., 2018). Gold nanoparticles are known to have high reactivity and efficiency as metallic colloids. However, as colloids tend to lose their stability over time, there need to be solid matrices to avoid

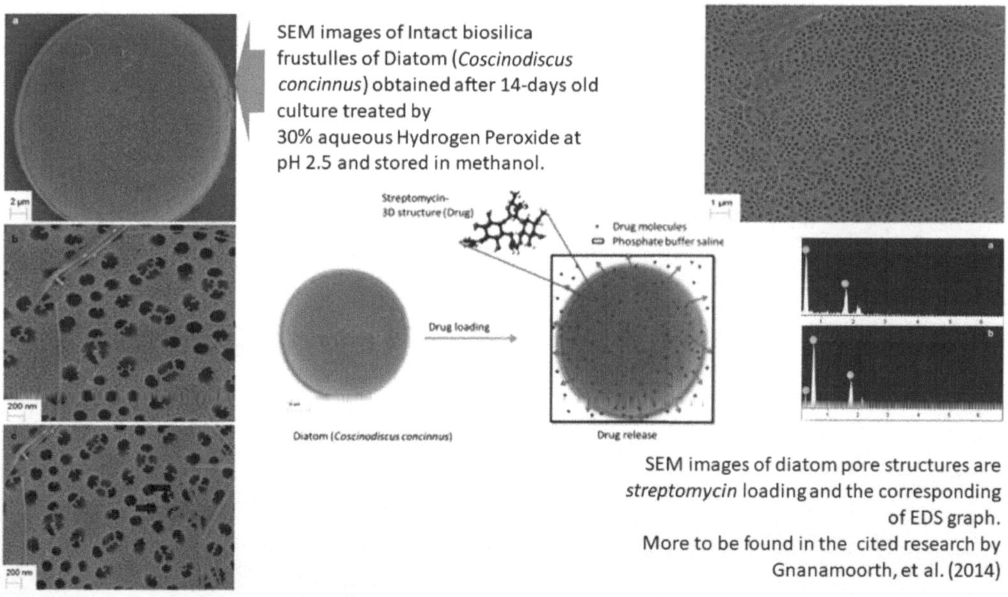

FIGURE 9.4 Frustule of a diatom (*Coscinodiscus* sp.) (modified from Gnanamoorth et al., 2014).

aggregation that often leads to a reduction of catalytic activity (Irfan et al., 2018; Yang et al., 2019). Green synthesis is preferable for sustainability. Greener biosynthesis of gold nanoparticles has been done by applying novel synthesis of gold nanoparticles by a biologically driven process employing two diatom strains, *Navicula atomus* and *Diadesmis gallica* (Schröfel et al., 2011). Fabrication of gold nanoparticles to functionalize diatom frustules through a bottom-up self-assembled procedure has been carried out (Shen et al., 2017). The research showed that its complex structure allows trapping, scattering, and local enchantment of diatom frustules, which allows the generation of more hot electrons which, through Landau damping, can serve in many applications, such as photodetectors, photocatalysts, photoelectrochemical cells, photovoltaics, biosensors, and chemical sensors (Tang et al., 2020; Pawolski et al., 2018). For this purpose, the obtaining of silica is done by cleaning the diatoms by soaking their surface with a solution of NH_3:H_2O_2:H_2O at a ratio of 1:1:4 for 6 hours to functionalize their surface with hydroxyl which will seed gold nanoparticles (Lee et al., 2007).

Biosilica obtained from diatoms also found application in gene therapy in the form of engineered diatomite nanoparticles (DNPs) as nucleic-acid delivery systems carrying short-interfering RNA (Martucci et al., 2016). The synthesized diatomite particles were found to be not cytotoxic, and the downregulation of gene expression effectivity was improved and evaluated by modifying their surface with amino-salinization and covalently conjugated Id-peptide (pA2036) which can be recognized by B-cell receptors (BC) on A20 cell line. The similar usage of DNP biosilica in gene therapy was also found to deliver short-interfering RNA to epidermoid cancer cells, H1355, using the same method of obtaining the diatomite biosilicate by suspending the separated by centrifugation diatomic solid in piranha solution (2M H_2SO_4, 10% H_2O_2) for 30 min at 80°C (Rea et al., 2014).

As diatomic silica nanostructures have drawn attention as high-value materials, more research into removal of organic matter is needed. The latest recovery of biosilica from diatoms was done by plasma technology which had a least damaging effect on the structure or disturbed pore geometry as revealed in FESEM and BET analysis as reported, and which showed better surface area and quality the longer the process was applied in comparison to the conventional method, as can be seen in

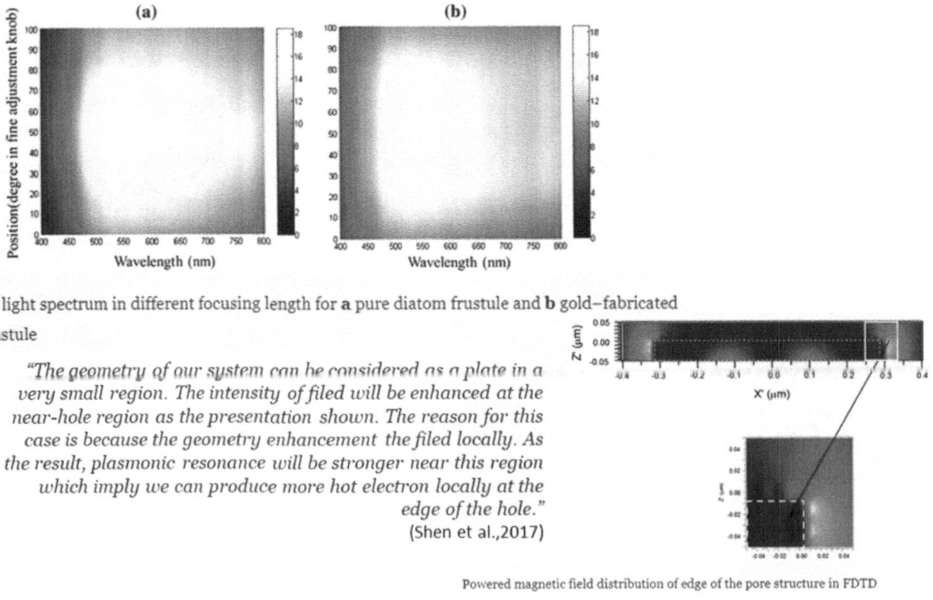

FIGURE 9.5 Hot electrons of a diatom's frustule and gold-fabricated frustule (modified from Shen et al., 2017).

Figure 9.5 (Jain et al., 2022). The recovery was done by uniformly spreading 100 mg of *Denticula* sp. biomass on a glass plate then exposing it to oxygen plasma asher at 500 watts with an oxygen flowrate of 60 sscm and a temperature of 25°C (Watanabe et al., 2010).

To conclude, biosilica is one of many valuable bioproducts that diatoms have that pave the way for technological advantages. Diatoms naturally have this material secreted even without natural stress. The abundance of diatoms also plays a great role in obtaining biosilicate for greater use in the future.

9.6 CONCLUSIONS AND PERSPECTIVES

The potential for wastewater remediation using diatoms has been evaluated with the presented research on diatom bioremediation capabilities. Diatoms' rigidity and biological signature played a role in their possible bioremediatory uses. Diatoms' rigidity lies in their 3D-hierarchical structure, which gives them a large surface-area-to-volume ratio, which provides more area for the substrate, or in this case contaminants, to attach to. Wastewater treatment by diatoms, just like microalgae, still has a many steps to take to make it a reality but, for bioremediating, diatoms' natural sturdiness as well as existence in every area of surface water and marine water should be good proof that it can serve uses in treating wastewater, especially that with extreme conditions.

REFERENCES

Agusti, S., González-Gordillo, J. I., Vaqué, D., Estrada, M., Cerezo, M. I., Salazar, G., Gasol, J. M., & Duarte, C. M. (2015). Ubiquitous healthy diatoms in the deep sea confirm deep carbon injection by the biological pump. *Nature Communications* 6. https://doi.org/10.1038/ncomms8608

Agwanda, P. O., & Iqbal, M. M. (2019). Engineering control of eutrophication: Potential impact assessment of wastewater treatment plants around winam gulf of lake Victoria in Kenya. *Journal of Coastal Research* 91(sp1), 221–225. https://doi.org/10.2112/SI91-045.1

Aitken, Z. H., Luo, S., Reynolds, S. N., Thaulow, C., & Greer, J. R. (2016). Microstructure provides insights into evolutionary design and resilience of Coscinodiscus sp. frustule. *Proceedings of the National Academy of Sciences of the United States of America* 113(8), 2017–2022. https://doi.org/10.1073/pnas.1519790113

Akita, T., & Kamo, M. 2015. Theoretical lessons for increasing algal biofuel: Evolution of oil accumulation to avert carbon starvation in microalgae. *Journal of Theoretical Biology* 380, 183–191. https://doi.org/http://dx.doi.org/10.1016/j.jtbi.2015.05.027

Al-Jabri, H., Das, P., Khan, S., Thaher, M., & Abdulquadir, M. (2021). Treatment of wastewaters by microalgae and the potential applications of the produced biomass—a review. In: *Water (Switzerland)* (Vol. 13, Issue 1). MDPI AG. https://doi.org/10.3390/w13010027

Allen, A. E., Dupont, C. L., Oborník, M., Horák, A., Nunes-Nesi, A., Mccrow, J. P., Zheng, H., Johnson, D. A., Hu, H., Fernie, A. R., & Bowler, C. (2011). Evolution and metabolic significance of the urea cycle in photosynthetic diatoms. *Nature*. https://doi.org/10.1038/nature10074

Amin, S. A., Parker, M. S., & Armbrust, E. V. (2012). Interactions between diatoms and bacteria. *Microbiology and Molecular Biology Reviews* 76(3), 667–684. https://doi.org/10.1128/mmbr.00007-12

Arthur, E. L., Rice, P. J., Rice, P. J., Anderson, T. A., Baladi, S. M., Henderson, K. L. D., & Coats, J. R. (2005). Phytoremediation – An overview. *Critical Reviews in Plant Sciences* 24(2), 109–122. https://doi.org/10.1080/07352680590952496

Bai, X., Sun, C., Xie, J., Song, H., Zhu, Q., Su, Y., Qian, H., & Fu, Z. (2015). Effects of atrazine on photosynthesis and defense response and the underlying mechanisms in Phaeodactylum tricornutum. *Environmental Science and Pollution Research* 22(22), 17499–17507. https://doi.org/10.1007/s11356-015-4923-7

Bilbao, J., Muñiz, O., Rodríguez, J. G., Revilla, M., Laza-Martínez, A., & Seoane, S. (2021). Assessment of a sheltered euhaline area of the southeastern Bay of Biscay to sustain bivalve production in terms of phytoplankton community composition. *Oceanologia* 63(1), 12–26. https://doi.org/10.1016/j.oceano.2020.08.007

Bowler, C., Karl, D. M., & Colwell, R. R. (2009). Microbial oceanography in a sea of opportunity. *Nature* 459(7244), 180–184. https://doi.org/10.1038/nature08056

Brennan, L., & Owende, P. (2010). Biofuels from microalgae—A review of technologies for production, processing, and extractions of biofuels and co-products. *Renewable and Sustainable Energy Reviews* 14(2), 557–577. https://doi.org/http://dx.doi.org/10.1016/j.rser.2009.10.009

Brianskaia, A. V., Namsaraev, Z. B., Kalashnikova, O. M., Barkhutova, D. D., Namsaraev, B. B., & Gorlenko, V. M. (2006). Biogeochemical processes in the algal-bacterial mats of the Urinskii alkaline hot spring. *Mikrobiologiia* 75(5), 702–712.

Butler, T., Kapoore, R. V., & Vaidyanathan, S. (2020). Phaeodactylum tricornutum: a diatom cell factory. *Trends in Biotechnology*. https://doi.org/10.1016/j.tibtech.2019.12.023

Cerón García, M. C., Sánchez Mirón, A., Fernández Sevilla, J. M., Molina Grima, E., & García Camacho, F. (2005). Mixotrophic growth of the microalga *Phaeodactylum tricornutum*: Influence of different nitrogen and organic carbon sources on productivity and biomass composition. *Process Biochemistry* 40(1), 297–305. https://doi.org/http://dx.doi.org/10.1016/j.procbio.2004.01.016

Cerón-García, M. C., Fernández-Sevilla, J. M., Sánchez-Mirón, A., García-Camacho, F., Contreras-Gómez, A., & Molina-Grima, E. (2013). Mixotrophic growth of Phaeodactylum tricornutum on fructose and glycerol in fed-batch and semi-continuous modes. *Bioresource Technology* 147, 569–576. https://doi.org/https://doi.org/10.1016/j.biortech.2013.08.092

Chauton, M. S., Olsen, Y., & Vadstein, O. (2013). Biomass production from the microalga *Phaeodactylum tricornutum*: Nutrient stress and chemical composition in exponential fed-batch cultures. *Biomass and Bioenergy* 58, 87–94. https://doi.org/http://dx.doi.org/10.1016/j.biombioe.2013.10.004

Chawla, P., Gola, D., Dalvi, V., Sreekrishnan, T. R., Ariyadasa, T. U., & Malik, A. (2022). Design and development of mini-photobioreactor system for strategic high throughput selection of optimum microalgae-wastewater combination. *Bioresource Technology Reports* 17. https://doi.org/10.1016/j.biteb.2022.100967

Chen, G. Q., Jiang, Y., & Chen, F. (2008). Salt-induced alterations in lipid composition of diatom *Nitschia laevis* (Bacillariophyceae) under heterotrophic culture condition. *Journal of Phycology* 44(5), 1309–1314. https://doi.org/10.1111/j.1529-8817.2008.00565.x

Chetia, L., Kalita, D., & Ahmed, G. A. (2017). Enhanced photocatalytic degradation by diatom templated mixed phase titania nanostructure. *Journal of Photochemistry and Photobiology A: Chemistry* 338, 134–145. https://doi.org/10.1016/J.JPHOTOCHEM.2017.01.035

Choi, H-G., Hyoung, M., Joo, W., Jung, S., Hong, J-S., Kang, Boo, S. M., & Kang, S-H. (2008). Morphology and phylogenetic relationships of some psychrophilic polar diatoms (Bacillariophyta). *Nova Hedwigia, Beiheft* 133: 7–30.

Choi, P. M., Tscharke, B., Samanipour, S., Hall, W. D., Gartner, C. E., Mueller, J. F., Thomas, K. v., & O'Brien, J. W. (2019). Social, demographic, and economic correlates of food and chemical consumption measured by wastewater-based epidemiology. *Proceedings of the National Academy of Sciences of the United States of America* 116(43), 21864–21873. https://doi.org/10.1073/pnas.1910242116

Conradi, M. (2013). Nanosilica-reinforced polymer composites. *Materiali in Tehnologije*, 47(3), 285–293.

Cunningham, C., & Gharipour, M. (2018). Pipe dreams: urban wastewater treatment for biodiversity protection. *Urban Science* 2(1), 10. https://doi.org/10.3390/urbansci2010010

de Castro Araújo, S., & Garcia, V. M. T. (2005). Growth and biochemical composition of the diatom *Chaetoceros* cf. *wighamii* brightwell under different temperature, salinity and carbon dioxide levels. I. Protein, carbohydrates and lipids. *Aquaculture* 246(1–4), 405–412. https://doi.org/http://dx.doi.org/10.1016/j.aquaculture.2005.02.051

Debenest, T., Pinelli, E., Coste, M., Silvestre, J., Mazzella, N., Madigou, C., & Delmas, F. (2009). Sensitivity of freshwater periphytic diatoms to agricultural herbicides. *Aquatic Toxicology* 93(1), 11–17. https://doi.org/10.1016/j.aquatox.2009.02.014

Ehrlich, H., & Witkowski, A. (2015). Biomineralization in diatoms: The organic templates. In: C. Hamm (Ed.), *Evolution of Lightweight Structures: Analyses and Technical Applications* (pp. 39–58). Springer: Netherlands. https://doi.org/10.1007/978-94-017-9398-8_3

Ehrlich, H., Bailey, E., Wysokowski, M., & Jesionowski, T. (2021). *Biomimetics Forced Biomineralization: A Review*. https://doi.org/10.3390/biomimetics6030046

Esteves, S. M., Almeida, S. F. P., Gonçalves, S., Rimet, F., Bouchez, A., & Figueira, E. (2018). Sensitive vs. tolerant *Nitzschia palea* (Kützing) W. Smith strains to atrazine: a biochemical perspective. *Ecotoxicology* 27, 860–870. https://doi.org/10.1007/s10646-018-1953-1

Fazey, F. M. C., & Ryan, P. G. (2016). Biofouling on buoyant marine plastics: An experimental study into the effect of size on surface longevity. *Environmental Pollution* 210, 354–360. https://doi.org/10.1016/j.envpol.2016.01.026

Frankland, P. F. (1888). The micro-organisms of air and water. *Nature* 38(975), 232–236. https://doi.org/10.1038/038232a0

Gao, X., Wang, Y., Ji, G., Cui, R., & Liu, Z. (2018). One-pot synthesis of hierarchical-pore metal-organic frameworks for drug delivery and fluorescent imaging. *CrystEngComm* 20(8), 1087–1093. https://doi.org/10.1039/c7ce02053h

Gautam, R. K., Sharma, S. K., Mahiya, S., & Chattopadhyaya, M. C. (2014). *Contamination of Heavy Metals in Aquatic Media: Transport, Toxicity and Technologies for Remediation*. In: *Heavy Metals in Water* (pp. 1–24). https://doi.org/10.1039/9781782620174-00001

Gerhardt, A., de Bisthoven, L. J., Guhr, K., Soares, A. M., & Pereira, M. J. (2008). Phytoassessment of acid mine drainage: Lemna gibba bioassay and diatom community structure. *Ecotoxicology* 17(1), 47–58. https://doi.org/10.1007/s10646-007-0175-8

Gholami, P., Khataee, A., & Bhatnagar, A. (2020a). Environmentally superior cleaning of diatom frustules using sono-Fenton process: Facile fabrication of nanoporous silica with homogeneous morphology and controlled size. *Ultrasonics Sonochemistry* 64. https://doi.org/10.1016/j.ultsonch.2020.105044

Gholami, P., Khataee, A., & Bhatnagar, A. (2020b). Photocatalytic degradation of antibiotic and hydrogen production using diatom-templated 3D WO3-x@mesoporous carbon nanohybrid under visible light irradiation. *Journal of Cleaner Production* 275. https://doi.org/10.1016/j.jclepro.2020.124157

Gnanamoorthy, P., Anandhan, S., & Prabu, V. A. (2014). Natural nanoporous silica frustules from marine diatom as a biocarrier for drug delivery. *Journal of Porous Materials* 21(5), 789–796. https://doi.org/10.1007/s10934-014-9827-2

Gordon, R., Losic, D., Tiffany, M. A., Nagy, S. S., & Sterrenburg, F. A. S. (2009). The glass menagerie: diatoms for novel applications in nanotechnology. *Trends in Biotechnology* 27(2), 116–127. https://doi.org/10.1016/j.tibtech.2008.11.003

Griffiths, M. J., & Harrison, S. T. L. (2009). Lipid productivity as a key characteristic for choosing algal species for biodiesel production. *Journal of Applied Phycology* 21(5), 493–507. https://doi.org/10.1007/s10811-008-9392-7

Grmela, J., Kopp, R., & Hadašová, L. (2014). Eutrophication potential of wastewater treatment plants in the upper reaches of svratka river. *Acta Universitatis Agriculturae et Silviculturae Mendelianae Brunensis* 62(3), 469–475. https://doi.org/10.11118/actaun201462030469

Guckert, J. B., & Cooksey, K. E. (1990). Triglyceride accumulation and fatty acid profile changes in *Chlorella* (Chlorophyta) during high pH-induced cell cycle inhibition. *Journal of Phycology* 26(1), 72–79. https://doi.org/https://doi.org/10.1111/j.0022-3646.1990.00072.x

Hamm, C. E., Merkel, R., Springer, O., Jurkojc, P., Maiert, C., Prechtelt, K., & Smetacek, V. (2003). Architecture and material properties of diatom shells provide effective mechanical protection. *Nature* 421(6925), 841–843. https://doi.org/10.1038/nature01416

Hariz, H. B., Takriff, M. S., Mohd Yasin, N. H., Ba-Abbad, M. M., & Mohd Hakimi, N. I. N. (2019). Potential of the microalgae-based integrated wastewater treatment and CO_2 fixation system to treat palm oil mill effluent (POME) by indigenous microalgae; *Scenedesmus* sp. and *Chlorella* sp. *Journal of Water Process Engineering* 32(July). https://doi.org/10.1016/j.jwpe.2019.100907

Hecky, R. E., & Kilham, P. (1973). Diatoms in alkaline, saline lakes: ecology and geochemical implications. *Limnology and Oceanography* 18(1), 53–71. https://doi.org/https://doi.org/10.4319/lo.1973.18.1.0053

Hellebust, J. (2008). Glucose uptake by *Cyclotella cryptica*: Dark induction and light inactivation of transport system. *Journal of Phycology* 7, 345–349. https://doi.org/10.1111/j.1529-8817.1971.tb01530.x

Hevia-Orube, J., Orive, E., David, H., Díez, A., Laza-Martínez, A., Miguel, I., & Seoane, S. (2016). Molecular and morphological analyses of solitary forms of brackish Thalassiosiroid diatoms (Coscinodiscophyceae), with emphasis on their phenotypic plasticity. *European Journal of Phycology* 51(1), 11–30. https://doi.org/10.1080/09670262.2015.1077394

Heydarizadeh, P., Boureba, W., Zahedi, M., Huang, B., Moreau, B., Lukomska, E., Couzinet-Mossion, A., Wielgosz-Collin, G., Martin-Jézéquel, V., Bougaran, G., Marchand, J., & Schoefs, B. B. (2017). Response of CO_2-starved diatom *Phaeodactylum tricornutum* to light intensity transition. *Philosophical Transactions B* https://doi.org/10.1098/rstb.2016.0396

Hildebrand, M., York, E., Kelz, J. I., Davis, A. K., Frigeri, L. G., Allison, D. P., & Doktycz, M. J. (2006). Nanoscale control of silica morphology and three-dimensional structure during diatom cell wall formation. https://doi.org/10.1557/JMR.2006.0333

Hu, Q., Sommerfeld, M., Jarvis, E., Ghirardi, M., Posewitz, M., Seibert, M., & Darzins, A. (2008). Microalgal triacylglycerols as feedstocks for biofuel production: perspectives and advances. *Plant Journal* 54(4), 621–639. https://doi.org/10.1111/j.1365-313X.2008.03492.x

International Year of Plant Health–Final Report. (2021). In: *International Year of Plant Health–Final report.* FAO. https://doi.org/10.4060/cb7056en

Irfan, M., Moniruzzaman, M., Ahmad, T., Osman, O. Y., Mandal, P. C., Bhattacharjee, S., & Hussain, M. (2018). Stability, interparticle interactions and catalytic performance of gold nanoparticles synthesized through ionic liquid mediated oil palm leaves extract. *Journal of Environmental Chemical Engineering* 6(4), 5024–5031. https://doi.org/10.1016/j.jece.2018.07.031

Jain, R., Dhali, S., Nigam, H., Malik, A., Malik, H. K., & Satyakam, R. (2022). Recovery of diatom bio-silica using chemical, thermal, and plasma treatment. *Bioresource Technology Reports* 18, 101035. https://doi.org/10.1016/j.biteb.2022.101035

Janech, M. G., Krell, A., Mock, T., Kang, J. S., & Raymond, J. A. (2006). Ice-binding proteins from sea ice diatoms (Bacillariophyceae). *Journal of Phycology* 42(2), 410–416. https://doi.org/10.1111/j.1529-8817.2006.00208.x

Javalkote, V. S., Pandey, A. P., Puranik, P. R., & Deshmukh, P. K. (2015). Magnetically responsive siliceous frustules for efficient chemotherapy. *Materials Science and Engineering: C* 50, 107–116. https://doi.org/10.1016/J.MSEC.2015.01.079

Jones, E. R., Van Vliet, M. T. H., Qadir, M., & Bierkens, M. F. P. (2021). Country-level and gridded estimates of wastewater production, collection, treatment and reuse. *Earth System Science Data* 13(2), 237–254. https://doi.org/10.5194/essd-13-237-2021

José, J., Karlusich, P., Ibarbalz, F. M., & Bowler, C. (2019). *Phytoplankton in the Tara Ocean.* https://doi.org/10.1146/annurev-marine-010419

Kamyab, H. (2018). *Palm Oil Mill Effluent as an Environmental Pollutant* (S. Chelliapan, Ed.; Ch. 2). IntechOpen. https://doi.org/10.5772/intechopen.75811

Khan, M. I., Cheema, S. A., Anum, S., Niazi, N. K., Azam, M., Bashir, S., Ashraf, I., & Qadri, R. (2020). *Phytoremediation of Agricultural Pollutants BT–Phytoremediation: In-situ Applications* (B. R. Shmaefsky, Ed.; pp. 27–81). Springer International Publishing. https://doi.org/10.1007/978-3-030-00099-8_2

Khan, M. J., Singh, R., Shewani, K., Shukla, P., Bhaskar, P. v, Ballabh Joshi, K., & Vinayak, V. (2020). Exopolysaccharides directed embellishment of diatoms triggered on plastics and other marine litter. *Scientific Reports* 10, 18448. https://doi.org/10.1038/s41598-020-74801-7

Kim, M., Gwak, Y., Jung, W., & Jin, E. S. (2017). Identification and characterization of an isoform antifreeze protein from the antarctic marine diatom, chaetoceros neogracile and suggestion of the core region. *Marine Drugs* 15(10). https://doi.org/10.3390/md15100318

Kiran Marella, T., Saxena, A., & Tiwari, A. (2020). Diatom mediated heavy metal remediation: A review. *Bioresource Technology* 305. https://doi.org/10.1016/j.biortech.2020.123068

Knothe, G. (2008). "Designer" biodiesel: optimizing fatty ester composition to improve fuel properties. *Energy & Fuels* 22(2), 1358–1364. https://doi.org/10.1021/ef700639e

Kooistra, W. H. C. F., Gersonde, R., Medlin, L. K., & Mann, D. G. (2007). The origin and evolution of the diatoms: their adaptation to a planktonic existence. *Evolution of Primary Producers in the Sea* 207–249. https://doi.org/10.1016/B978-012370518-1/50012-6

Krämer, L. C., Wasser, D., Haitz, F., Sabel, B., & Büchel, C. (2022). Heterologous expression of HUP1 glucose transporter enables low-light mediated growth on glucose in Phaeodactylum tricornutum. *Algal Research* 64, 102719. https://doi.org/10.1016/j.algal.2022.102719

Kuczynska, P., Jemiola-Rzeminska, M., & Strzalka, K. (2015). Photosynthetic pigments in diatoms. *Marine Drugs* 13, 5847–5881. https://doi.org/10.3390/md13095847

Kumar, B. R. , Mathimani, T., Sudhakar, M. P., Rajendran, K., Nizami, A-S., Brindhadevi, K., & Pugazhendhi, A. (2021). A state of the art review on the cultivation of algae for energy and other valuable products: Application, challenges, and opportunities. *Renewable and Sustainable Energy Reviews* 138, 110649. https://doi.org/https://doi.org/10.1016/j.rser.2020.110649

Lee, K. H., Huang, K. M., Tseng, W. L., Chili, T. C., Lin, Y. W., & Chang, H. T. (2007). Manipulation of the growth of gold and silver nanomaterials on glass by seeding approach. *Langmuir* 23(3), 1435–1442. https://doi.org/10.1021/la061880j

Lewin, J. C., & Guillard, R. R. (1963). Diatoms. *Annual Review of Microbiology* 17, 373–414. https://doi.org/10.1146/annurev.mi.17.100163.002105

Leyland, B., Boussiba, S., & Khozin-Goldberg, I. (2020). A review of diatom lipid droplets. *Biology* 9(2), 38. www.mdpi.com/2079-7737/9/2/38

Li, A., Zhao, X., Duan, G., Anderson, S., & Zhang, X. (2019). Diatom frustule-inspired metamaterial absorbers: the effect of hierarchical pattern arrays. *Advanced Functional Materials* 29(22). https://doi.org/10.1002/adfm.201809029

Li, J., Holze, R., Moyo, S., Wang, S., Li, S., Tang, T., & Chen, X. (2021). Three-dimensional hierarchical porous carbon derived from natural resources for highly efficient treatment of polluted water. *Environmental Sciences Europe* 33(1). https://doi.org/10.1186/s12302-021-00527-6

Litchfield, C. (2005). Thirty years and counting: bioremediation in its prime? *BioScience* 55(3), 273–279.

Liyanage, C. P., & Yamada, K. (2017). Impact of population growth on the water quality of natural water bodies. *Sustainability (Switzerland)* 9(8). https://doi.org/10.3390/su9081405

Lozano, P., Trombini, C., Crespo, E., Blasco, J., & Moreno-Garrido, I. (2014). ROI-scavenging enzyme activities as toxicity biomarkers in three species of marine microalgae exposed to model contaminants (copper, Irgarol and atrazine). *Ecotoxicology and Environmental Safety* 104(1), 294–301. https://doi.org/10.1016/j.ecoenv.2014.03.021

Luo, Y., Ahmad, M., Schug, A., & Tsotsalas, M. (2019). Rising up: hierarchical metal–organic frameworks in experiments and simulations. *Advanced Materials* 31(26). https://doi.org/10.1002/adma.201901744

Ma, J., Zhou, B., Duan, D., Wei, Y., & Pan, K. (2018). Silicon limitation reduced the adsorption of cadmium in marine diatoms. *Aquatic Toxicology* 202, 136–144. https://doi.org/10.1016/j.aquatox.2018.07.011

Malavin, S., Shmakova, L., Claverie, J.-M., & Rivkina, E. (2018). Frozen zoo: A collection of permafrost samples containing viable protists and their viruses. *Biodiversity Data Journal* 8, e51586–e51586. https://doi.org/10.3897/BDJ.8.e51586

Mann, D. G., & Vanormelingen, P. (2013). An inordinate fondness? the number, distributions, and origins of diatom species. *Journal of Eukaryotic Microbiology* 60(4), 414–420. https://doi.org/10.1111/jeu.12047

Mansour, M. P., Frampton, D. M. F., Nichols, P. D., Volkman, J. K., & Blackburn, S. I. 2005. Lipid and fatty acid yield of nine stationary-phase microalgae: Applications and unusual C24–C28 polyunsaturated fatty acids. *Journal of Applied Phycology* 17(4), 287–300. https://doi.org/10.1007/s10811-005-6625-x

Martucci, N. M., Migliaccio, N., Ruggiero, I., Albano, F., Calì, G., Romano, S., Terracciano, M., Rea, I., Arcari, P., & Lamberti, A. (2016). Nanoparticle-based strategy for personalized B-cell lymphoma therapy. *International Journal of Nanomedicine* 11, 6089–6101. https://doi.org/10.2147/IJN.S118661

Matsuda, Y., Hopkinson, B. M., Nakajima, K., Dupont, C. L., & Tsuji, Y. (2017). Mechanisms of carbon dioxide acquisition and CO_2 sensing in marine diatoms: a gateway to carbon metabolism. *Philosophical Transaction B*. https://doi.org/10.1098/rstb.2016.0403

Mayzel, B., Aram, L., Varsano, N., Wolf, S. G., Gal, A., & Lior, A. (2021). Structural evidence for extracellular silica formation by diatoms. *Nature Communications*. https://doi.org/10.1038/s41467-021-24944-6

Mazumder, N., Gogoi, A., Buragohain, A. K., Ahmed, G. A., & Choudhury, A. (2014). Structural and optical characterization of freshwater diatoms (*Cyclotella* sp.): nature's nanoporous silica manufacturing plant. *Quantum Dots and Nanostructures: Synthesis, Characterization, and Modeling XI* 8996, 89960Y. https://doi.org/10.1117/12.2039793

Milligan, A. J., & M. Morel, F. M. (2001). A proton buffering role for silica in diatoms. *Acta Crystallographica D. Biology Crystallography* 276. www.sciencemag.org/cgi/content/full/297/5588/1837/

Minggat, E., Roseli, W., & Tanaka, Y. (2021). Nutrient absorption and biomass production by the marine diatom *Chaetoceros muelleri*: effects of temperature, salinity, photoperiod, and light intensity. *Journal of Ecological Engineering* 22(1), 231–240. https://doi.org/10.12911/22998993/129253

Mishra, M., Arukha, A. P., Bashir, T., Yadav, D., & Prasad, G. B. K. S. (2017). All new faces of diatoms: Potential source of nanomaterials and beyond. *Frontiers in Microbiology* 8(July), 1–8. https://doi.org/10.3389/fmicb.2017.01239

Mock, T., & Hoch, N. (2005). Long-term temperature acclimation of photosynthesis in steady-state cultures of the polar diatom *Fragilariopsis cylindrus*. *Photosynthesis Research* 85(3), 307–317. https://doi.org/10.1007/s11120-005-5668-9

Muylaert, K., & Sabbe, K. (1996). The diatom genus *Thalassiosira* (Bacillariophyta) in the estuaries of the Schelde (Belgium/The Netherlands) and the Elbe (Germany). *Botanica Marina* 39(2), 103–115. https://doi.org/10.1515/botm.1996.39.1-6.103

Neori, A., & Holm-Hansen, O. (1982). Effect of temperature on rate of photosynthesis in Antarctic phytoplankton. *Polar Biology* 1(1), 33–38. https://doi.org/10.1007/BF00568752

Ngatiman, M., Jami, M. S., Abu Bakar, M. R., Subramaniam, V., & Loh, S. K. (2021). Investigation of struvite crystals formed in palm oil mill effluent anaerobic digester. *Heliyon* 7(1), e05931. https://doi.org/10.1016/j.heliyon.2021.e05931

Not, F., Siano, R., Kooistra, W. H. C. F., Simon, N., Vaulot, D., & Probert, I. (2012). Diversity and ecology of eukaryotic marine phytoplankton. *Advances in Botanical Research* 64, 1–53. https://doi.org/10.1016/B978-0-12-391499-6.00001-3

Nur, M. M. A., Muizelaar, W., Boelen, P., & Buma, A. G. J. (2019). Environmental and nutrient conditions influence fucoxanthin productivity of the marine diatom *Phaeodactylum tricornutum* grown on palm oil mill effluent. *Journal of Applied Phycology* 31(1), 111–122. https://doi.org/10.1007/s10811-018-1563-6

Panneerselvam, K., Manimaran, K., Sampathkumar, P., & Rameshkumar, L. (2013). Growth and nutrient removal properties of the diatoms, *Chaetoceros curvisetus* and *C. simplex* under different nitrogen sources. *Applied Water Science* 3(1), 49–55. https://doi.org/10.1007/s13201-012-0056-z

Pawolski, D., Heintze, C., Mey, I., Steinem, C., & Kröger, N. (2018). Reconstituting the formation of hierarchically porous silica patterns using diatom biomolecules. *Journal of Structural Biology* 204(1), 64–74. https://doi.org/10.1016/j.jsb.2018.07.005

Pierella Karlusich, J. J., Bowler, C., & Biswas, H. (2021). Carbon dioxide concentration mechanisms in natural populations of marine diatoms: insights from Tara Oceans. *Frontiers in Plant Science* 12. https://doi.org/10.3389/fpls.2021.657821

Pruvost, J., Van Vooren, G., Le Gouic, B., Couzinet-Mossion, A., & Legrand, J. (2011). Systematic investigation of biomass and lipid productivity by microalgae in photobioreactors for biodiesel application. *Bioresource Technology* 102(1), 150–158. https://doi.org/http://dx.doi.org/10.1016/j.biortech.2010.06.153

Puy-Alquiza, M. J., Luna, B. N., Miranda-Aviles, R., Hernández, Ma. M. S., & Aguilera, G. C. (2021). Diatom of the Bacillariophyceae class in thermophilic microbial mats present in sulphurous hot springs and their possible biotechnological application. *Global Journal of Earth Science and Engineering* 8, 80–96. https://doi.org/10.15377/2409-5710.2021.08.7

Pytlik, N., Butscher, D., Machill, S., & Brunner, E. (2018). Diatoms – a "green" way to biosynthesize gold-silica nanocomposites? *Zeitschrift Für Physikalische Chemie* 232(9–11), 1353–1368. https://doi.org/doi:10.1515/zpch-2018-1141

Radakovits, R., Eduafo, P. M., & Posewitz, M. C. (2011). Genetic engineering of fatty acid chain length in *Phaeodactylum tricornutum*. *Metabolic Engineering* 13(1), 89–95. https://doi.org/10.1016/j.ymben.2010.10.003

Rea, I., Martucci, N. M., de Stefano, L., Ruggiero, I., Terracciano, M., Dardano, P., Migliaccio, N., Arcari, P., Taté, R., Rendina, I., & Lamberti, A. (2014). Diatomite biosilica nanocarriers for siRNA transport inside cancer cells. *Biochimica et Biophysica Acta (BBA)–General Subjects* 1840(12), 3393–3403. https://doi.org/10.1016/J.BBAGEN.2014.09.009

Rehman, W., Zeb, A., Noor, N., & Nawaz, M. (2008). Heavy metal pollution assessment in various industries of Pakistan. *Environmental Geology* 55(2), 353–358. https://doi.org/10.1007/s00254-007-0980-7

Reyes, J. G., Sánchez-Saavedra, Md. P., Viana, Mt., Flores-Acevedo, N., & Peláez, C. 2009. Effect of eight benthic diatoms as feed on the growth of red abalone (*Haliotis rufescens*) postlarvae. *Journal of Applied Phycology* 21, 387–393. https://doi.org/10.1007/s10811-008-9381-x

Roessler, P. (2008). Effects of silicon deficiency on lipid composition and metabolism in the diatom *Cyclotella cryptica*. *Journal of Phycology* 24, 394–400. https://doi.org/10.1111/j.1529-8817.1988.tb04482.x

Rojas, S., Colinet, I., Cunha, D., Hidalgo, T., Salles, F., Serre, C., Guillou, N., & Horcajada, P. (2018). Toward understanding drug incorporation and delivery from biocompatible metal-organic frameworks in view of cutaneous administration. *ACS Omega* 3(3), 2994–3003. https://doi.org/10.1021/acsomega.8b00185

Round C. R., Mann D. G., & F. Eric. (2000). *The Diatoms: Biology & Morphology of the Genera*. Cambridge University Press.

Samantaray, S. S., Sangeetha, V., Abinaya, S., & Ramaprabhu, S. (2020). Diatom frustule-graphene based nanomaterial for room temperature hydrogen storage. *International Journal of Hydrogen Energy* 45(1), 764–773. https://doi.org/10.1016/j.ijhydene.2019.10.155

Schoefs, B., Hu, H., & Kroth, P. G. (2017). The peculiar carbon metabolism in diatoms. *Philosophical Transactions of the Royal Society B: Biological Sciences* 372(1728). Royal Society Publishing. https://doi.org/10.1098/rstb.2016.0405

Schröfel, A., Kratošová, G., Bohunická, M., Dobročka, E., & Vávra, I. (2011). Biosynthesis of gold nanoparticles using diatoms-silica-gold and EPS-gold bionanocomposite formation. *Journal of Nanoparticle Research* 13(8), 3207–3216. https://doi.org/10.1007/s11051-011-0221-6

Seth, K., Kumar, A., Rastogi, R. P., Meena, M., Vinayak, V., & Harish. (2021). Bioprospecting of fucoxanthin from diatoms — challenges and perspectives. *Algal Research* 60. https://doi.org/10.1016/j.algal.2021.102475

Shah, M., Hashmi, H. N., Ali, A., & Ghumman, A. R. (2014). Performance assessment of aquatic macrophytes for treatment of municipal wastewater. *Journal of Environmental Health Science & Engineering* 12, 106. https://doi.org/10.1186/2052-336X-12-106

Sharma, N., Simon, D. P., Diaz-Garza, A. M., Fantino, E., Messaabi, A., Meddeb-Mouelhi, F., Germain, H., Desgagné-Penix, I. (2021). Diatoms biotechnology: various industrial applications for a greener tomorrow. *Frontiers in Marine Science* 8. https://doi.org/10.3389/fmars.2021.636613

Sheehan, J., Terri Dunahay, T., Benemann, J., & Roessler, P. (1998). A look back at the U.S. Department of Energy's Aquatic Species Program—Biodiesel from algae. Prepared for U.S. Department of Energy's Office of Fuels Development and the National Renewable Energy Laboratory. www.nrel.gov/docs/legosti/fy98/24190.pdf

Shen, T.-W., Tsai, M.-H., Lai, Y.-S., & Su, Y.-H. (2017). Surface plasmon of gold nanoparticle on diatom template as light scattering center for local field enhancer. *Optical and Quantum Electronics* https://doi.org/10.1007/s11082-017-1133-1

Shifrin, N., & Chisholm, S. (2008). Phytoplankton lipids: interspecific differences and effects of nitrate, silicate and light–dark cycles. *Journal of Phycology* 17, 374–384. https://doi.org/10.1111/j.1529-8817.1981.tb00865.x

Siciliano, A., Limonti, C., Curcio, G. M., & Molinari, R. (2020). Advances in struvite precipitation technologies for nutrients removal and recovery from aqueous waste and wastewater. *Sustainability (Switzerland)* 12(18). https://doi.org/10.3390/su12187538

Silva, B. F., Wendt, E. V., Castro, J. C., Oliveira, A. Ed., Carrim, A. J. I., Vieira, J. D. G., Sassi, R., Sassi, C. Fd. C., Silva, A. L. Vd., Barboza, G. Fd. O., & Antoniosi Filho, N. R. (2015). Analysis of some chemical elements in marine microalgae for biodiesel production and other uses. *Algal Research* 9, 312–321. https://doi.org/http://dx.doi.org/10.1016/j.algal.2015.04.010

Souza-Egipsy, V., Altamirano, M., Amils, R., & Aguilera, A. (2011). Photosynthetic performance of phototrophic biofilms in extreme acidic environments. *Environmental Microbiology* 13(8), 2351–2358. https://doi.org/10.1111/j.1462-2920.2011.02506.x

Suresh Kumar, K., Dahms, H. U., Lee, J. S., Kim, H. C., Lee, W. C., & Shin, K. H. (2014). Algal photosynthetic responses to toxic metals and herbicides assessed by chlorophyll a fluorescence. *Ecotoxicology and Environmental Safety* 104(1), 51–71). https://doi.org/10.1016/j.ecoenv.2014.01.042

Tang, H., Chen, C. J., Huang, Z., Bright, J., Meng, G., Liu, R. S., & Wu, N. (2020). Plasmonic hot electrons for sensing, photodetection, and solar energy applications: A perspective. *Journal of Chemical Physics* 152(22). https://doi.org/10.1063/5.0005334

Thomas, D. N., & Dieckmann, G. S. (2002). Antarctic Sea ice—a habitat for extremophiles. *Science* 295(5555), 641–644. https://doi.org/10.1126/science.1063391

Thompson, G. A. (1996). Lipids and membrane function in green algae. *Biochimica et Biophysica Acta (BBA)–Lipids and Lipid Metabolism* 1302(1), 17–45. https://doi.org/https://doi.org/10.1016/0005-2760(96)00045-8

Tolomio, C., Ceschi-Berrini, C., Moschin, E., & Galzigna, L. (1999). Colonization by diatoms and antirheumatic activity of thermal mud. *Cell Biochemistry and Function* 17(1), 29–33. https://doi.org/10.1002/(sici)1099-0844(199903)17:1<29::aid-cbf808>3.0.co;2-4

Tréguer, P., Bowler, C., Moriceau, B., Dutkiewicz, S., Gehlen, M., Aumont, O., Bittner, L., Dugdale, R., Finkel, Z., Iudicone, D., Jahn, O., Guidi, L., Lasbleiz, M., Leblanc, K., Levy, M., & Pondaven, P. (2018).

Influence of diatom diversity on the ocean biological carbon pump. *Nature Geoscience* 11(1), 27–37. Nature Publishing Group. https://doi.org/10.1038/s41561-017-0028-x

UNESCO (2017). *The United Nations World Water Development Report, 2017: Wastewater: The Untapped Resource*. https://unesdoc.unesco.org/notice?id=p::usmarcdef_0000247153

United Nations. (2021). *Progress on Wastewater Treatment–2021 Update*. www.unwater.org/publications/progress-on-wastewater-treatment-631-2021-update/#print_all

Vasani, R. B., Losic, D., Cavallaro, A., & Voelcker, N. H. (2015). Fabrication of stimulus-responsive diatom biosilica microcapsules for antibiotic drug delivery. *Journal of Material Chemistry B* 3, 4325. https://doi.org/10.1039/c5tb00648a

Wang, J. K., & Seibert, M. (2017). Prospects for commercial production of diatoms. *Biotechnology for Biofuels* 10(1). BioMed Central Ltd. https://doi.org/10.1186/s13068-017-0699-y

Wang, J., Li, Y., Lu, L., Zheng, M., Zhang, X., Tian, H., Wang, W., & Ru, S. (2019). Polystyrene microplastics cause tissue damages, sex-specific reproductive disruption and transgenerational effects in marine medaka (*Oryzias melastigma*). *Environmental Pollution* 254. https://doi.org/10.1016/j.envpol.2019.113024

Wang, J.-K., & Seibert, M. (2017). Prospects for commercial production of diatoms. *Biotechnology for Biofuels* 10, 16. https://doi.org/10.1186/s13068-017-0699-y

Wang, S., Wang, Y., Liang, Y., Cao, W., Sun, C., Ju, P., & Zheng, L. (2020). The interactions between microplastic polyvinyl chloride and marine diatoms: Physiological, morphological, and growth effects. *Ecotoxicology and Environmental Safety* 203. https://doi.org/10.1016/j.ecoenv.2020.111000

Watanabe, T., Kodama, Y., & Mayama, S. (2010). Application of a novel cleaning method using low-temperature plasma on tidal flat diatoms with heterovalvy or delicate frustule structure. *Proceedings of the Academy of Natural Sciences of Philadelphia* 160(1), 83–87. https://doi.org/10.1635/053.160.0109

Weyer, K. M., Bush, D. R., Darzins, A., & Willson, B. D. (2009). *Theoretical Maximum Algal Oil Production*. https://doi.org/10.1007/s12155-009-9046-x

Wu, J., Wang, T., Shi, N., Min, F., & Pan, W. P. (2022). Hierarchically porous biochar templated by in situ formed ZnO for rapid Pb^{2+} and Cd^{2+} adsorption in wastewater: Experiment and molecular dynamics study. *Environmental Pollution* 302. https://doi.org/10.1016/j.envpol.2022.119107

Yang, X., Du, C., Guo, Y., Guo, Y., Wang, L., Wang, Y., & Zhan, W. (2019). Al_2O_3 supported hybrid Pd–CeO_2 colloidal spheres and its enhanced catalytic performances for methane combustion. *Journal of Rare Earths* 37(7), 714–719. https://doi.org/10.1016/j.jre.2018.10.005

Yeh, Y. Q., Su, C. J., Wang, C. A., Lai, Y. C., Tang, C. Y., Di, Z., Frielinghaus, H., Su, A. C., Jeng, U. S., & Mou, C. Y. (2021). Diatom-inspired self-assembly for silica thin sheets of perpendicular nanochannels. *Journal of Colloid and Interface Science* 584, 647–659. https://doi.org/10.1016/j.jcis.2020.10.114

Yona, D., Febriana, R., & Handayani, M. (2018). The comparison of heavy metals (Pb and Cd) in the water and sediment during spring and neap tide tidal periods in Popoh Bay, Indonesia. *IOP Conference Series: Earth and Environmental Science* 137(1). https://doi.org/10.1088/1755-1315/137/1/012037

Yoshida, D., Raebiger, H., Shudo, K., & Ohno, K. (2018). Core electron topologies in chemical compounds: case study of carbon versus silicon. *Angewandte Chemie (International Ed.*, 57(24), 7012–7018. https://doi.org/10.1002/anie.201713108

Yu, E. T., Zendejas, F. J., Lane, P. D., Gaucher, S., Simmons, B. A., & Lane, T. W. (2009). Triacylglycerol accumulation and profiling in the model diatoms *Thalassiosira pseudonana* and *Phaeodactylum tricornutum* (Baccilariophyceae) during starvation. *Journal of Applied Phycology* 21(6), 669. https://doi.org/10.1007/s10811-008-9400-y

Zhu, J., Du, J., Pei, X., Qin, X., & Liu, B. (2017). Cultivation of microalgae in wastewater: A review. *Proceedings of the 2016 4th International Conference on Renewable Energy and Environmental Technology (ICREET 2016)*, 112(Icreet 2016), 303–310. https://doi.org/10.2991/icreet-16.2017.52

Zielinski, U, Gersonde, R. (1997). Diatom distribution in Southern Ocean surface sediments (Atlantic sector): Implications for paleoenvironmental reconstructions. *Palaeogeography, Palaeoclimatology, Palaeoecology* 129(3), 213–250. https://doi.org/https://doi.org/10.1016/S0031-0182(96)00130-7

10 Cultivation of Diatoms for Wastewater Remediation

G. Saranya[a,b] *and T. V. Ramachandra*[a,b,c]
[a]Energy & Wetlands Research Group, Centre for Ecological Sciences, Indian Institute of Science, Bangalore, Karnataka, India
[b]Centre for Sustainable Technologies, Indian Institute of Science, Bangalore, Karnataka, India
[c]Centre for Infrastructure, Sustainable Transportation and Urban Planning, Indian Institute of Science, Bangalore, Karnataka, India

10.1 INTRODUCTION

Wastewaters of domestic, industrial, and agricultural origins mainly consist of water (~ 99%) with a minute percentage (~ 1%) of solids, comprising both organics and inorganics (Saranya and Ramachandra, 2020). Wastewater generated in households (domestic) is a mixture of organic (carbohydrates, fats, lignin) and chemical (aliphatic and aromatic) compounds. In contrast, aquaculture wastewater comprises inorganic constituents with a higher total suspended solids (TSS) content, and the organic load of unutilized aquaculture feed and detritus matters. The quantum of wastewater generation in domestic and industrial sectors has been increasing subsequent to unregulated urbanization and industrialization. Untreated or partially treated wastewater being discharged to waterbodies (rivers, lakes, etc.) has been posing severe challenges of water pollution in aquatic ecosystems in recent times, and remediation is pivotal to minimizing human health hazards (Khan et al., 2021). Therefore, regular monitoring of water bodies is essential for prudent management and protecting inhabiting aquatic biota. Remediation of wastewaters by growing microalgae or phyco-remediation is the most viable and economical method of treating wastewater. It has emerged as a promising economical option to treat wastewater from industrial and anthropogenic origins, including agricultural, dairy, and slaughterhouse wastes. Microalgal treatment of wastewaters is based on the uptake of nutrients through the mechanism of bacterial and algal symbiotic interactions. The bacteria help in the degradation of organic nutrients into inorganic forms that are then assimilated by microalgae, while microalgae, in turn, release the oxygen required by the bacteria for organic compound degradation (Shah et al., 2021). Microalgae, especially diatoms, possess unique, unaltered, robust, and transparent silica that does not perish even after cell death, thus aiding in ecological health assessment as bioindicators and in reconstructing the past environmental status of water bodies (Gautam et al., 2017). Phytoplankton are being used as bioindicators. In this context, periphytons consisting of a consortium of microorganisms (such as bacteria and algae) in an embedded exopolysaccharide are a better choice, as these matrices are long-lasting in water bodies (Larned, 2010).

In recent decades, integrated approaches to wastewater treatment with microalgae in the production of bioenergy are gaining popularity, and research attention as biological waste treatment systems contribute toward zero-waste with a reduction in carbon and water footprints (Vishwakarma et al., 2022). Diatoms' ability to photosynthesize atmospheric CO_2 into valuable biomolecules such as proteins, lipids, carbohydrates, and pigments has widened their scope for value-added

DOI: 10.1201/9781003436553-10

products with phyco-remediation (Tiwari and Marella, 2019). Diatoms uptake inorganic nutrients, nitrates, phosphates, and silica and absorb/adsorb micronutrients like iron, molybdenum, and copper in addition to toxic heavy metals such as cadmium chromium, mercury, and lead (Kiran et al., 2015). Diatoms absorb heavy metals from wastewater and undergo morphological alterations with the enhanced production of stress-responsive secondary metabolites (Pandey et al., 2014). These eukaryotic algae dominate under both nutrient-rich and nutrient-limiting conditions. Their ability to utilize silica to form frustules, the biological protection rendered by hard silica frustules against grazer attacks, efficient carbon transport mechanism, and higher cell division rate are critical factors that help to dominate diverse inland and coastal water bodies (Madhavi et al., 2014). Diatoms are known for their better carbon sequestering and nutrient uptake ability than other microalgae. Higher nutrient storing capacity, faster doubling rates, ability to grow in light-limiting and more turbulent water flow conditions make diatoms superior to other microalgae for remediation of wastewaters.

10.2 PHYSIOLOGICAL RESPONSE OF DIATOMS TO VARIOUS STRESSES

Diatoms undergo several physiological and morphological changes while remediating wastewater through biosorption of heavy metals with the electrostatic interaction that facilitates biosorption of heavy metals on the cell surface, leading to the degradation of toxic environmental pollutants. Heavy metals with a positive charge are attracted through electrostatic forces to the outermost layer of the diatom cell wall, which is negatively charged due to carboxyl, phosphoryl, and amine groups. The heavy metals thus attracted undergo biosorption. As the diatom grows, heavy metal accumulation increases with cell growth. Detoxification of these accumulated heavy metals in the cell wall takes place as a stress response to impairment of routine cell metabolism by producing reactive oxygen species (ROS), which detoxifies the heavy metals through dehalogenation and denitrification processes (Kiran et al., 2020). For instance, diatoms accumulate high-energy molecules like carotenoids and lipids under extreme salinity, silica- and other nutrient-deficiency conditions, high temperature, and solar irradiance (Ramachandra et al., 2009).

Silicon starvation in diatoms acts as a critical stress that ceases the production of the frustule exoskeleton, leading to higher levels of lipid accumulation, evident in *Cyclotella cryptica*, *Cylindrotheca fusiformis*, and *Phaeodactylum tricornutum* with enhanced lipid contents from 27.6% to 54.1%, 27% to 32% (dcw), and 18% to 28%, respectively (Griffiths et al., 2012). Biofilm cultivation of the diatom *Nitzschia palea* with silicon deficiency and nitrogen limitation outperformed the green microalga *Scenedesmus obliquus* with a higher lipid productivity of 0.45 $g\ m^{-2}\ d^{-1}$ (Schnurr et al., 2013). Phosphorus limitation in *Phaeodactylum tricornutum* doubled the lipid content but lowered the biomass productivity (Xia et al., 2013). Diatoms were tested for their ability to modulate lipid content during photo-oxidative stress and UV exposure. *Nitzschia palea* exhibited rapid lipid accumulation within an hour (Arts et al., 2000) under UV-A and UV-B exposure. Table 10.1 details the responses of diatoms to different stressors.

10.3 DIATOM CULTIVATION IN WASTEWATER

The open/raceway ponds are the most widely employed large-scale diatom cultivation systems that are often constructed using concrete, fiberglass, or polycarbonate materials (Edenhofer et al., 2014). These ponds are equipped with mixers or powered paddles for constant mixing of microalgae, and facilitate uniform exposure of the algal cells to natural sunlight. The open pond systems vary regarding shape, size, type of mixers (agitators), angle of inclination, and construction materials (Medipally et al., 2015). These cultivation systems have been widely used for aquaculture feed production (Mishra et al., 2017), wastewater bioremediation (Graham et al., 2012), and biofuel production (Matsumoto et al., 2017). The open pond commercial cultivation of *Chaetoceros* sp. for feeding

TABLE 10.1
Diatoms' Responses to Different Stressors

Diatom strain	Nature of stress	Stress response	References
Phaeodactylum tricornutum	N depletion	Increment in neutral lipid from 69% to 75%	(Alonso et al., 2000)
Gomphonema pseudoaugur	Heavy metal stress	Induction of lipid bodies and cell deformities	(Gautam et al., 2017)
Halamphora veneta	Heavy metal (Cu, Hg) toxicity	High accumulation of chlorophyll a, shape deformity	(Mu et al., 2017)
Amphora coffeaeformis	Cu, Cd stress	Reduction in growth rate, carbohydrate, lipid, amino acids, and protein	
Cyclotella cryptica	Si starvation	Increase in total lipids from 27.6% to 54.1%	(Roessler, 1988)
Phaeodactylum tricornutum	P limitation	Shift in relative content of C16:1 to C18:1	(Reitan et al., 1994)
Cylindrotheca fusiformis	N limitation	Total lipid content increased from 27% to 32%	(Griffiths et al., 2012)
Phaeodactylum tricornutum	N limitation	Total lipid content increased from 18% to 28%	
Cyclotella cryptica	N limitation	Increment in lipid content from 14% to 42%	(Sheehan, 1998)
Thalassiosira weisflogii	N limitation	Lipid content increase from 22% to 24%	(Fields and Kociolek, 2015)
Navicula saprophila		Lipid content increase from 26% to 44%	(Sheehan, 1998)
Melosira nummuloides	N limitation	Increase in lipid content from 18.50% to 33.02%	(Chen, 2012)

shrimps and bivalves has been practiced since the 1990s in Hawaii (US PATENT US6673592B1, 2004). In India, *Chaetoceros* sp., *Skeletonema* sp., and *Thalassiosira* sp. are cultured in large-scale open pond production systems as live feeds for brood stock rearing in hatcheries (Dash et al., 2016). The unialgal cultures of *Phaeodactylum tricornutum*, grown in corrugated raceway ponds, were tested for their ability to bioremediate primary influent sewage diluted with seawater (Wang et al., 2019). The recent ventures of biofuel production from oleaginous diatoms include a year-long outdoor mass cultivation of *Mayamaea* sp. JPCC CTDA0820 and *Fistulifera solaris* in open-pond bioreactors (Matsumoto et al., 2017). Figure 10.1 illustrates the different unit operations involved in diatom cultivation.

The photobioreactors are closed-loop biomass production systems that are artificially illuminated at a lab scale, mainly by two modes (internal and external). The various configurations of photobioreactors include stirred-tank (Sobczuk et al., 2006), vertical flat-plate (Zhang et al., 2002), tubular horizontal/serpentine (Molina et al., 2001), airlift tubular (Acién et al., 2012), and bubble column (Merchuk et al., 2000). The biomass productivities that have been reported range between 0.06–1.50 g L^{-1} d^{-1} in outdoor photobioreactors of varying configurations (Ugwu et al., 2008); there were different hydrodynamic and mass transfer characteristics such as rate of mixing, residence time, gas bubble velocity, gas holdup volume, type of sparger used for agitation, shear stress, and superficial fluid velocity experienced by the algal cultures inside the reactor and these act as critical factors in determining the efficiency of biomass production in such closed photobioreactor systems (Kaewpintong et al., 2007). Earlier estimates reported that nearly 12–2000 kg of water is required to

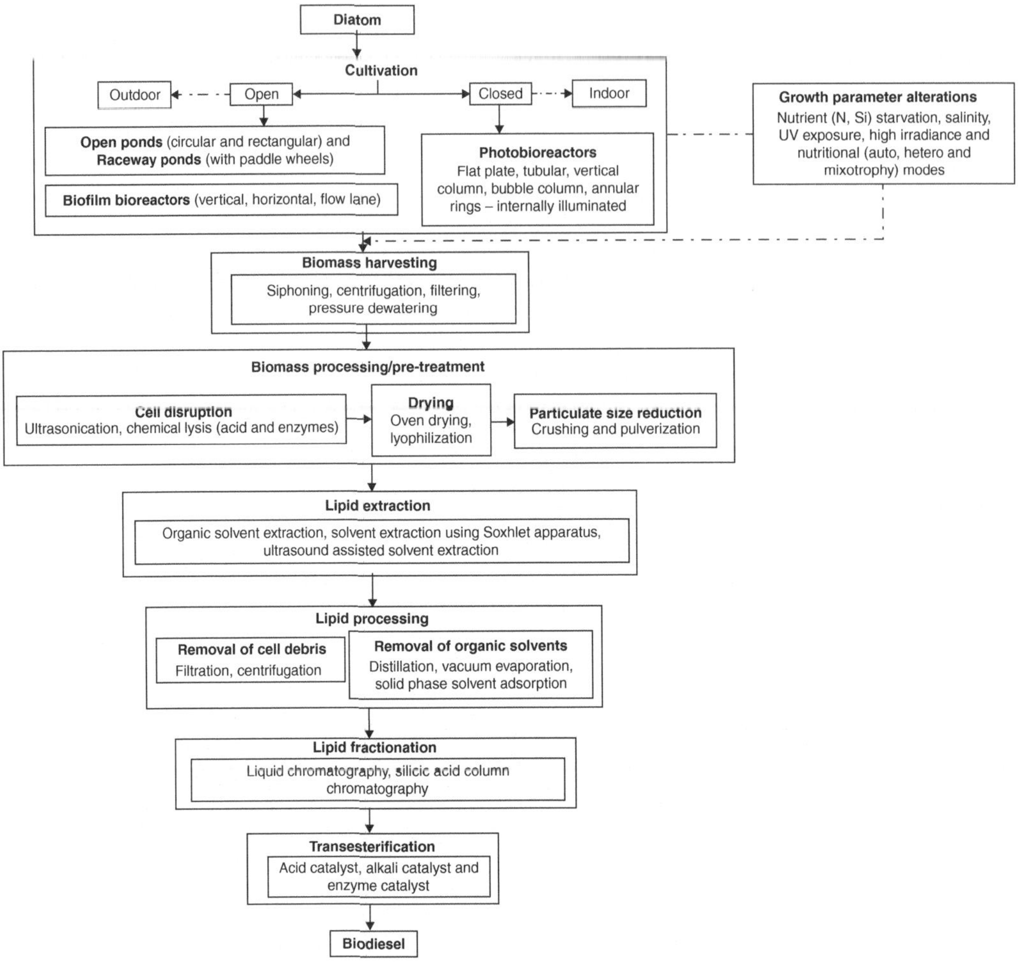

FIGURE 10.1 Schematics of the processes involved in diatom cultivation.

produce 1 kg of dry microalgal biomass (Berner et al., 2015). In vertical bubble column bioreactors, *Fistulifera* sp. yielded a maximum biomass concentration of 0.57 g L^{-1} with an oil content of 37.7% of the dry cell weight (Sato et al., 2014). Table 10.2 lists the various diatom cultivation practices in different wastewaters.

10.4 SUBSTRATE-BASED ATTACHED CULTIVATION OF DIATOMS

The last decade has witnessed significant efforts toward substrate-based diatom cultivation to exploit the biofilm-forming potential of diatoms (Mantzorou et al., 2018). Biofilms are composed of symbiotically interacting diverse microbes such as bacteria, protozoa, larvae, microalgae, microzooplankton, and macroalgae (Katarzyna et al., 2015). The biofilm formation process involves the adherence of microbial cells to a solid substratum through adsorption (usually reversible) followed by irreversible adherence with the help of EPS secretion (Shen, Xu, et al., 2014). The biofilm cultivation systems can be broadly classified as either constantly or intermittently submerged and perfused systems in which the moisture is provided by a porous substrate, with the system exposed to an ambient gas phase (Berner et al., 2015). Different solid-surface materials used as substrates include polystyrene foam, polyurethane, cotton duct/rope (Christenson and Sims, 2012), concrete, stainless steel (Shen

TABLE 10.2
Various Diatom Cultivation Ventures in Wastewater

Diatoms	Type of wastewater	Productivity	Application	References
Corrugated raceways	Primary sewage effluent	–	Bioremediation	(Park et al., 2011)
Culture flasks	Domestic and industrial wastewater + Nualgi	37 mg L^{-1} d^{-1}	Wastewater treatment	(Marella et al., 2018)
Erlenmeyer flasks	Aquaculture wastewater + Nualgi	58.4 × 10^8 cells mL^{-1}	Bioremediation	(Li et al., 2017)
Culture flasks	Aquaculture wastewater	32–39 mg L^{-1} d^{-1}	Bioremediation and biofuel production	(Saranya and Ramachandra, 2020)
Erlenmeyer flasks	POME industrial wastewater	Biomass 34.1 mg L^{-1} and lipid 24.1 mg L^{-1}	Bioremediation and lipid synthesis	(Govindan et al., 2019)
Culture flasks	Municipal wastewater and seawater	Maximum lipid productivity 33.2 mg L^{-1} d^{-1} obtained with 1:1 SW and WW ratio	Bioremediation and lipid synthesis	(Wang et al., 2019)
Chaetoceros gracillis	Secondary sewage	0.768 µg mL^{-1} chlorophyll	Larval feed nourishment	(Da Costa et al., 1999)
Thalassiosira sp.	Secondary municipal sewage	0.883 µg mL^{-1} chlorophyll	Larval feed nourishment	

et al., 2014), nylon sponge, loofah sponge (Johnson, 2009), nylon mesh, polyethylene fiber, polycarbonate, glass fiber reinforced plastic (GFRP) (Shen et al., 2014), nylon nets and plastic scrubber screens, Perspex, titanium, stainless steel 316-L, glass, copper, aluminum brass and admiralty brass (Sekar et al., 2004), polypropylene, high-density polyethylene, polyester, aluminum and stainless steel (Gross, 2015), cellulose acetate (Cheng et al., 2017), flexible fiber bundles (Gao et al., 2015), cotton, linen, and mohair (Zhang et al., 2017), glass (Schnurr and Allen, 2015), and printing paper (Naumann et al., 2013).

Substrate-based biofilm cultivation offers innumerable advantages at multiple dimensions: it has a lower carbon footprint area, and is economical, readily available with a re-useable supporting surface (Gifuni et al., 2019); reduced harvesting costs, easy scalability, higher biomass productivity, lower water requirement with bioremediation capability, increased light availability owing to the wider exposed surface area (Roostaei et al., 2018), and better control of the cell growth area when cultivated in lagoons and open oceans (Shen et al., 2018). The diatom strain *Haslea ostrearia*, grown in immobilized cell photobioreactors using agar gel layers, yielded a two-fold higher cell growth (Lebeau et al., 2002). The colonizing tendency of benthic diatoms with EPS secretions has been exploited for successful biofuel production in biofilm bioreactors. Under natural conditions, the influence of substrates on water quality assessment revealed the diatom–nutrient relationship in substrates like mud and ropes. Ropes were found to be the ideal substrate for assessing the sensitivity and nutrient status of the streams (Richards et al., 2020).

10.5 INTEGRATED DIATOM CULTIVATION IN AQUACULTURE WASTEWATER

Microalgae, especially diatoms, can produce a plethora of products that find application in diverse industrial sectors (Chew et al., 2017). In a conventional refinery, the raw material (e.g., crude petroleum) undergoes a series of production processes with a potential output of diverse energy sources (gasoline, diesel, LPG, and ethanol) and an array of complex and valuable chemicals (volatile acids, fine chemicals, detergents, pharmaceuticals, waxes, and asphalt). Similarly, in the diatom integrated system, renewable raw materials are used at low cost to produce value-added products with inherent energy potential (Laurens et al., 2017). Hence, integrating diatom cultivation with wastewater (Ramachandra and Saranya, 2022) would provide various products and cumulative benefits, proving microalgal biofuel generation to be economically viable and sustainable. Figure 10.2 depicts an integrated algal cultivation system using aquaculture wastewater for diverse bioproduct production with the added advantage of wastewater remediation.

10.6 VALORIZATION OF WASTEWATER-GROWN DIATOM BIOMASS

Cell walls of Bacillariophyceae (diatoms) are protected by a rigid inorganic wall of silica or calcium carbonate (Ferreira et al., 2018), and the thickness and chemical composition of the microalgal cell wall depends on the growth environment (Praveenkumar et al., 2015). Disruption of cells is required to extract accumulated lipid, which is done either by physical or chemical pre-treatment methods to improve the amenability of cell constituents by organic solvents. Cell disruption methods include bead milling, osmotic shock, pulsed electric field, microwave, ultrasound, and freezing/thawing. The different pre-treatment and bioenergy conversion processes used for bioenergy production from microalgal biomass are illustrated in Figure 10.3. Among the cell disruption methods, it is reported that ultrasound treatment (sonication) improves the hydrolysis of carbohydrates by inducing structural modifications to the crystalline structures of sweet potato starch (Zheng et al., 2013) and improves the cell disruption efficiency of microalgae (Ramachandran et al., 2013; Ramachandra et al., 2011). Microwave treatment is a rapid process with 80% of the cell lysis (Abbassi et al., 2014). After cell disruption, either thermochemical or biological processes are carried out to derive secondary products like bioethanol, bio-oil, biogas, biobutanol, volatile fatty acids, biohydrogen,

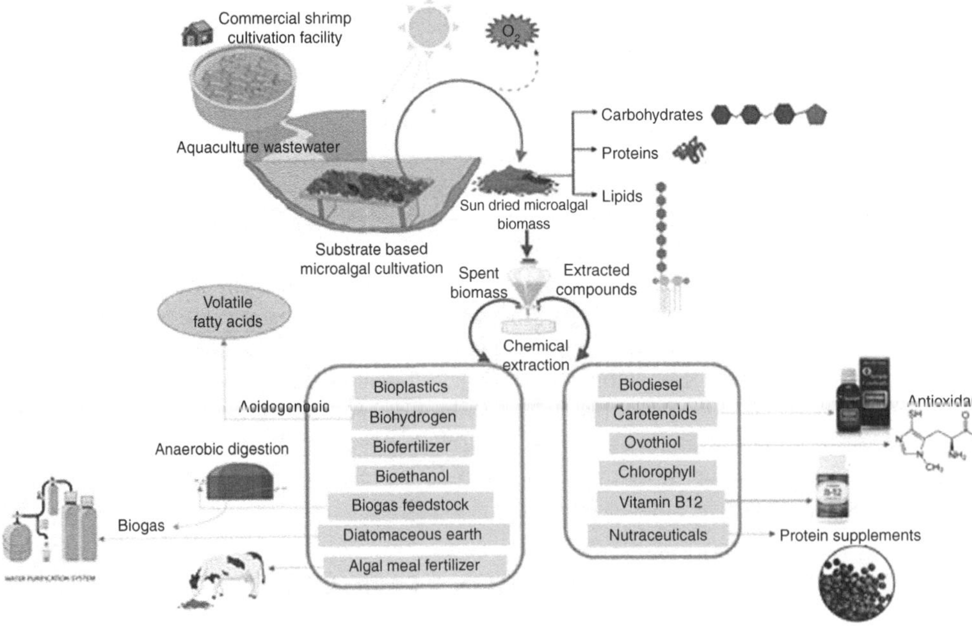

FIGURE 10.2 Integrated diatom cultivation system using aquaculture wastewater for bioproducts production.

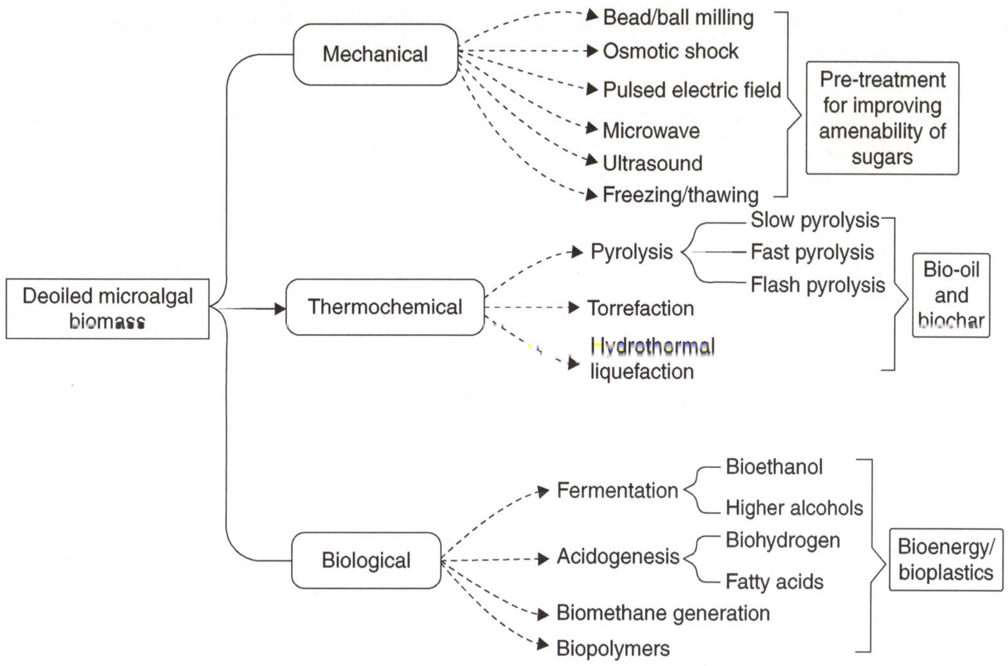

FIGURE 10.3 Different conversion processes for the production of various energy products from diatoms.

and biopolymers (Venkata Mohan et al., 2020). Gasification, pyrolysis, hydrogenation, liquefaction, and combustion are thermochemical conversion processes that involve applying heat energy to obtain end-products. The energy obtained from such thermochemical processes is mainly gaseous forms or energy-rich biocrude that must be further upgraded to bioenergy products. Biological treatments such as anaerobic digestion, fermentation, and heterotrophic fermentation result in biogas, bioethanol, alcohols, and liquid hydrocarbons, whereas physicochemical conversion using extraction and transesterification leads to biodiesel (Jankowska et al., 2017). Possible bioenergy components from microalgal biomass from an energy perspective are methane, biodiesel, biocrude, with possible refinement into gasoline and green diesel, biobutanol and ethanol, biohydrogen, and bioelectricity from microalgal carbohydrates.

10.6.1 Diatom Biomass into Biogas

Anaerobic fermentation (digestion) of biomass is emerging as a promising technology where anaerobic digestion (AD) involves anaerobic bacteria in the decomposition of diatom biomass to produce biogas. AD is a four-step process involving hydrolysis, acidogenesis, acetogenesis, and methanogenesis (Ward et al., 2008). The main components of biogas are methane and carbon dioxide. The biogas quality is usually determined by the relative amount of methane, which in turn is dependent on substrate and operating conditions for anaerobic fermentation (Sialve et al., 2009). Several studies have shown the scope for using spent diatom biomass as a feedstock for biogas production (Ruiz-Domínguez et al., 2022; González-González et al., 2021; Solé-Bundó et al., 2018; Zabed et al., 2020).

The diatom biomass constitutes a high quantity of carbohydrates (51–68%), followed by proteins (6–21%) and lipids (7–16) (Jankowska et al., 2017). The biogas potential reported for different diatom species varies between 0.09–0.54 L CH_4 g^{-1} (Ruiz-Domínguez et al., 2022). However, the efficiency of biogas production is highly species-specific as the anaerobic digestion of diatoms is hindered by its complex cell wall structure and composition. Recalcitrant cell walls of diatoms showed resistance to being penetrated by bacteria (anaerobic biodegradability) during anaerobic digestion, resulting in low methane yields. Hence, various pre-treatment methods for disrupting cell walls include thermal, microwave, ultrasonic, chemical, and mechanical methods (Kwietniewska and Tys, 2014). The marine diatom *Phaeodactylum tricornutum*, when subjected to anaerobic digestion after super-critical extraction of fucoxanthin from the fresh biomass, yielded 56.6 L_{STP} CH_4/kg VS added (Ruiz-Domínguez et al., 2022). The bio-methanation potential of *Chaetoceros mulleri* tested in an integrated biogas and biodiesel production plant revealed an average methane yield of 52 L CH_4/kg VS added (González-González et al., 2021). Techno-economic feasibility studies have demonstrated a cost reduction of about 35% by integrating a methane production plant into a biodiesel production plant (Harun et al., 2011). Overall improvement in the energy balance of biodiesel was seen with biogas production from the waste biomass (Francisco et al., 2010; Poveda-Giraldo et al., 2021; Zhao et al., 2014). Energy recovery of 80% was recorded in *Isochrysis galbana* through an integrated approach of biodiesel and biogas production (Sánchez-Bayo et al., 2020).

10.6.2 Fermentation into Bioethanol

Microalgae, especially diatoms, are receiving attention as a carbohydrate (~ 50–68%) feedstock for their effective fermentation into bioethanol. In the biorefinery approach, the de-oiled biomass composed of large fractions of starch and cellulose after lipid extraction has been used effectively for bioethanol production (Shokrkar and Ebrahimi, 2018). Simultaneous saccharification and fermentation of microalgal biomass using enzyme cocktails has shown higher yields of reducing sugars with prospects for cost-effective fermentation of microalgal biomass into bioethanol (Shokrkar and Ebrahimi, 2018). The marine diatom *Amphora* sp. assessed as a bioethanol feedstock using the brewer's yeast *Saccharomyces cerevisiae* yielded ~ 18 µL/mL of biomass hydrolysate (Hogan et al., 2021).

10.6.3 Fermentation to Biobutanol

Microalgal residues after lipid extraction are being used as feedstocks for selective fermentation into higher alcohol such as biobutanol, etc. These include microalgae's ability to accumulate readily digestible polysaccharides like starch in a major portion of their dry weight. The biological production of biobutanol is achieved by acetone-butanol-ethanol (ABE) fermentation of microalgal biomass using a solventogenic anaerobic bacterium belonging to the genus *Clostridium*. This bacterium is known for its capability to convert a wide range of organic carbon sources, including glucose, cellobiose, arabinose, galactose, xylose, and mannose, by secreting numerous polymers degrading enzymes such as α-amylase, β-amylase, and β-glucosidase, glucoamylase, amylopullulanase, and pullulanase (Ezeji et al., 2007). With incorporation of these bacterial strains into carbohydrate-rich feedstocks under anaerobic conditions, the bacterium breaks down complex polysaccharides into butyric and acetic acid through an acidogenic process followed by solventogenesis with the synthesis of acetone, ethanol, and butanol (Lee et al., 2008). Thus, the species-specific carbohydrate composition of microalgae determines the efficiency of ABE fermentation. Green microalgae have been extensively studied for their ability to produce biobutanol, and, for example, marine microalgae *Nannochloropsis gaditana* showed a biobutanol production of ~ 2.9 g L^{-1} (Onay, 2020).

10.6.4 Diatom Biomass Conversion to Biocrude using Hydrothermal Liquefaction

Hydrothermal liquefaction (HTL) is the process of converting microalgal biomass into liquid biocrude by subjecting the biomass to a high temperature (280–370°C) and pressure (10–25 MPa).

The HTL process utilizes wet algal feedstock circumventing higher energy costs for biomass drying associated with centrifugation for water removal. Earlier studies have shown that for an algal concentration ranging between 5–20%, HTL treatment can be carried out with just < 5% of the energy costs required for drying (Xu et al., 2011). The biocrude synthesized possesses an energy value close to fossil petroleum (Jena and Das, 2011), which can be fractionated into different energy products. Biocrude is a dark, viscous, energy-rich liquid produced due to hydrothermal degradation of microalgal biochemical constituents (carbohydrates, lipids, proteins) (López Barreiro et al., 2013). In comparison with other thermo-chemical conversion technologies of pyrolysis and gasification, HTL possesses prominent characteristics such as (i) complete conversion of whole algal biomass into biocrude and other chemicals, thus higher oil yields are possible, (ii) elimination of drying process leading to considerable cost reduction, (iii) enhanced HTL efficiency due to the enthalpy of phase change of water at higher pressure is reduced chiefly, and (iv) non-requirement for additional rectification or extraction, and the principal product can be self-regulated (Tian et al., 2014).

10.6.5 Diatom Feedstock for Biopolymer Production

Bio-based polymers are gaining attention over petroleum-based polymers due to their biodegradability and benign environmental characteristics (Garrison et al., 2016). Various bio-based raw materials such as resins, lignin derivatives, polysaccharides sourced from multiple feedstocks, proteins, and vegetable oils have been investigated for their suitability in bio-polymer production (Gandini, 2008). The current primary production of bioplastics is from terrestrial plant-based starch and poly-lactic acid (PLA) polymers derived from corn and sugar beets (Sreedevi et al., 2014). However, the production of PLA competes with arable land, which makes it a less attractive process. Several bacterial strains can produce a range of exopolysaccharides and polyhydroxyalkanoates (PHA) that are already commercially used as bioplastics precursors (Rehm, 2010). However, as the world's plastic demand is increasing, conventional plastics pose more significant threats to the environment, especially to marine ecosystems (Rahman and Miller, 2017). Microalgae secrete extracellular polymeric substances (EPSs) comprising polysaccharides, proteins, lipids, and uronic and nucleic acids (Venkata Mohan et al., 2020). These EPSs are converted into bioplastics (polyhydroxyalkanoates [PHA], polylactic acid [PLA], or thermoplastic starch [TPS]) through the thermochemical conversion process by using glycerol as a co-substrate (Jerez et al., 2007). Biological routes of bioplastic production from microalgae are realized either through direct utilization and conversion of microalgal biomass or by utilizing the aqueous phase (hydrolysate), the spent biomass that is left out after lipid extraction as a source of nutrient for growing recombinant *E. coli* which produces the bioplastic–polyhydroxy butyrate (PHB) (Rahman et al., 2015).

10.6.6 Diatoms as Precursors of Microbial Fuel Cells

Microalgae/microbial fuel cells (mMFCs) convert solar energy into electricity through a bioelectrochemical process by combining live and dead microalgae in the respective cathode and anode chambers (Lee et al., 2015). Live microalgae are placed in a cathode chamber (photoreactor) that acts as a biocathode in a microbial fuel cell. The biocathode produces O_2 during photosynthesis and is an electron acceptor from the external circuit (Wang et al., 2010). The anode chamber consisting of organics undergoes digestion by releasing CO_2 and excess protons to the cathode chamber and acts as an electron donor resulting in the generation of electricity (Juang et al., 2012). Live green algae *Chlorella vulgaris* grown in the cathodic chamber was found to photosynthesize and increase the dissolved oxygen levels of the chamber. This DO increase positively correlates with voltage output (Juang et al., 2012). The influence of light intensity on the cathodic resistance was investigated (Wu et al., 2014) and it was found that the light intensity increased the rate of photosynthesis of *Desmodesmus* sp., resulting in higher O_2 levels, which enhanced the mMFC's cathodic resistance to induce bioelectricity. Similar research on diatom biomass would open new avenues for substantial improvements in diatom-coupled microbial fuel cell processes.

Diatoms' ability to uptake organic and inorganic carbon, nitrogen, and phosphorus in addition to trace/heavy metals from wastewater could circumvent water contamination with sustainable and eco-friendly wastewater treatment. Moreover, the diatom biomass produced during wastewater treatment acts as a reservoir for a plethora of bioactive compounds such as pigments, proteins, sterols, and flavonoids (Kuppusamy et al., 2017). Diatoms also produces docosahexaenoic acid (DHA), eicosapentaenoic acid (EPA), and minute traces of squalene, which can make good-quality nutritional supplements for aquaculture production. However, from a whole-cell utilization perspective to avert the loss of valuable biocomponents and enable transition toward a zero-waste circular economy, the above-mentioned thermochemical routes of diatom biomass valorization would render complete conversion of biomass into low-value high-volume energy products such as biogas, bioethanol, biopolymer, and biocrude, while providing higher energy returns on investment as against conventional culture/media-grown diatoms cultivated in suspended forms.

10.7 DIATOM BIOMASS AND WASTEWATER REMEDIATION—POTENTIAL COMMERCIAL APPLICATIONS

Diatoms are gaining attention in diverse disciplines such as energy, nanotechnology, materials engineering, geology, oceanography, climate studies, environment, health, and forensic sciences. The significance of diatoms in the field of renewable energy has risen greatly, especially after the proven hypothesis of diatoms being responsible as progenitor elements of the current crude petroleum reserves, third-generation biofuel precursors, silicon nanofabrication, bio-optical sensors in solar photovoltaic panels, and eminent sources of bio-active compounds and pigments.

The Indian coast has a length of 7517 km (Jong, 2016) with diverse land uses for agriculture, aquaculture, salt pans, forest plantations, mangroves, mudflats, scrublands, built-up areas, and water bodies (Jayanthi et al., 2018). Among the coastal water bodies, estuaries are known for their higher productivity (Ramachandra and Shwetmala, 2012), with an estimated mean annual primary productivity of 1500 grams of dry matter/m^2, which is approximately 10 times higher than the open ocean's primary productivity (Malviya et al., 2016). Hence, estuaries are categorized as highly productive ecotone regions; here, the nutrients from land-based discharges and salinity variations (influenced by regular tidal action) with an established mangrove ecosystem conflux favor increased population densities of microorganisms, benthos, and plankton (Kennish, 2002). Diatoms that dominate the global oceans and estuaries (Guo et al., 2013) are broadly classified as planktonic (the ones that drift with the water currents) and benthic (those attached to substrates such as stones, sand, and aquatic vegetation).

The benthic diatoms are photoautotrophic organisms inhabiting intertidal sediments, rocks, and other biotic substrates (sedges, aquatic plants, and oyster beds) and exhibiting high rates of primary production (Malviya et al., 2016). They are functionally crucial in sequestration and transforming organic nutrients into inorganic forms. The benthic diatoms, also known as microphytobenthos (MPB), which form a predominant portion of primary producers of the estuaries, contribute up to 50% of the total estuarine carbon budget (McKew et al., 2011), even though their habitat is restricted only to photic zones (~ 400 μm depth) with limited periods of light exposure. The carbon fixation rates of benthic microalgae in the top 5 mm of the sediment layer are estimated to be 17.38 mg C m^{-2} h^{-1} (Jacobs et al., 2021).

The prominent biotechnological applications of diatoms are: (i) aquaculture feed (Kiatmetha et al., 2011), (ii) intracellular metabolites (lipids, fucoxanthin, and eicosapentaenoic acid [EPA]) that are beneficial in the pharmaceutical and cosmetic industries (Wen and Chen, 2002), (iii) extracellular metabolites such as exopolysaccharides (EPS) (Gügi et al., 2015; Zhou et al., 2018), amino acids, and proteins as high-value molecules, and (iv) silicate frustules for nanobiotechnological

TABLE 10.3
Research Institutions across the Globe Working on Diatoms

Species	Research Institution	Targeted Biomolecules	References
Navicula cincta, Nitzschia punctata, Amphiprora sp., *Chaetoceros* spp., *Cyclotella* sp.	Indian Institute of Science, Bangalore	Biodiesel, EPA	(Ramachandra and Saranya, 2022)
Amphora coffeaeformis	Centre for Advanced Studies in Botany, University of Madras, Chennai	Biofuel and essential fatty acids	(Rajaram et al., 2018)
Nitzschia paea, Gomphonema parvulum, Nitzschia inconspicua, Diadesmis confervaceae, Sellaphora sp., *Placoneis elginensis*	Harish Singh Gour University, Madhya Pradesh	Diatom lipids and diatom solar panels	(Gautam et al., 2016)
Pinnularia sp.	School of Chemical, Biological & Environmental Engineering, Oregon State University, USA	Solar cells, batteries and electroluminescent devices, diatom solar panels	(Jeffryes et al., 2011)
Coscinodiscus walesii	National Council of Research, Institute for Microelectronics and Microsystems, Department of Naples	Diatom solar panels	(De Stefano et al., 2007)
Synedra sp., *Diatom consortium*	Noida International University, Uttar Pradesh, India	Fatty acids and triglycerides	(Tiwari and Marella, 2019)
Achnanthes sp.	Scripps Institution of Oceanography	Biofuel	(Hildebrand et al., 2012)
Halamphora coffeaformis, Navicula cincta	Bahia Blanca, Argentina	Biodiesel and essential fatty acids	(Martín et al., 2018)
Diatoms	University of Colorado, Boulder, USA	Diatom based taxonomy and limnological studies	(Andrejić et al., 2018)
Diatoms	Rutgers University, New Brunswick, USA	Physiological and molecular aspects of diatoms	(Kranzler et al., 2019)
Phaeodatylum tricornutum	Shandong University, China	Seasonal dynamics studies	(Zhang et al. 2019)
P. tricornutum, Skeletonema costatum	University of Iceland, Reykjavik, Iceland	Anti-cancer compounds	(Hussein and Abdullah, 2020)
Thalassiosira weisflogii, Cyclotella cryptica	Istituto di Chimica Biomolecolare (ICB)–CNR, Via Campi Flegrei, Italy	Biofuel	(d'Ippolito et al., 2015)
Phaeodatylum tricornutum	Swansea University, UK	Diatom biorefinery	(Butler, Kapoore, and Vaidyanathan, 2020)
Thalassiosira pseudonana	University of Sheffield, UK	Bioactive compounds	(Sethi et al., 2020)
Thalassiosira weissflogi	International Crop Research Institute for Semi-arid Tropics (ICRISAT), Telangana, India	Diatom biorefinery	(Marella and Tiwari, 2020)
Diatoms	Centre for rural development and technology, IIT Delhi	Bioactive compounds, biorefinery	(Vishwakarma et al., 2022)

applications such as electrochemical devices and solar cells. Table 10.3 lists the research institutions across the globe that are currently working on diverse commercial aspects of diatoms. Diatom cultivation in open ponds and closed photobioreactors has been in practice for more than 50 years (Levitan et al., 2014). However, the commercial-scale production of diatoms in India and across the globe has been restricted to feeding shrimps and molluscans in aquaculture mass-production facilities (López-Elías et al., 2005). Different species of diatoms belonging to the genera *Chaetoceros* and *Skeletonema* are commonly used in intensive *Penaeid* shrimp cultivation systems. *Chaetoceros muelleri* was mass cultivated both indoors and outdoors in tubular photobioreactors as live feed for zooplankton in a Mexican hatchery, yielding a biomass productivity of 22.5–45.7 g m^{-3} d^{-1} (López-Elías et al., 2005). Pennate diatoms that have been commonly utilized as aquaculture feed include *Navicula* sp. and *Nitzschia* sp. (Thompson et al., 2002). Figure 10.4 illustrates diverse industrially relevant bioproducts that can be extracted from diatom biomass.

Another important commercial application of diatoms is the production of long-chain polyunsaturated fatty acids (LC–PUFAs) such as eicosapentaenoic acid (EPA). *Phaeodactylum tricornutum*, cultivated in a semi-continuous mode to produce EPA, yielded 3.9% of 4.0 g m^{-2} d^{-1} of biomass on an ash-free dry weight (AFDW) basis (Thierry et al., 2003). *Nitzschia laevis* was grown heterotrophically using glucose and yielded an EPA productivity of 73 mg L^{-1} d^{-1} (Wen and Chen, 2002). Studies are in progress toward devising nanostructure semiconductors made of diatom frustules by metabolic insertion or peptide-mediated deposition of titanium dioxide (TiO_2) into silica frustules; this is useful for photocatalytic and photovoltaic applications (Gautam et al., 2016; Jeffryes et al., 2011; Jeffryes et al., 2008; Ramachandra et al., 2009) from the perspective of nanobiotechnological solar cells and other electrochemical devices. Table 10.4 lists the diatoms that are being used in diverse commercial applications, including food, feed, and fuel (Levitan et al., 2014).

A substrate-based diatom cultivation system using low-cost substrates such as concrete slabs or gravel stones is integrated with constructed wetland to optimize bioremediation. Diatom cultivation systems, when integrated with macrophyte plants, would effectively help in the reduction of nutrient concentrations from raw waters of domestic or industrial origin. The raw wastewaters, when let into the constructed wetlands comprising of macrophytes, lead to an effective reduction of chemical oxygen demand (COD), total nitrogen (TN), and total phosphorus (TP) from the wastewaters (Ramachandra and Saranya, 2022). Thus, when entering the diatom cultivation system, the partially

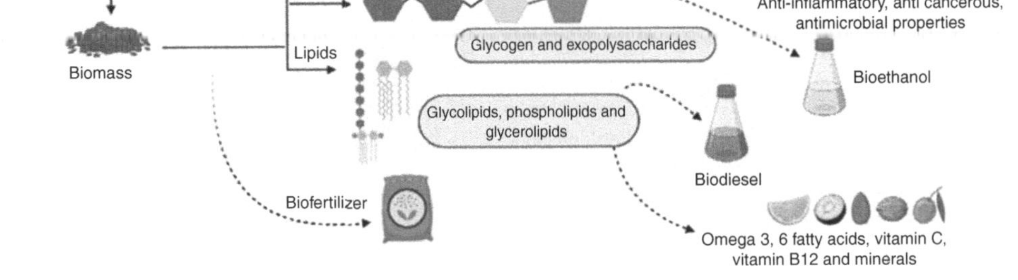

FIGURE 10.4 Different industrially relevant bio-products extractable from diatoms.

TABLE 10.4
Diverse Commercial Applications of Diatoms

Diatoms	Reactor used	Nutritional mode	Organic/inorganic substrate used	Productivity	Field of application	References
			Aquaculture feed			
Chaetoceros calcitrans	Culture flasks	M	Interstitial water	6×10^5 cells mL^{-1}	Fish larvae feed	(Yusoff et al., 2001)
Chaetoceros calcitrans	Polyethylene carboys	P	Conway medium	7–13 cells mL^{-1} d^{-1}	Live feed in hatchery	(Kaspar et al., 2014)
Skeletonema costatum	Plastic flasks	P	Provasoli medium	61% PUFA of total fatty acids	Diet for mollusks	(Berge et al., 1995)
Chaetoceros muelleri	300L transparent fiber glass cylinders	P	–	22.5–45.7 g m^{-3} d^{-1}	Shrimp larval feed	(López-Elías et al., 2005)
Phaeodactylum tricornutum	Twin layer PBR	P	Modified F/2 media	1.8 g m^{-2} d^{-1} (B)	Live feed in hatcheries	(Naumann et al., 2013)
Chaetoceros gracillis	Culture flasks	M	Secondary sewage	0.768 µg mL^{-1} chlorophyll	Larval feed nourishment	(Da Costa et. al., 1999)
Thalassiosira sp.	Culture flasks	M	Secondary municipal sewage	0.883 µg mL^{-1} chlorophyll	Larval feed nourishment	
			Fucoxanthin and EPA production			
Nitzschia inconspicua	Conical flasks	M	Glucose and acetate	15.6–48.9 mg g^{-1} dw	PUFA extraction	(Chu et al., 1996)
Nitzschia laevis	2.4 L fermenters	H	Glucose	46.5 mg L^{-1} d^{-1} (EPA)	Bioactive compound production	(Wen and Chen, 2002)
Chaetoceros gracilis	Culture flasks	P	f/2 silicate Guillard medium	283 pg $cell^{-1}$	PUFA synthesis	(Pratiwi et al., 2009)
Nitzschia saprophila	Erlenmeyer flasks	M	Acetic acid	10.6 mg L^{-1} EPA	Food and allied applications	(Kitano et al., 1997)
Chaetoceros brevis	Erlenmeyer flasks	P	F2 medium	1.06 fg EPA $µm^{-3}$	Nutraceutical industry	(Boelen et al., 2013)
Thalassiosira weisflogii	Erlenmeyer flasks	P	F2 medium	2.31 fg EPA $µm^{-3}$	Nutraceutical industry	
Chaetoceros muelleri	Flat plate glass reactors	P	f/2 enriched medium	2.29 EPA (pg $cell^{-1}$)	EPA analysis	(Zhang and Richmond, 2003)
Nitzschia sp.	Erlenmeyer flasks	P	Guillard F medium	0.18 EPA L^{-1} D^{-1}	Food and allied applications	(Kitano et al., 1997)

(Continued)

TABLE 10.4 (Continued)
Diverse Commercial Applications of Diatoms

Diatoms	Reactor used	Nutritional mode	Organic/inorganic substrate used	Productivity	Field of application	References
Nitzschia laevis	Culture flasks	M	LDM medium	60.12% fucoxanthin and 50.67% EPA	High value products	(Lu et al., 2019)
Nitzschia laevis	Fed batch fermenter	M	LDM medium	16.5 mg L^{-1} d^{-1} (fucoxanthin)	Fucoxanthin production	
P. tricornutum	Shake flasks	M	Glycerol	2.40% on biomass dw	Metabolite production	(Cerón García et al., 2006)
Skeletonema sp.	1 L Roux flasks	P	f/2 medium	1.52 pg EPA $cell^{-1}$	Nutraceuticals	(Blanchemain and Grizeau, 1996)
			Biofuel			
Nitzschia palea	Biofilm bioreactor	P	CHU 10 diatom medium	2.8 g m^{-2} d^{-1} (B)	Biodiesel	(Schnurr et al., 2013)
Halamphora coffeaeformis	Indoor PVC raceway ponds	P	f/2 medium	33.4% total lipids of AFDW	Biodiesel	(Martín et al., 2018)
Navicula gregaria	Erlenmeyer flasks	P	f/2 media	20.83 pg lipids $cell^{-1}$	Biodiesel	(Popovich et al., 2012)
Synedra acus	Erlenmeyer flasks	M	glycerol	~ 71% of SFAs and MUFAs	Fatty acids and chrysolaminarin production	(Shishlyannikov et al., 2014)
Navicula cincta	Erlenmeyer flasks	P	f/2 media	41% of AFDW	Biodiesel and EPS	(Bielsa et al., 2016)
Skeletonema costatum	Erlenmeyer flasks	P	f/2 media	19.2 pg lipids $cell^{-1}$	Biodiesel	
Navicula cincta	Photobioreactor	P	SWES + f/2 media	49.7% total lipid of AFDW	Biodiesel	(Popovich et al., 2019)
Synedra ulna	Erlenmeyer flasks	P	sD11 medium + 180 µg $Na_2 SiO_3$	16,480 cells mL^{-1}	Fatty acid composition	(Graham et al., 2012)
Cyclotella meneghiniana	Erlenmeyer flasks	P	sD11 medium + 180 µg $Na_2 SiO_3$	7.71 × 10^5 cells mL^{-1}	Fatty acid composition	
Aulacoseira granulata	Erlenmeyer flasks	P	sD11 medium + 180 µg $Na_2 SiO_3$	15,650 cells mL^{-1}	Fatty acid composition	
Thalassiosira weisflogii	Conical flasks	P	f/2 medium	4.12% lipids	Biodiesel	(Borges, Montes, and Abreu, 2011)

			Extracellular polysaccharides (EPS) production			
Cylindrotheca closterium	Erlenmeyer flasks	P	Enriched f/2 medium	2.56 μmol C 10^{-6} cells	EPS production	(Urbani et al., 2005)
Skeletonema costatum	Erlenmeyer flasks	P	Enriched f/2 medium	1.18 μmol C 10^{-6} cells	EPS production	
Thalassiosira pseudonana	Erlenmeyer flasks	P	Enriched f/2 medium	28.4 μmol C 10^{-6} cells	EPS production	
Achnanthes longipes	Conical flasks	P	Enriched f/2 medium	Sulphated galactosyl, t–2-GlucUA residues	Composition of EPS	(Wustman et al., 1997)
Pinnularia viridis	Pyrex conical flasks	P	K–medium	19 sugars identified in EPS	Bioactive compound extraction	(Chiovitti et al., 2003)
Amphora sp.	Erlenmeyer flasks	P	F/2 Guillard's medium	13 monomers including Gal, Glc, and Xyl	High value molecules	(Khan et al., 2018)
Navicula cincta	Erlenmeyer flasks	P	F/2 Guillard's medium	9–18 mg L^{-1}	Valuable co-production extraction	(Bielsa et al., 2016)
Stauroneis sp.	Erlenmeyer flasks	P	F/2 Guillard's medium	10 monomers (Glc, Xyl and Man)	High value molecules	(Jiang et al., 2019)
			Nanobiotechnology and solar cell applications			
Pinnularia sp.	II stage bioreactor	P	Harrison's seawater medium substituted with f/2 nutrients	Metabolic insertion of TiO_2 into silica frustules	Diatom solar panels	(Jeffryes et al., 2008)
Nitzschia palea	II stage bioreactor	P	Modified f/2 media	Metabolic insertion of TiO_2 into silica frustules	Diatom nanotubes for solar cell applications	(Gautam et al., 2016)
Pinnularia sp.	500 mL flasks	P	Artificial seawater medium	Peptide mediates TiO_2 deposition on frustule biosilica	Photocatalytic and photovoltaic applications	(Jeffryes et al., 2008)

P, photoautotrophic; H, heterotrophic; M, mixotrophic cultivation; B, biomass; AFDW, ash-free dry weight; SW, seawater; WW, wastewater.

treated wastewaters from macrophytes would effectively act as a tertiary biological treatment of wastewater. A similar study that investigated the performance of algae integration to improve the nitrogenous waste management from an aquaculture production system showed the hydraulic retention time (HRT) and surface area of the algal reactor (Ramli et al., 2020).

10.8 CONCLUSIONS AND PERSPECTIVES

Diatom cultivation in wastewaters would undoubtedly address the most pressing issue of efficient resource recovery from waste. The potential of diatoms to transform waste into wealth would aid in the economical treatment of wastewater, lowering the carbon footprint, and enhancing the quality of the aquatic ecosystem has recently gained significant research attention. Innovative downstream processing and optimal choice of biomass to energy conversion technologies applied on wastewater-grown diatom biomass would revolutionize the concept of a zero-waste circular economy. Bio-valorization of diatom biomass grown using wastewaters in substrate-based biofilm cultivation would redefine the economic feasibility bottlenecks faced by microalgal biofuel and bioproduct industries.

FUNDING

This research was financially supported by: (i) ISRO–IISc space technology cell [ISTC/BES/TVR/389] and (ii) Gajanana and Madhavi Hegde [CES/TVR/GJH001] endowment for biofuel research.

ACKNOWLEDGMENTS

The authors are grateful to (i) EIACP Division, The Ministry of Environment, Forests and Climate Change, Government of India (SP/DEOO-22-0001), (ii) the NRDMS division, Ministry of Science and Technology, Government of India (CES/TVR/DST1045), and iii) the Indian Institute of Science for the infrastructure support for ecological research.

REFERENCES

Abbassi, A., M. Ali, and I.A. Watson. (2014). Temperature dependency of cell wall destruction of microalgae with liquid nitrogen pretreatment and hydraulic pressing. *Algal Research* 5: 190–194. www.sciencedirect.com/science/article/pii/S2211926414000095

Acién, F.G., J.M. Fernández, J.J. Magán, and E. Molina. (2012). Production cost of a real microalgae production plant and strategies to reduce it. *Biotechnology Advances*. Elsevier.

Alonso, D.L., E.H. Belarbi, J.M. Fernández-Sevilla, J. Rodríguez-Ruiz, and E.M. Grima. (2000). Acyl lipid composition variation related to culture age and nitrogen concentration in continuous culture of the microalga *Phaeodactylum tricornutum*. *Phytochemistry* 54 (5): 461–471.

Andrejić, J.Z., S.A. Spaulding, K.M. Manoylov, and M.B. Edlund. (2018). Phenotypic plasticity in diatoms: Janus cells in four Gomphonema taxa. *Diatom Research* 33(4): 453–470. www.tandfonline.com/doi/abs/10.1080/0269249X.2019.1572652

Arts, M.T., H. Rai, and V.P. Tumber. (2000). Effects of artificial UV-A and UV-B radiation on carbon allocation in *Synechococcus elongatus* (cyanobacterium) and *Nitzschia palea* (diatom) . *SIL Proceedings, 1922–2010* 27(4): 2000–2007.

Berge, J., J. Gouygou, J. Dubacqt, and P. Durand. (1995). Reassessment of lipid composition of the diatom, Skeletonema costatum. *Phytochemistry* 39: 1017–1021.

Berner, F., K. Heimann, and M. Sheehan. (2015). Microalgal biofilms for biomass production. *Journal of Applied Phycology* 27(5): 1793–1804. http://link.springer.com/10.1007/s10811-014-0489-x

Bielsa, G.B., C.A. Popovich, M.C. Rodríguez, A.M. Martínez, L.A. Martín, M.C. Matulewicz, and P.I. Leonardi. (2016). Simultaneous production assessment of triacylglycerols for biodiesel and exopolysaccharides as valuable co-products in *Navicula cincta*. *Algal Research* 15: 120–128.

Blanchemain, A., and D. Grizeau. (1996). Eicosapentaenoic acid content of *Skeletonema costatum* as a function of growth and irradiance; relation with chlorophyll a content and photosynthetic capacity. *Journal of Experimental Marine Biology and Ecology* 196(1–2): 177–188.

Boelen, P., R. van Dijk, J.S.S. Damsté, W.I.C. Rijpstra, and A.G.J. Buma. (2013). On the potential application of polar and temperate marine microalgae for EPA and DHA production. *AMB Express* 3(1): 1–9.

Borges, L., M.G. Montes, and P.C. Abreu. (2011). Effects of flocculants on lipid extraction and fatty acid composition of the microalgae *Nannochloropsis oculata* and *Thalassiosira weissflogii.* Biosafety and Occupational Health in Aquaculture View Project Automation of Microalgae Cultivation Using Arduino Open-Source Platform View Project. *Biomass and Bioenergy*. www.researchgate.net/publication/251628949.

Butler, T., R.V. Kapoore, and S. Vaidyanathan. (2020). *Phaeodactylum tricornutum*: a diatom cell factory. *Trends in Biotechnology* 38(6): 606–622.

Cerón García, M.C., F. García Camacho, A. Sánchez Mirón, J.M. Fernández Sevilla, Y. Chisti, and E. Molina Grima. (2006). Mixotrophic production of marine microalga *Phaeodactylum tricornutum* on various carbon sources. *Journal of Microbiology and Biotechnology* 16(5): 689–694.

Chen, Y.-C. (2012). The biomass and total lipid content and composition of twelve species of marine diatoms cultured under various environments. *Food Chemistry* 131(11): 211–219. www.sciencedirect.com/science/article/pii/S0308814611012118

Cheng, P., Y. Wang, T. Liu, and D. Liu. (2017). Biofilm attached cultivation of *Chlorella pyrenoidosa* is a developed system for swine wastewater treatment and lipid production. *Frontiers in Plant Science* 8: 1594. http://journal.frontiersin.org/article/10.3389/fpls.2017.01594/full

Chew, K.W., J.Y. Yap, P.L. Show, N.H. Suan, J.C. Juan, T.C. Ling, D.J. Lee, and J.S. Chang. (2017). Microalgae biorefinery: high value products perspectives. *Bioresource Technology* 229: 53–62.

Chiovitti, A., M.J. Higgins, R.E. Harper, R. Wetherbee, and A. Bacic. (2003). The complex polysaccharides of the Raphid diatom *Pinnularia viridis* (Bacillariophyceae). *Journal of Phycology* 39(31): 543–554. http://doi.wiley.com/10.1046/j.1529-8817.2003.02162.x

Christenson, L.B., and R.C. Sims. (2012). Rotating algal biofilm reactor and spool harvester for wastewater treatment with biofuels by-products. *Biotechnology and Bioengineering* 109(71): 1674–1684. http://doi.wiley.com/10.1002/bit.24451

Chu, W.-L., S.-M. Phang, and S.-H. Goh. (1996). Environmental effects on growth and biochemical composition of *Nitzschia inconspicua* Grunow. *Journal of Applied Phycology* 8: 389–396.

Da Costa, R.A.A.M., M.L. Koening, and S.J. De Macêdo. (1999). Use of secondary sewage water as a culture medium for *Chaetoceros gracilis* and *Thalassiosira* sp (Chrysophyceae) in laboratory conditions. *Brazilian Archives of Biology and Technology* 42(2): 205–212.

D'Ippolito, G., A. Sardo, D. Paris, F.M. Vella, M.G. Adelfi, P. Botte, C. Gallo, and A. Fontana. (2015). Potential of lipid metabolism in marine diatoms for biofuel production. *Biotechnology for Biofuels* 8(1): 1–10. https://biotechnologyforbiofuels.biomedcentral.com/articles/10.1186/s13068-015-0212-4

Dash, B., R. Ranjan, B. Vamsi, and P. Shiva. (2016). *Training Manual on Live Feed for Marine Finfish and Shellfish Culture. CMFRI Manuel Customized Training Book.* ://eprints.cmfri.org.in/10857/

Edenhofer, O., R. Pichs-Madruga, Y. Sokona, S. Kadner, J. Minx, S. Brunner, S. Agrawala, et al. (2014.) *Technical Summary. Climate Change 2014: Mitigation of Climate Change. Contribution of Working Group III to the Fifth Assessment Report of the Intergovernmental Panel on Climate Change.* Jan Fuglestvedt Edgar Hertwich Elmar Kriegler. http://pure.iiasa.ac.at/id/eprint/11124/1/ipcc_wg3_ar5_technical-summary.pdf

Ezeji, T., N. Qureshi, and H.P. Blaschek. (2007). Butanol production from agricultural residues: impact of degradation products on *Clostridium beijerinckii* growth and butanol fermentation. *Biotechnology and Bioengineering* 97(6): 1460–1469. www.ncbi.nlm.nih.gov/pubmed/17724071

Ferreira, J.A., P. Brancoli, S. Agnihotri, K. Bolton, and M.J. Taherzadeh. (2018). A review of integration strategies of lignocelluloses and other wastes in 1st generation bioethanol processes. *Process Biochemistry* 75: 173–186.

Fields, F.J., and J.P. Kociolek. (2015). An evolutionary perspective on selecting high-lipid-content diatoms (Bacillariophyta). *Journal of Applied Phycology* 27(6): 2209–2220. http://link.springer.com/10.1007/s10811-014-0505-1

Francisco, É.C., D.B. Neves, E. Jacob-Lopes, and T.T. Franco. (2010). Microalgae as feedstock for biodiesel production: carbon dioxide sequestration, lipid production and biofuel quality. *Journal of Chemical Technology & Biotechnology* 85(3): 395–403. http://doi.wiley.com/10.1002/jctb.2338

Gandini, A. (2008). Polymers from renewable resources: a challenge for the future of macromolecular materials. *Macromolecules* 41(24): 9491–9504. https://pubs.acs.org/sharingguidelines

Gao, F., Z.-H. Yang, C. Li, G.-M. Zeng, D.-H. Ma, and L. Zhou. (2015). A novel algal biofilm membrane photobioreactor for attached microalgae growth and nutrients removal from secondary effluent. *Bioresource Technology* 179: 8–12. www.sciencedirect.com/science/article/pii/S0960852414017295

Garrison, T.F., A. Murawski, and R.L. Quirino. (2016). Bio-based polymers with potential for biodegradability. *Polymers* 8(7): 1–22.

Gautam, S., M. Kashyap, S. Gupta, V. Kumar, B. Schoefs, R. Gordon, C. Jeffryes, K.B. Joshi, and V. Vinayak. (2016). Metabolic engineering of TiO_2 nanoparticles in *Nitzschia palea* to form diatom nanotubes: an ingredient for solar cells to produce electricity and biofuel. *RSC Advances* 6(99): 97276–97284. http://xlink.rsc.org/?DOI=C6RA18487A.

Gautam, S., L.K. Pandey, V. Vinayak, and A. Arya. (2017). Morphological and physiological alterations in the diatom *Gomphonema pseudoaugur* due to heavy metal stress. *Ecological Indicators* 72: 67–76.

Gifuni, I., A. Pollio, C. Safi, A. Marzocchella, and G. Olivieri. (2019). Current bottlenecks and challenges of the microalgal biorefinery. *Trends in Biotechnology* 37(3): 242–252.

González-González, L.M., S. Astals, S. Pratt, P.D. Jensen, and P.M. Schenk. (2021). Osmotic shock pre-treatment of *Chaetoceros muelleri* wet biomass enhanced solvent-free lipid extraction and biogas production. *Algal Research* 54: 102177.

Govindan, N., G.P. Maniam, M.M. Yusoff, M.H. Mohd, T. Chatsungnoen, R. Ramaraj, and Y. Chisti. (2019). Statistical optimization of lipid production by the diatom *Gyrosigma* sp. grown in industrial wastewater. *Journal of Applied Phycology*. https://doi.org/10.1007/s10811-019-01971-x

Graham, J.M., L.E. Graham, S.B. Zulkifly, B.F. Pfleger, S.W. Hoover, and J. Yoshitani. (2012). Freshwater diatoms as a source of lipids for biofuels. *Journal of Industrial Microbiology & Biotechnology* 39(319): 419–428. http://link.springer.com/10.1007/s10295-011-1041-5

Griffiths, M.J., R.P. van Hille, and S.T.L. Harrison. (2012). Lipid productivity, settling potential and fatty acid profile of 11 microalgal species grown under nitrogen replete and limited conditions. *Journal of Applied Phycology* 24(5): 989–1001. http://link.springer.com/10.1007/s10811-011-9723-y

Gross, M.A. (2015). *Development and Optimization of Biofilm Based Algal Cultivation*. https://lib.dr.iastate.edu/etd

Gügi, B., T. Le Costaouec, C. Burel, P. Lerouge, W. Helbert, and M. Bardor. (2015). Diatom-specific oligosaccharide and polysaccharide structures help to unravel biosynthetic capabilities in diatoms. *Marine Drugs* 13(9): 5993–6018.

Guo, Z., Y. Liu, H. Guo, S. Yan, and J. Mu. (2013). Microalgae cultivation using an aquaculture wastewater as growth medium for biomass and biofuel production. *Journal of Environmental Sciences* 25: S85–S88. www.sciencedirect.com/science/article/pii/S100107421460632X

Harun, R., M. Davidson, M. Doyle, R. Gopiraj, M. Danquah, and G. Forde. (2011). Technoeconomic analysis of an integrated microalgae photobioreactor, biodiesel and biogas production facility. *Biomass and Bioenergy* 35(1): 741–747.

Hildebrand, M., A.K. Davis, S.R. Smith, J.C. Traller, and R. Abbriano. (2012). The place of diatoms in the biofuels industry. *Biofuels* 3(2): 221–240. www.tandfonline.com/doi/full/10.4155/bfs.11.157

Hogan, P., P. Otero, P. Murray, and S.K. Saha. (2021). Effect of biomass pre-treatment on supercritical CO_2 extraction of lipids from marine diatom *Amphora* sp. and its biomass evaluation as bioethanol feedstock. *Heliyon* 7(11): e05995.

Hussein, H.A., and M.A. Abdullah. (2020). Anticancer compounds derived from marine diatoms. *Marine Drugs* 18(7): 356.

Jacobs, P., J. Pitarch, J.C. Kromkamp, and C.J.M. Philippart. (2021). Assessing biomass and primary production of microphytobenthos in depositional coastal systems using spectral information. *PLOS ONE* 16(7): e0246012. https://journals.plos.org/plosone/article?id=10.1371/journal.pone.0246012

Jankowska, E., A.K. Sahu, and P. Oleskowicz-Popiel. (2017). Biogas from microalgae: review on microalgae's cultivation, harvesting and pretreatment for anaerobic digestion. *Renewable and Sustainable Energy Reviews* 75©: 692–709.

Jayanthi, M., S. Thirumurthy, M. Muralidhar, and P. Ravichandran. (2018). Impact of shrimp aquaculture development on important ecosystems in India. *Global Environmental Change* 52: 10–21. www.sciencedirect.com/science/article/pii/S0959378018302954#fig0010

Jeffryes, C., J. Campbell, H. Li, J. Jiao, and G. Rorrer. (2011). The potential of diatom nanobiotechnology for applications in solar cells, batteries, and electroluminescent devices. *Energy & Environmental Science* 4(10): 3930. http://xlink.rsc.org/?DOI=c0ee00306a.

Jeffryes, C., T. Gutu, J. Jiao, and G.L. Rorrer. (2008). Metabolic insertion of nanostructured TiO_2 into the patterned biosilica of the diatom *Pinnularia* sp. by a two-stage bioreactor cultivation process. *ACS Nano* 2(10): 2103–2112. https://s3.amazonaws.com/academia.edu.documents/3469809/Jeffryes_2008_Ti.pdf?AWSAccessKeyId=AKIAIWOWYYGZ2Y53UL3A&Expires=1524474891&Signature=YYBoOvhmNg0ZsAWX4u25j%2BRRYKM%3D&response-content-disposition=inline%3B filename%3Dmetabolic_insertion_of_nanostr.

Jena, U., and K.C. Das. (2011). Comparative evaluation of thermochemical liquefaction and pyrolysis for bio-oil production from microalgae. *Energy and Fuels* 25(11): 5472–5482.

Jerez, A., P. Partal, I. Martínez, C. Gallegos, and A. Guerrero. (2007). Protein-based bioplastics: effect of thermo-mechanical processing. *Rheologica Acta* 46: 711–720. https://link.springer.com/article/10.1007/s00397-007-0165-z

Jiang, Y., C. Ba, D. Wang, S. Chen, M. Yang, H. Zou, and Y. Zhang. (2019). Effect of mixed polymer-coated urea reduction on soil NH_3 volatilization and N_2O emission. *Research of Environmental Sciences* 32(1): 159–165.

Johnson, M.B. (2009). *Microalgal Biodiesel Production through a Novel Attached Culture System and Conversion Parameters*. https://vtechworks.lib.vt.edu/handle/10919/32034

Jong, J. De. (2016). *Aquaculture in Indiano*. November 2015: 15. www.rvo.nl/sites/default/files/2017/04/aquaculture-in-india-report-2017.pdf

Juang, D.F., C.H. Lee, and S.C. Hsueh. (2012). Comparison of electrogenic capabilities of microbial fuel cell with different light power on algae grown cathode. *Bioresource Technology* 123: 23–29.

Kaewpintong, K., A. Shotipruk, S. Powtongsook, and P. Pavasant. (2007). Photoautotrophic high-density cultivation of vegetative cells of *Haematococcus pluvialis* in airlift bioreactor. *Bioresource Technology* 98(2): 288–295. www.ncbi.nlm.nih.gov/pubmed/16516464

Kaspar, H.F., E.F. Keys, N. King, K.F. Smith, A. Kesarcodi-Watson, and M.R. Miller. (2014). Continuous production of *Chaetoceros calcitrans* in a system suitable for commercial hatcheries. *Aquaculture* 420: 1–9.

Katarzyna, L., G. Sai, and O.A. Singh. (2015). Non-enclosure methods for non-suspended microalgae cultivation: literature review and research needs. *Renewable and Sustainable Energy Reviews* 42: 1418–1427. www.sciencedirect.com/science/article/pii/S1364032114009605

Kennish, M.J. (2002). Environmental threats and environmental future of estuaries. *Environmental Conservation* 29(1): 78–107. www.cambridge.org/core/services/aop-cambridge-core/content/view/830CE2DF3F482B87AA60ADAB773906D6/S037689290200061a.pdf/environmental_threats_and_environmental_future_of_estuaries.pdf.

Khan, M.I., J.H. Shin, and J.D. Kim. (2018). The promising future of microalgae: current status, challenges, and optimization of a sustainable and renewable industry for biofuels, feed, and other products. *Microbial Cell Factories* 17(1): 36.

Khan, S., M. Naushad, J. Iqbal, C. Bathula, and G. Sharma. (2021). Production and harvesting of microalgae and an efficient operational approach to biofuel production for a sustainable environment. *Fuel* 122543. https://linkinghub.elsevier.com/retrieve/pii/S0016236121024121

Kiatmetha, P., W. Siangdang, B. Bunnag, S. Senapin, and B. Withyachumnarnkul. (2011). Enhancement of survival and metamorphosis rates of *Penaeus monodon* larvae by feeding with the diatom *Thalassiosira weissflogii*. *Aquaculture International* 19(4): 599–609. https://link.springer.com/article/10.1007/s10499-010-9375-y

Kiran Marella, T., A. Saxena, and A. Tiwari. (2020). Diatom mediated heavy metal remediation: a review. *Bioresource Technology* 305: 123068.

Kiran, T., A. Tiwari, and M.V. Bhaskar. (2015). A new novel solution to grow diatom algae in large natural water bodies and its impact on CO_2 capture and nutrient removal. *Journal of Algal Biomass Utilization* 6(2): 22–27.

Kitano, M., R. Matsukawa, and I. Karube. (1997). Changes in eicosapentaenoic acid content of *Navicula saprophila, Rhodomonas salina* and *Nitzschia* sp. under mixotrophic conditions. *Journal of Applied Phycology* 9(6): 559–563.

Kranzler, C.F., J.W. Krause, M.A. Brzezinski, B.R. Edwards, W.P. Biggs, M. Maniscalco, J.P. McCrow, et al. (2019). Silicon limitation facilitates virus infection and mortality of marine diatoms. *Nature Microbiology*. Nature Publishing Group. www.nature.com/articles/s41564-019-0502-x

Kuppusamy, P., I. Soundharrajan, S. Srigopalram, M.M. Yusoff, G. Pragas Maniam, N. Govindan, and K.C. Choi. (2017). Potential pharmaceutical and biomedical applications of diatoms microalgae—an overview. *Indian Journal of Geo-Marine Sciences* 46(4): 663–667.

Kwietniewska, E., and J. Tys. (2014). Process characteristics, inhibition factors and methane yields of anaerobic digestion process, with particular focus on microalgal biomass fermentation. *Renewable and Sustainable Energy Reviews* 34: 491–500.

Larned, S.T. (2010). A prospectus for periphyton: recent and future ecological research. https://Doi.Org/10.1899/08-063.1 29, no. 1: 182–206. https://bioone.org/journals/journal-of-the-north-american-benthological-society/volume-29/issue-1/08-063.1/A-prospectus-for-periphyton-recent-and-future-ecological-research/10.1899/08-063.1.full

Laurens, L.M.L., J. Markham, D.W. Templeton, E.D. Christensen, S. Van Wychen, E.W. Vadelius, M. Chen-Glasser, T. Dong, R. Davis, and P.T. Pienkos. (2017). Development of algae biorefinery concepts for biofuels and bioproducts; a perspective on process-compatible products and their impact on cost-reduction. *Energy and Environmental Science* 10(8): 1716–1738.

Lebeau, T., P. Gaudin, R. Moan, and J.M. Robert. (2002). A new photobioreactor for continuous marennin production with a marine diatom: influence of the light intensity and the immobilised-cell matrix (alginate beads or agar layer). *Applied Microbiology and Biotechnology* 59(2–3): 153–159.

Lebeau, Thierry, and J.-M. Robert. (2003). Diatom cultivation and biotechnologically relevant products. Part I: Cultivation at various scales. *Applied Microbiology and Biotechnology* 60: 612–623.

Lee, D.J., J.S. Chang, and J.Y. Lai. (2015). Microalgae-microbial fuel cell: a mini review. *Bioresource Technology* 198: 891–895.

Lee, S.Y., J.H. Park, S.H. Jang, L.K. Nielsen, J. Kim, and K.S. Jung. (2008). Fermentative butanol production by clostridia. *Biotechnology and Bioengineering*. www.ncbi.nlm.nih.gov/pubmed/18727018

Levitan, O., J. Dinamarca, G. Hochman, and P.G. Falkowski. (2014). Diatoms: a fossil fuel of the future. *Trends in Biotechnology* 32(3): 117–124.

Li, X.L., T.K. Marella, L. Tao, L. Peng, C.F. Song, L.L. Dai, A. Tiwari, and G. Li. (2017). A novel growth method for diatom algae in aquaculture waste water for natural food development and nutrient removal. *Water Science and Technology* 75(12): 2777–2783.

López-Elías, J.A., D. Voltolina, F. Enríquez-Ocaña, and G. Gallegos-Simental. (2005). Indoor and outdoor mass production of the diatom *Chaetoceros muelleri* in a Mexican commercial hatchery. *Aquacultural Engineering* 33(31): 181–191.

López Barreiro, D., W. Prins, F. Ronsse, and W. Brilman. (2013). Hydrothermal liquefaction (HTL) of microalgae for biofuel production: state of the art review and future prospects. *Biomass and Bioenergy* 53: 113–127.

Lu, X., B. Liu, Y. He, B. Guo, H. Sun, and F. Chen. (2019). Novel insights into mixotrophic cultivation of *Nitzschia laevis* for co-production of fucoxanthin and eicosapentaenoic acid. *Bioresource Technology* 294: 122145.

Madhavi, K., G. Gowda, E.G. Jayaraj, M.T. Lakshmipathi, and C.S. Sree. (2014). Distribution of diatoms in riverine, estuarine and coastal waters off Mangalore, Karnataka. *Journal of Academy and Industrial Research* 3(3). 142–147.

Malviya, S., E. Scalco, S. Audic, F. Vincent, A. Veluchamy, J. Poulain, P. Wincker, et al. (2016). Insights into global diatom distribution and diversity in the world's ocean. *Proceedings of the National Academy of Sciences of the United States of America* 113(11): E1516–E1525.

Mantzorou, A., E. Navakoudis, K. Paschalidis, and F. Ververidis. (2018). Microalgae: a potential tool for remediating aquatic environments from toxic metals. *International Journal of Environmental Science and Technology* 15(8): 1815–1830. http://link.springer.com/10.1007/s13762-018-1783-y

Marella, T.K., N.R. Parine, and A. Tiwari. (2018). Potential of diatom consortium developed by nutrient enrichment for biodiesel production and simultaneous nutrient removal from waste water. *Saudi Journal of Biological Sciences* 25(4): 704–709. http://linkinghub.elsevier.com/retrieve/pii/S1319562X17301493

Marella, T.K., and A. Tiwari. (2020). Marine diatom *Thalassiosira weissflogii* based biorefinery for co-production of eicosapentaenoic acid and fucoxanthin. *Bioresource Technology* 307: 123245.

Martín, L.A., C.A. Popovich, A.M. Martínez, P.G. Scodelaro Bilbao, M.C. Damiani, and P.I. Leonardi. (2018). Hybrid two-stage culture of *Halamphora coffeaeformis* for biodiesel production: growth phases, nutritional stages and biorefinery approach. *Renewable Energy* 118: 984–992.

Matsumoto, M., D. Nojima, T. Nonoyama, K. Ikeda, Y. Maeda, T. Yoshino, and T. Tanaka. (2017). Outdoor cultivation of marine diatoms for year-round production of biofuels. *Marine Drugs* 15(4). www.ncbi.nlm.nih.gov/pubmed/28346334

McKew, B.A., J.D. Taylor, T.J. McGenity, and G.J.C. Underwood. (2011). Resistance and resilience of benthic biofilm communities from a temperate saltmarsh to desiccation and rewetting. *The ISME Journal* 5(1): 30–41. www.ncbi.nlm.nih.gov/pubmed/20596071

Medipally, S.R., F.M. Yusoff, S. Banerjee, and M. Shariff. (2015). Microalgae as sustainable renewable energy feedstock for biofuel production. *BioMed Research International* 2015: 519513. www.ncbi.nlm.nih.gov/pubmed/25874216

Merchuk, J., M. Gluz, and I. Mukmenev. (2000). Comparison of photobioreactors for cultivation of the red microalga *Porphyridium* sp. *Journal of Chemical Technology & Biotechnology* 75(12): 1119–1126. http://doi.wiley.com/10.1002/1097-4660%28200012%2975%3A12%3C1119%3A%3AAID-JCTB329%3E3.0.CO%3B2-G

Mishra, M., A.P. Arukha, T. Bashir, D. Yadav, and G.B.K.S. Prasad. (2017). All new faces of diatoms: potential source of nanomaterials and beyond. *Frontiers in Microbiology* 8(July): 1239.

Molina, E., J. Fernández, F.G. Acién, and Y. Chisti. (2001). Tubular photobioreactor design for algal cultures. *Journal of Biotechnology* 92(2): 113–131. https://linkinghub.elsevier.com/retrieve/pii/S0168165601003534

Mu, W., K. Jia, Y. Liu, X. Pan, and Y. Fan. (2017). Response of the freshwater diatom *Halamphora veneta* (Kützing) Levkov to copper and mercury and its potential for bioassessment of heavy metal toxicity in aquatic habitats. *Environmental Science and Pollution Research* 24(34): 26375–26386. https://link.springer.com/article/10.1007/s11356-017-0225-6

Naumann, T., Z. Çebi, B. Podola, and M. Melkonian. (2013). Growing microalgae as aquaculture feeds on twin-layers: a novel solid-state photobioreactor. *Journal of Applied Phycology* 25(5): 1413–1420. http://link.springer.com/10.1007/s10811-012-9962-6

Onay, M. (2020). Enhancing carbohydrate productivity from *Nannochloropsis gaditana* for bio-butanol production. *Energy Reports* 6: 63–67.

Pandey, L.K., D. Kumar, A. Yadav, J. Rai, and J.P. Gaur. (2014). Morphological abnormalities in periphytic diatoms as a tool for biomonitoring of heavy metal pollution in a river. *Ecological Indicators* 36: 272–279.

Park, J.B.K., R.J. Craggs, and A.N. Shilton. (2011). Wastewater treatment high rate algal ponds for biofuel production. *Bioresource Technology* 102(1): 35–42. http://linkinghub.elsevier.com/retrieve/pii/S0960852410011636

Popovich, C.A., M. Pistonesi, P. Hegel, D. Constenla, G.B. Bielsa, L.A. Martín, M.C. Damiani, and P.I. Leonardi. (2019). Unconventional alternative biofuels: quality assessment of biodiesel and its blends from marine diatom *Navicula cincta*. *Algal Research* 39: 101438. www.sciencedirect.com/science/article/pii/S2211926418308361

Popovich, Cecilia A., C. Damiani, D. Constenla, and P.I. Leonardi. (2012). Lipid quality of the diatoms *Skeletonema costatum* and *Navicula gregaria* from the South Atlantic coast (Argentina): evaluation of its suitability as biodiesel feedstock. *Journal of Applied Phycology* 24(1): 1–10.

Poveda-Giraldo, J.A., S. Piedrahita-Rodríguez, and C.A. Cardona-Alzate. (2021). Life cycle analysis of biotechnological processes based on the composition of the raw material. Eucalyptus, avocado, and plantain cases in a biorefinery system. *Chemical Engineering Transactions* 83: 397–402.

Pratiwi, A., D. Syah, L. Hardjito, L.M.G. Panggabean, and M.T. Suhartono. (2009). Fatty acid synthesis by Indonesian marine diatom, *Chaetoceros gracilis*. *HAYATI Journal of Biosciences* 16(41): 151–156.

Praveenkumar, R., K. Lee, J. Lee, and Y.K. Oh. (2015). Breaking dormancy: an energy-efficient means of recovering astaxanthin from microalgae. *Green Chemistry* 17(2): 1226–1234. https://pubs.rsc.org/en/content/articlehtml/2015/gc/c4gc01413h

Rahman, A., and C.D. Miller. (2017). Microalgae as a source of bioplastics. In: *Algal Green Chemistry: Recent Progress in Biotechnology*, 121–138. Elsevier.

Rahman, Asif, R.J. Putman, K. Inan, F.A. Sal, A. Sathish, T. Smith, C. Nielsen, R.C. Sims, and C.D. Miller. (2015). Polyhydroxybutyrate production using a wastewater microalgae based media. *Algal Research* 8: 95–98.

Rajaram, M.G., S. Nagaraj, M. Manjunath, A.B. Boopathy, C. Kurinjimalar, R. Rengasamy, T. Jayakumar, J.R. Sheu, and J.Y. Li. (2018). Biofuel and biochemical analysis of *Amphora coffeaeformis* RR03, a novel marine diatom, cultivated in an open raceway pond. *Energies* 11(6): 1–12.

Ramachandra, T.V., D.M. Mahapatra, K. B, and R. Gordon. (2009). Milking diatoms for sustainable energy: biochemical engineering versus gasoline-secreting diatom solar panels. *Industrial & Engineering Chemistry Research* 48(19): 8769–8788. http://pubs.acs.org/doi/abs/10.1021/ie900044j

Ramachandra, T.V., and Shwetmala. (2012). Decentralised carbon footprint analysis for opting climate change mitigation strategies in India. *Renewable and Sustainable Energy Reviews*. Pergamon. www.sciencedirect.com/science/article/pii/S1364032112003693

Ramachandra, T.V., B. Alakananda, and G. Supriya. (2011). Biofuel prospects of microalgal community in urban wetlands. *World* 1(2): 54–61.

Ramachandra, T.V., and G. Saranya. (2022). Sustainable bioeconomy prospects of diatom biorefineries in the Indian West coast. *Renewable and Sustainable Energy Reviews* 161: 112399. https://linkinghub.elsevier.com/retrieve/pii/S1364032122003094

Ramachandran, K., T. Suganya, N.N. Gandhi, and S. Renganathan. (2013). Recent developments for biodiesel production by ultrasonic assist transesterification using different heterogeneous catalyst: a review. *Renewable and Sustainable Energy Reviews* 22(C): 410–418. http://dx.doi.org/10.1016/j.rser.2013.01.057

Ramli, N.M., J.A.J. Verreth, F.M. Yusoff, K. Nurulhuda, N. Nagao, and M.C.J. Verdegem. (2020). Integration of algae to improve nitrogenous waste management in recirculating aquaculture systems: a review. *Frontiers in Bioengineering and Biotechnology* 8: 1004.

Rehm, B.H.A. (2010). Bacterial polymers: biosynthesis, modifications and applications. *Nature Reviews Microbiology*. Nature Publishing Group. www.nature.com/reviews/micro

Reitan, K.I., J.R. Rainuzzo, and Y. Olsen. (1994). Effect of nutrient limitation on fatty acid and lipid content of marine microalgae. *Journal of Phycology* 30(6): 972–979. http://doi.wiley.com/10.1111/j.0022-3646.1994.00972.x

Richards, J., J. Tibby, C. Barr, and P. Goonan. (2020). Effect of substrate type on diatom-based water quality assessments in the Mount Lofty Ranges, South Australia. *Hydrobiologia* 847(14): 3077–3090. https://link.springer.com/article/10.1007/s10750-020-04316-9

Roessler, P.G. (1988). Effects of silicon deficiency on lipid composition and metabolism in the diatom *Cyclotella cryptica*. *Journal of Phycology* 24(3): 394–400. http://doi.wiley.com/10.1111/j.1529-8817.1988.tb04482.x

Roostaei, J., Y. Zhang, K. Gopalakrishnan, and A.J. Ochocki. (2018). Mixotrophic microalgae biofilm: a novel algae cultivation strategy for improved productivity and cost-efficiency of biofuel feedstock production. *Scientific Reports* 8(121): 12528. www.ncbi.nlm.nih.gov/pubmed/30131525

Ruiz-Domínguez, M.C., F. Salinas, E. Medina, B. Rincón, M. de los Á. Martín, M.D.C. Gutiérrez, and P. Cerezal-Mezquita. (2022). Supercritical fluid extraction of fucoxanthin from the diatom *Phaeodactylum tricornutum* and biogas production through anaerobic digestion. *Marine Drugs* 20(2): 127. www.mdpi.com/1660-3397/20/2/127/htm.

Sánchez-Bayo, A., D. López-Chicharro, V. Morales, J.J. Espada, D. Puyol, F. Martínez, S. Astals, G. Vicente, L.F. Bautista, and R. Rodríguez. (2020). Biodiesel and biogas production from *Isochrysis galbana* using dry and wet lipid extraction: a biorefinery approach. *Renewable Energy* 146: 188–195.

Saranya, G., and T.V. Ramachandra. (2020). Biofuel and bioremediation prospects of estuarine diatoms. *International Journal of Renewable Energy Technology* 11(3): 248. www.inderscience.com/link.php?id=112006

Sato, R., Y. Maeda, T. Yoshino, T. Tanaka, and M. Matsumoto. (2014). Seasonal variation of biomass and oil production of the oleaginous diatom *Fistulifera* sp. in outdoor vertical bubble column and raceway-type bioreactors. *Journal of Bioscience and Bioengineering* 117(6): 720–724.

Schnurr, P.J., and D.G. Allen. (2015). Factors affecting algae biofilm growth and lipid production: a review. *Renewable and Sustainable Energy Reviews* 52: 418–429. www.sciencedirect.com/science/article/pii/S1364032115007376

Schnurr, P.J., G.S. Espie, and D.G. Allen. (2013). Algae biofilm growth and the potential to stimulate lipid accumulation through nutrient starvation. *Bioresource Technology* 136: 337–344.

Sekar, R., V.P. Venugopalan, K.K. Satpathy, K.V.K. Nair, and V.N.R. Rao. (2004). Laboratory studies on adhesion of microalgae to hard substrates. In: *Asian Pacific Phycology in the 21st Century: Prospects and Challenges*, 109–116. Dordrecht: Springer Netherlands. http://link.springer.com/10.1007/978-94-007-0944-7_14

Sethi, D., T.O. Butler, F. Shuhaili, and S. Vaidyanathan. (2020). Diatoms for carbon sequestration and bio-based manufacturing. *Biology* 9(8): 1–29.

Shah, A.V., V.K. Srivastava, S.S. Mohanty, and S. Varjani. (2021). Municipal solid waste as a sustainable resource for energy production: state-of-the-art review. *Journal of Environmental Chemical Engineering* 9(4): 105717.

Sheehan, J. (1998). A look back at the U.S. Department of Energy's Aquatic Species Program—Biodiesel from Algae. *Program* 328(July): 1–294.

Shen, Y., C. Chen, W. Chen, and X. Xu. (2014). Attached culture of *Nannochloropsis oculata* for lipid production. *Bioprocess and Biosystems Engineering* 37(9): 1743–1748. www.ncbi.nlm.nih.gov/pubmed/24554084

Shen, Y., X. Xu, Y. Zhao, and X. Lin. (2014). Influence of algae species, substrata and culture conditions on attached microalgal culture. *Bioprocess and Biosystems Engineering* 37(3): 441–450. http://link.springer.com/10.1007/s00449-013-1011-6

Shen, Ying, S. Wang, S.-H. Ho, Y. Xie, and J. Chen. (2018). Enhancing lipid production in attached culture of a thermotolerant microalga *Desmodesmus* sp. F51 using light-related strategies. *Biochemical Engineering Journal* 129: 119–128. www.sciencedirect.com/science/article/pii/S1369703X17302486

Shishlyannikov, S.M., I.V. Klimenkov, Y.D. Bedoshvili, I.S. Mikhailov, and A.G. Gorshkov. (2014). Effect of mixotrophic growth on the ultrastructure and fatty acid composition of the diatom *Synedra acus* from Lake Baikal. *Journal of Biological Research-Thessaloniki* 21(1): 15. https://jbiolres.biomedcentral.com/articles/10.1186/2241-5793-21-15

Shokrkar, H., and S. Ebrahimi. (2018). Evaluation of different enzymatic treatment procedures on sugar extraction from microalgal biomass, experimental and kinetic study. *Energy* 148: 258–268.

Sialve, B., N. Bernet, and O. Bernard. (2009). Anaerobic digestion of microalgae as a necessary step to make microalgal biodiesel sustainable. *Biotechnology Advances* 27(4): 409–416.

Sobczuk, T.M., F.G. Camacho, E.M. Grima, and Y. Chisti. (2006). Effects of agitation on the microalgae *Phaeodactylum tricornutum* and *Porphyridium cruentum*. *Bioprocess and Biosystems Engineering* 28(4): 243–250. www.ncbi.nlm.nih.gov/pubmed/16247611

Solé-Bundó, M., H. Salvadó, F. Passos, M. Garfí, and I. Ferrer. (2018). Strategies to optimize microalgae conversion to biogas: co-digestion, pretreatment and hydraulic retention time. *Molecules* 23(9): 2096. www.mdpi.com/1420-3049/23/9/2096/htm

Sreedevi, S., K.N. Unni, S. Sajith, P. Priji, M.S. Josh, and S. Benjamin. (2014). *Bioplastics: Advances in Polyhydroxybutyrate Research.* 1–30. Springer, Berlin, Heidelberg. https://link.springer.com/chapter/10.1007/12_2014_297

De Stefano, L., I. Rea, I. Rendina, M. De Stefano, and L. Moretti. (2007). Lensless light focusing with the centric marine diatom *Coscinodiscus walesii*. *Optics Express* 15(26): 18082. www.osapublishing.org/oe/abstract.cfm?uri=oe-15-26-18082

Thompson, F.L., P.C. Abreu, and W. Wasielesky. (2002). Importance of biofilm for water quality and nourishment in intensive shrimp culture. *Aquaculture* 203(3–4): 263–278.

Tian, C., B. Li, Z. Liu, Y. Zhang, and H. Lu. (2014). Hydrothermal liquefaction for algal biorefinery: a critical review. *Renewable and Sustainable Energy Reviews* 38(C): 933–950.

Tiwari, A., and T.K. Marella. (2019). Potential and application of diatoms for industry-specific wastewater treatment. *Application of Microalgae in Wastewater Treatment* 321–339.

Ugwu, C.U., H. Aoyagi, and H. Uchiyama. (2008). Photobioreactors for mass cultivation of algae. *Bioresource Technology* 99(10): 4021–4028.

Urbani, R., E. Magaletti, P. Sist, and A.M. Cicero. (2005). Extracellular carbohydrates released by the marine diatoms *Cylindrotheca closterium*, *Thalassiosira pseudonana* and *Skeletonema costatum*: effect of P-depletion and growth status. *Science of the Total Environment* 353(1–3):300–306.

Venkata Mohan, S., M. Hemalatha, D. Chakraborty, S. Chatterjee, P. Ranadheer, and R. Kona. (2020). Algal biorefinery models with self-sustainable closed loop approach: trends and prospective for blue-bioeconomy. *Bioresource Technology* 295: 122128.

Vishwakarma, R., V. Dhaka, T.U. Ariyadasa, and A. Malik. (2022). Exploring algal technologies for a circular bio-based economy in rural sector. *Journal of Cleaner Production* 354(April): 131653. https://doi.org/10.1016/j.jclepro.2022.131653

Wang, X., Y. Feng, J. Liu, H. Lee, C. Li, N. Li, and N. Ren. (2010). Sequestration of CO_2 discharged from anode by algal cathode in microbial carbon capture cells (MCCs). *Biosensors and Bioelectronics* 25(12): 2639–2643.

Wang, X.W., L. Huang, P.Y. Ji, C.P. Chen, X.S. Li, Y.H. Gao, and J.R. Liang. (2019). Using a mixture of wastewater and seawater as the growth medium for wastewater treatment and lipid production by the marine diatom *Phaeodactylum tricornutum*. *Bioresource Technology* 289: 121681.

Ward, A.J., P.J. Hobbs, P.J. Holliman, and D.L. Jones. (2008). Optimisation of the anaerobic digestion of agricultural resources. *Bioresource Technology* 99(17): 7928–7940.

Wen, Z.Y., and F. Chen. (2002). Continuous cultivation of the diatom *Nitzschia laevis* for eicosapentaenoic acid production: physiological study and process optimization. *Biotechnology Progress* 18(1): 21–28.

Wu, Y., Cheng, Z., Jie Wang, Y., Zheng, Y., Xiao, Z., Hui Yang, and F. Zhao. (2014). Light intensity affects the performance of photo microbial fuel cells with *Desmodesmus* sp. A8 as cathodic microorganism. *Applied Energy* 116: 86–90.

Wustman, B.A., M.R. Gretz, and K.D. Hoagland. (1997). Extracellular matrix assembly in diatoms (Bacillariophyceae): I. A model of adhesives based on chemical characterization and localization of polysaccharides from the marine diatom *Achnanthes longipes* and other diatoms. *Plant Physiology* 113.

Xia, S., L. Wan, A. Li, M. Sang, and C. Zhang. (2013). Effects of nutrients and light intensity on the growth and biochemical composition of a marine microalga *Odontella aurita*. *Chinese Journal of Oceanology and Limnology* 31(6): 1163–1173.

Xu, L., D.W.F. Wim Brilman, J.A.M. Withag, G. Brem, and S. Kersten. (2011). Assessment of a dry and a wet route for the production of biofuels from microalgae: energy balance analysis. *Bioresource Technology* 102(8): 5113–5122. www.ncbi.nlm.nih.gov/pubmed/21334889

Yusoff, F.M., H.B. Matias, Z.A. Khalid, and S.M. Phang. (2001). Culture of microalgae using interstitial water extracted from shrimp pond bottom sediments. *Aquaculture* 201(3–4): 263–270.

Zabed, H.M., S. Akter, J. Yun, G. Zhang, Y. Zhang, and X. Qi. (2020). Biogas from microalgae: technologies, challenges and opportunities. *Renewable and Sustainable Energy Reviews* 117: 109503.

Zhang, C.W., and A. Richmond. (2003). Sustainable, high-yielding outdoor mass cultures of *Chaetoceros muelleri* var. *subsalsum* and *isochrysis* Galbana in vertical plate reactors. *Marine Biotechnology* 5(3): 302–310.

Zhang, K., N. Kurano, and S. Miyachi. (2002). Optimized aeration by carbon dioxide gas for microalgal production and mass transfer characterization in a vertical flat-plate photobioreactor. *Bioprocess and Biosystems Engineering* 25(2): 97–101. http://link.springer.com/10.1007/s00449-002-0284-y

Zhang, Q., C. Liu, Y. Li, Z. Yu, Z. Chen, T. Ye, X. Wang, et al. (2017). Cultivation of algal biofilm using different lignocellulosic materials as carriers. *Biotechnology for Biofuels* 10: 115. www.ncbi.nlm.nih.gov/pubmed/28484515

Zhang, S., H. Pei, J. Wei, Y. Zhu, Y. Wang, and Z. Yang. (2019). The seasonal and spatial variations in diatom communities and the influence of environmental factors on three temperate reservoirs in Shandong Province, China. *Environmental Science and Pollution Research* 26(24): 24503–24515. https://pubmed.ncbi.nlm.nih.gov/31230246/

Zhao, J., X. Ge, J. Vasco-Correa, and Y. Li. (2014). Fungal pretreatment of unsterilized yard trimmings for enhanced methane production by solid-state anaerobic digestion. *Bioresource Technology* 158: 248–252.

Zheng, J., Q. Li, A. Hu, L. Yang, J. Lu, X. Zhang, and Q. Lin. (2013). Dual-frequency ultrasound effect on structure and properties of sweet potato starch. *Starch–Stärke* 65(7–8): 621–627. http://doi.wiley.com/10.1002/star.201200197

Zhou, X., J. Meng, Z. Yu, L. Miao, and C. Jin. (2018). The alterations of biofilm formation and EPS characteristics of a diatom by a sponge-associated bacterium *Psychrobacter* Sp. *Scientifica* 1892520. doi: 10.1155/2018/1892520.

Index

Note: Page numbers in *italic* refers to Tables.